COLLEGE ALGEBRA AND TRIGONOMETRY

COLLEGE ALGEBRA AND TRIGONOMETRY

Margaret F. Willerding
San Diego State College

Stephen Hoffman
Bates College

JOHN WILEY & SONS, INC. NEW YORK · LONDON · SYDNEY · TORONTO

Library of Congress Catalogue Card Number: 70-127672

ISBN 0-471-94658-3

Printed in the United States of America

10 9 8 7 6 5 4 3 2 1

PREFACE

The material in this book can be covered in a five-hour, one-semester course for students who have completed two years of high school algebra and one year of high school geometry or equivalent college courses. The book presents topics in college algebra and trigonometry in a form that will be useful to those students who will later study analytic geometry and calculus as well as for those students for whom it is their final mathematics course.

The approach is modern and reflects the recommendations of various mathematics curriculum study groups. Emphasis is on the algebraic structure of the ordered field of real numbers. In the treatment of the topics of trigonometry the emphasis is on the study of trigonometric functions as functions.

Chapter 1 illustrates how the manipulative aspects of elementary algebra are based on the field axioms, which are stated for the system of real numbers. This chapter provides rigorous proofs for simple algebraic theorems.

A concentrated review of elementary algebra, including work in finding solution sets of linear equations and inequalities, quadratic equations and inequalities, factoring, exponents, radicals, and rational expressions, is found in Chapters 2 and 3. For the well-prepared student these chapters may be treated as review. For students who are not adequately prepared or who have forgotten most of their elementary algebra, these chapters should be studied in detail.

Chapters 4 to 6 examine the function concept. Chapter 4 introduces the concept of a function as a set of ordered pairs of real numbers. Here the concepts of linear and quadratic functions are presented. Chapter 5 introduces the exponential function, and logarithmic functions are discussed as inverses of appropriate exponential functions. Chapter 6 introduces the trigonometric functions that are considered initially as functions of real numbers.

Chapters 6 to 10 are concerned entirely with the trigonometric functions. Chapter 11 presents determinants and their properties as well as the solution of systems of equations. Chapter 12 presents the complex numbers as an algebraic extension of the real numbers and includes sections on the trigonometric form of a complex number and DeMoivre's theorem. Chapter 13 introduces topics from the theory of equations. Mathematical induction is discussed in Chapter 14. Sequences and series are treated in this chapter as are the binomial theorem and binomial series. Chapter 15 includes the usual topics on permutations and combinations in addition to an elementary introduction to probability that includes conditional probability.

Adequate exercises are provided for each section and a chapter review is included at the end of each chapter. Answers to odd-numbered exercises are provided at the back of the book. Answers to even-numbered exercises are available to the instructor in a separate instructor's manual.

Margaret F. Willerding
Stephen Hoffman

CONTENTS

Review   Sept 5

COLLEGE ALGEBRA AND TRIGONOMETRY

CHAPTER 1

The Real Numbers and Their Properties

1.1 ◆ Definitions and Symbols of Set Theory

A fundamental idea of mathematics is the concept of a set. A **set** is simply a collection of things or objects. These objects are called **elements** or **members** of the set and are said to **belong to** the set or **be contained in** it. For example, Friday is a member of the set of days of the week; December is a member of the set of months of the calendar year.

We denote a set in such a way that there is no doubt about which elements belong to it. One way to do this is to list its members. If we consider the set of the first three months of the calendar year, we may use the symbol

$$\{\text{January, February, March}\}$$

to denote this set. The symbol {January, February, March} is read: "The set whose elements are January, February, and March." When we use the above symbol to denote a set, the elements are enclosed in **braces,** { }, with commas separating them.

We customarily use capital letters as names for sets. For example,

$$A = \{\text{January, February, March}\}$$

When all the elements of a set are tabulated, as in set A above, we say that we have **listed** the elements. If there are a great many elements in a set we often abbreviate when we list its members. For example,

$$B = \{\text{a, b, c, d}, \ldots, \text{x, y, z}\}$$

is the set of letters of the English alphabet. The three dots indicate that the letters e through w are also elements of B but have not been listed.

The set of numbers used in counting is called the set of **natural numbers,** namely,

$$N = \{1, 2, 3, 4, 5, \ldots\}$$

The set of natural numbers is a nonending set. Notice that to designate this set we again use three dots. In this case, the three dots mean "and so on in the same manner and continuing indefinitely."

Another way of designating a set is by specifying a characteristic that each element of the set must have. Let us consider the set of natural numbers less than 6. To be a member of this set an object must be a natural number and be less than 6. We may denote this set by

$$C = \{x \mid x \text{ is a natural number and } x \text{ is less than } 6\}$$

We read this symbol "the set of all x such that x is a natural number and x is less than 6." This notation for denoting sets is called **set-builder notation.** Notice that the bar in set-builder notation is read "such that."

The letter used in set-builder notation, for example, x in the notation for set C, is called a **variable.** In set C, x may be replaced by any one of the natural numbers 1, 2, 3, 4, or 5. The set from which we select replacements for x (in this case the set of natural numbers) is called the **replacement set,** the **domain of the variable,** or simply the **domain.**

When two sets have exactly the same elements, as for example,

$$P = \{1, 2, 3, 4\} \qquad \text{and} \qquad Q = \{x \mid x \text{ is a natural number less than } 5\}$$

they are said to be equal and we write $P = Q$.

DEFINITION 1.1. Two sets, A and B, are **equal,** denoted by $A = B$, if and only if they have exactly the same elements.

The phrase "if and only if" used in Definition 1.1 is a way of saying two statements at once. Thus, Definition 1.1 is equivalent to the compound statement: If two sets A and B are equal, then they have exactly the same elements, and if two sets A and B have exactly the same elements, then they are equal.

We use the symbol $\in$ to mean "is an element of" or "belongs to" when referring to an element of a set. Thus, if

$$C = \{1, 2, 3, 4\}$$

we see that 2 is an element of C and we write

$$2 \in C$$

The symbol $\notin$ is used to mean "is not an element of" or "does not belong to" when referring to an element of a set. Since 7 is not a member of C, we write

$$7 \notin C$$

Sometimes a set has no elements. For example, the set of months of the calendar year with exactly 52 days contains no elements. A set that has no elements is called an **empty set.** We use the symbol $\varnothing$ to denote the empty set.

A set may contain many elements. For example,

$$R = \{x \mid x \text{ is a natural number less than } 1{,}000{,}001\}$$

contains one million elements. If there is any doubt about this we could list all of the elements and count them. A set whose elements can be arranged in some systematic fashion and counted until the last element is reached is called a **finite set.** Set R above is a finite set. It should be noted that a set may have a fantastically large number of elements and still be a finite set. For example, the set of grains of sand on all the beaches in Texas at a particular moment has a very large number of elements but, nevertheless, is a finite set. Some examples of finite sets are:

The set of natural numbers less than 7,000,000.
The set of months of the year.
The set of persons residing in the United States at this moment.

A set that is not a finite set is called an **infinite set.** An example of an infinite set is the set of natural numbers. Since there is no last or greatest natural number, it is impossible to count the elements of the set of natural numbers with the counting coming to an end. Some examples of other infinite sets are:

The set of odd natural numbers.
The set of squares of the natural numbers.
The set of natural numbers greater than 1,000,000.

If every member of set B is also a member of set A, then B is called a **subset** of A. For example,

$$B = \{2, 4\}$$

is a subset of

$$A = \{1, 2, 3, 4, 5\}$$

since every member of B is a member of A. We use the symbol $\subseteq$ to mean "is a subset of." Thus

$$B \subseteq A$$

DEFINITION 1.2. If every member of set B is also a member of set A, then B is a **subset** of A.

Using Definition 1.2, can we say that set A is a subset of itself? Certainly every element of set A is a member of A, hence Definition 1.2 is satisfied and A is a subset of A. *Every set is a subset of itself.*

The empty set is a subset of every set. Every element of $\varnothing$ is a member of set A, since $\varnothing$ contains no elements. In this sense, $\varnothing$ is "contained" in every set.

EXAMPLE 1. List the members of the set of natural numbers less than 10 between braces.

SOLUTION $\{1, 2, 3, 4, 5, 6, 7, 8, 9\}$

EXAMPLE 2. Denote the set of natural numbers less than 10 using set-builder notation.

SOLUTION $\{x \mid x$ is a natural number and x is less than $10\}$

The set of all elements that are involved in any particular discussion is called the **universal set** or, simply, the **universe** and is denoted by the capital letter U. For example, the countries of the world are frequently classified into sets, such as the set of European countries, the set of American countries, the set of African countries, and so forth. Each of these sets is a subset of the universal set, which in this particular case, is the set composed of all of the countries of the world. As the discussion changes, so does the universe. Thus, if we consider the St. Louis Cardinals, the Atlanta Braves, the New York Mets, and the San Francisco Giants, all of whose members are professional baseball players, the universe for this particular discussion is the set of professional baseball players.

EXERCISES 1.1

List the members of the following sets in braces (Exercises 1–10).

1. The set of the days of the week.
2. The set of natural numbers between 1 and 15.
3. The set of natural numbers less than 30 that are divisible by 3.
4. The set of natural numbers greater than 11.
5. $\{x \mid x$ is a month of the year$\}$.

6. $\{x \mid x$ is a season of the year$\}$.
7. $\{x \mid x$ is a natural number less than 12$\}$.
8. $\{x \mid x$ is a natural number and x is greater than 10 and less than 20$\}$.
9. $\{x \mid x$ is a state of the United States and x borders the Pacific Ocean$\}$.
10. $\{x \mid x$ is an even natural number$\}$.

Denote the following sets by using set-builder notation (Exercises 11–20).

11. $\{a, b, c, d\}$.
12. $\{$January, June, July$\}$.
13. $\{1, 2, 3, 4, 5\}$.
14. $\{1, 2, 3, 4, \ldots\}$.
15. $\{2, 4, 6, 8, \ldots\}$.
16. $\{$Tuesday, Thursday$\}$.
17. $\{11, 12, 13, 14, 15, 16\}$.
18. $\{5, 10, 15, 20, \ldots\}$.
19. $\{15, 16, 17, \ldots\}$.
20. $\{2, 3, 4, 5, 6, 7, 8\}$.
21. Which of the following sets are empty?
    (a) The set of natural numbers.
    (b) The set of five-sided squares.
    (c) The set of states of the United States whose names begin with the letter A.
    (d) $\{x \mid x$ is a solution of $x + 3 = 7$ and x is a natural number$\}$.
    (e) $\{x \mid x$ is a solution of $x^2 + 1 = 0$ and x is a natural number$\}$.
22. Write all the subsets of $A = \{1, 2, 3\}$.
23. Write all the subsets of $A = \{1, 2, 3, 4, 5\}$ that have three elements.
24. Write all the subsets of $A = \{2, 4, 6, 8, 10\}$ that have two elements.
25. Which of the following are true statements?
    (a) $A \subseteq A$ (c) $U \subseteq A$
    (b) $\varnothing \subseteq A$ (d) $\varnothing \subseteq U$

1.2 ◆ Set Operations

We shall now define the complement of a set and the set operations of union and intersection. We begin our discussion by defining the union of two sets, A and B.

DEFINITION 1.3. The **union** of two sets, A and B, denoted by $A \cup B$ and read "A union B," is the set of all elements belonging to A or B or to both A and B.

The union of A and B may be denoted using set-builder notation:

$$A \cup B = \{x \mid x \in A \text{ or } x \in B\}$$

The "or" in the definition of union is an example of the "inclusive" use of "or." It means one or the other or both. In everyday language we often use the word "or" in the "exclusive" sense, meaning one or the other but *not* both. Thus, when we say "Today is Tuesday or Thursday" we are using the exclusive sense of the word "or," since we mean that today is Tuesday or Thursday but not both. In this book we shall always use "or" in the inclusive sense.

The two examples below will help toward a better understanding of the union of sets.

EXAMPLE 1. Given $A = \{1, 2, 3, 4, 5, 6\}$ and $B = \{1, 3, 5, 7, 9\}$, find $A \cup B$.

SOLUTION $A \cup B = \{1, 2, 3, 4, 5, 6, 7, 9\}$

Notice that although some elements belong to both A and B, they are listed only once in $A \cup B$. Thus, in forming $A \cup B$, we need only adjoin to the elements of A, those elements of B which are not already elements of A.

EXAMPLE 2. Given $P = \{a, b, c, d\}$ and $Q = \{r, s, t\}$, find $P \cup Q$.

SOLUTION $P \cup Q = \{a, b, c, d, r, s, t\}$

We now define the intersection of two sets A and B.

DEFINITION 1.4. The **intersection** of two sets A and B, denoted by $A \cap B$ and read "A intersection B," is the set of all elements that belong to both A and B.

The intersection of A and B may be denoted using set-builder notation:

$$A \cap B = \{x \mid x \in A \text{ and } x \in B\}$$

EXAMPLE 3. Given $A = \{1, 2, 3, 4, 5, 6\}$ and $B = \{2, 4, 6, 8, 10\}$, find $A \cap B$.

SOLUTION $A \cap B = \{2, 4, 6\}$

EXAMPLE 4. Given $C = \{a, b, c, d, e\}$ and $D = \{p, i, g\}$, find $C \cap D$.

SOLUTION. Since sets C and D have no elements in common, their intersection is the empty set, $\varnothing$:

$$C \cap D = \varnothing$$

Two sets that have no elements in common are called **disjoint sets.** The intersection of two disjoint sets is the empty set, $\emptyset$.

We now define the complement of a set A which is a subset of a universe U.

DEFINITION 1.5. The **complement** of a set A, $A \subseteq U$, denoted by A' and read "the complement of A," is the set of all elements of U which are not elements of A.

In set-builder notation we write:

$$A' = \{x \mid x \in U \text{ and } x \notin A\}$$

EXAMPLE 5. If $U = \{1, 2, 3, 4, 5, 6, 7, 8, 9, 10\}$ and $P = \{2, 4, 6\}$, find P'.

SOLUTION $P' = \{1, 3, 5, 7, 8, 9, 10\}$

EXERCISES 1.2

1. Let $A = \{1, 2, 3, 4, 5, 6, 7, 8\}$, $B = \{1, 3, 5, 7, 9\}$, and $C = \{2, 4, 6, 8, 10, 12\}$.
   List the elements of:
   (a) $A \cup B$ (d) $A \cap C$
   (b) $A \cup C$ (e) $B \cup C$
   (c) $A \cap B$ (f) $C \cap B$

2. Given $U = \{x \mid x \text{ is a natural number}\}$, and $A = \{x \mid x \text{ is an even number}\}$, $B = \{x \mid x \text{ is a number less than 10}\}$, and $C = \{x \mid x \text{ is a multiple of 3}\}$ all subsets of U. Find:
   (a) $A \cup B$ (d) A'
   (b) $A \cap C$ (e) B'
   (c) $B \cap C$ (f) C'

Use the following to list the members of the sets in exercises 3 through 12. Sets A, B, and C are subsets of U.

$$U = \{1, 2, 3, 4, 5, 6, 7, 8, 9, 10\}$$
$$A = \{2, 4, 6, 8, 10\}$$
$$B = \{3, 6, 9\}$$
$$C = \{1, 2, 3, 4, 5\}$$

3. $A \cup B$ 8. $A \cup B'$
4. A' 9. $C' \cap B$
5. $B \cap C$ 10. $(B \cap C)'$
6. $B \cup C$ 11. $A' \cup C'$
7. B' 12. $(C' \cup B)'$

13. Suppose $A \cup B = \emptyset$. What conclusion can you draw about A and B?

14. If $A \cup B = B$, what conclusion can you draw about A and B?

15. If $A \cap B = A$, what conclusion can you draw about A and B?

16. Which of the following are true statements?

    (a) $A \cup U = U$ (e) $A \cap B = B \cap A$

    (b) $A \cap U = U$ (f) $A \cup B = B \cup A$

    (c) $A \cap \emptyset = \emptyset$ (g) $A' = A$

    (d) $A \cup \emptyset = \emptyset$ (h) $(A')' = A$

Complete each of the following (Exercises 20–24).

17. $A \cup A = ?$ 21. $A \cap A = ?$

18. $\emptyset \cap A = ?$ 22. $A \cup \emptyset = ?$

19. $U \cap U = ?$ 23. $A' \cap A = ?$

20. $A \cup A' = ?$ 24. $U \cup \emptyset = ?$

25. Explain why $A \cup B = B \cup A$.

26. Explain why $A \cap B = B \cap A$.

27. Which of the following are true statements?

    (a) $(A \cap B) \subseteq A$.

    (b) $(A \cap B) \subseteq B$.

    (c) $(A \cap A') \subseteq A$.

    (d) $(A \cap B') \subseteq B$.

    (e) $(A' \cap B') \subseteq A'$.

1.3 ◆ Sets of Numbers

Throughout this book we shall be considering sets of numbers, most of which are familiar to you from your past studies of mathematics. We shall define these sets now.

DEFINITION 1.6. The set of **natural numbers,** denoted by N, is the set consisting of the numbers used in counting.

We may write:

$$N = \{1, 2, 3, 4, 5, \ldots\}$$

DEFINITION 1.7. The set of **whole numbers,** denoted by W, is the set consisting of the natural numbers and zero.

We may write:

$$W = \{0, 1, 2, 3, 4, \ldots\}$$

DEFINITION 1.8. The set of **integers,** denoted by I, is the set consisting of the natural numbers, their negatives, and zero.

We may write:

$$I = \{\ldots, -3, -2, -1, 0, 1, 2, 3, \ldots\}$$

DEFINITION 1.9. The set of **rational numbers,** denoted by Q, is the set of numbers that can be represented as the quotient of two integers a and b, $\dfrac{a}{b}$ or $a \div b$, where $b \neq 0$.

Using set-builder notation, we may write

$$Q = \left\{ \frac{a}{b} \;\middle|\; a \in I, b \in I, \text{ and } b \neq 0 \right\}$$

Some elements of Q are $\frac{1}{2}$, $-\frac{3}{4}$, $\frac{7}{7}$, and $-\frac{8}{4}$.

We see that rational numbers are denoted by fractions. It should be noted that although all rational numbers may be denoted by fractions, not all fractions denote rational numbers. For example, the fractions $\dfrac{\sqrt{2}}{2}$ and $\dfrac{4\pi}{3}$ do not denote rational numbers because $\sqrt{2}$ and π are not integers.

We also recall that rational numbers may be denoted by terminating decimals or by repeating decimals. Thus

$$\frac{1}{2} = 0.5 \qquad\qquad \frac{1}{3} = 0.333\ldots$$
$$\frac{1}{4} = 0.25 \qquad\qquad \frac{1}{6} = 0.1666\ldots$$

We see that $\frac{1}{2}$ and $\frac{1}{4}$ are represented by terminating decimals. On the other hand, $\frac{1}{3}$ and $\frac{1}{6}$ are represented by repeating decimals. The three dots in $0.333\ldots$ mean that the 3 keeps repeating indefinitely.

Some numbers such as $\sqrt{2}$ can be approximated by a decimal, but no repeating pattern appears. Thus

$$\sqrt{2} \doteq 1.41$$
$$\doteq 1.414$$
$$\doteq 1.4142$$
$$\doteq 1.41421, \text{ and so forth}$$

(the symbol $\doteq$ means approximately). Such numbers are called **irrational numbers.** The set of irrational numbers may be defined as that set of numbers which may be approximated by nonrepeating, nonterminating decimals, but that cannot be represented by the quotient of two integers.

We recall that numbers may be represented by lengths of line segments. Thus in the right triangle in Figure 1.1, each leg represents 1 and the hypotenuse represents $\sqrt{2}$. Thus we can define irrational numbers as those numbers that may be represented by the lengths of line segments or their negatives, but cannot be represented as the quotient of two integers.

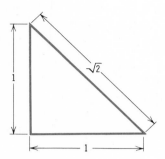

Figure 1.1

DEFINITION 1.10. The set of **irrational numbers,** denoted by S, is the set consisting of those numbers that may be represented by the lengths of line segments or their negatives but cannot be represented as the quotient of two integers.

Among the elements of S are such numbers as $\sqrt{2}$, π, $-\sqrt{5}$, and $\sqrt[3]{3}$.

DEFINITION 1.11. The set of **real numbers,** denoted by R, is the set consisting of all the rational numbers and all the irrational numbers.

We can write:

$$R = Q \cup S$$

DEFINITION 1.12. The set of **complex numbers,** denoted by C, is the set of all numbers that can be represented in the form $a + bi$, where a and b are real numbers and $i^2 = -1$.

Using set-builder notation we may write:

$$C = \{z \mid z = a + bi,\, a \in R,\, b \in R,\, \text{and } i^2 = -1\}$$

The following examples illustrate the use of set notation and various sets of numbers.

EXAMPLE 1. List the members of $\{x \mid x$ is an integer and x is divisible by 5$\}$.

SOLUTION. Since every element of the given set is an integer and a multiple of 5 (that is, every element is divisible by 5), each member is of the form $5k$ where k is an integer. Hence the set is

$$\{\ldots, -15, -10, -5, 0, 5, 10, 15, \ldots\}$$

EXAMPLE 2. Use set-builder notation to denote the set of real numbers greater than $\sqrt{2}$.

SOLUTION $\qquad \{x \mid x \in R \text{ and } x > \sqrt{2}\}$

EXAMPLE 3. Use set-builder notation to denote the set of real numbers greater than -6 and less than 6.

SOLUTION

$$\{x \mid x \in R \text{ and } -6 < x < 6\} = \{x \mid x \in R \text{ and } x > -6\} \cap \{x \mid x \in R \text{ and } x < 6\}$$

EXERCISES 1.3

List the members of the following sets (Exercises 1–10).

1. $\{x \mid x$ is a natural number greater than 6 and less than 18$\}$.
2. $\{x \mid x$ is an integer between -4 and 5$\}$.
3. $\{x \mid x$ is a natural number and x is greater than 6$\}$.
4. $\{x \mid x$ is a nonnegative integer$\}$.
5. $\{x \mid x$ is a negative integer$\}$.
6. $\{x \mid x$ is a natural number between 3 and 7$\}$.
7. $\{x \mid x$ is a natural number divisible by 3$\}$.
8. $\{x \mid x$ is an integer and x is divisible by 3$\}$.
9. $\{x \mid x$ is an even natural number$\}$.
10. $\{x \mid x$ is an odd integer$\}$.

Use set-builder notation to denote the following sets (Exercises 11–18).

11. The set of natural numbers greater than 5.
12. The set of integers between -4 and 4.
13. The set of positive integers.
14. The set of odd natural numbers.

15. The set of real numbers between -2 and 2.
16. The set of real numbers greater than -5 and less than 10.
17. The set of even integers.
18. The set of real numbers greater than 1.86.
19. Which of the following are true statements?
    (a) All real numbers are rational numbers.
    (b) All irrational numbers are real numbers.
    (c) All integers are either positive or negative.
    (d) All irrational numbers may be represented by $\frac{a}{b}$ where a and b are integers
       and $b \neq 0$.
20. Which of the following are true statements?
    (a) $\{x \mid x \text{ is a rational number}\} \subseteq \{x \mid x \text{ is a real number}\}$.
    (b) $\{x \mid x \text{ is a real number}\} \cap \{x \mid x \text{ is a rational number}\}$
       $= \{x \mid x \text{ is an irrational number}\}$.
    (c) $\{x \mid x \text{ is an integer}\} \subseteq \{x \mid x \text{ is a natural number}\}$.

1.4 ◆ The Real Number Axioms

Mathematical statements which we assume to be true without further justification are called **axioms** or **postulates.** The first axioms that we shall state are called the **axioms of equality.** We use the symbol $=$, read "equals," to denote the familiar relation of equality. When we write $a = b$, we mean that a and b are names for the same thing.

AXIOMS OF EQUALITY

In the following axioms, a, b, and c represent real numbers.
E-1. (*Reflexive Property*)   For all a, $a = a$.
E-2. (*Symmetric Property*)   For all a and b, if $a = b$, then $b = a$.
E-3. (*Transitive Property*)   For all a, b, and c, if $a = b$ and $b = c$, then $a = c$.
E-4. (*Substitution Property*)   Any quantity may be substituted for an equal quantity in any mathematical statement without changing the truth or falsity of the statement.
E-5. (*Addition Property*)   If $a = b$, then $a + c = b + c$ and $c + a = c + b$.
E-6. (*Multiplication Property*)   If $a = b$, then $ac = bc$ and $ca = cb$.

We now state the axioms of the real number system under the binary operations of addition and multiplication. A **binary operation** on the elements of a set S is a rule that assigns to each pair of elements a and b of S, taken in that order, a unique element of S. For example, the binary operation of addition assigns to the pair of real numbers 6 and 8 (in that order) the unique real number 14; the binary operation of multiplication assigns to the pair of real numbers 6 and 8 (in that order), the unique real number 48.

DEFINITION 1.13. A **binary operation** on a set S is a rule that assigns to each pair of elements a and b of S, taken in that order, a unique element of S.

The binary operation of addition is denoted by the symbol $+$. The unique real number which the operation of addition assigns to the pair of real numbers a and b is called their **sum** and is denoted by $a + b$. Thus the sum of 6 and 8 is denoted by $6 + 8$.

The binary operation of multiplication is denoted by the symbol $\times$, a raised dot, or by placing two numbers adjacent to one another. The unique real number that the operation of multiplication assigns to the pair of real numbers a and b is called their **product** and is denoted by $a \times b, a \cdot b$, or ab. Thus the product of 6 and 8 is denoted by $6 \times 8, 6 \cdot 8$, or $(6)(8)$.

In the axioms of the real number system stated below, a, b, and c are symbols for real numbers.

REAL NUMBER AXIOMS

R-1. (*Closure Property of Addition*)   For all a and b, their sum, denoted by $a + b$, is a unique element of R.

R-2. (*Commutative Property of Addition*)   For all a and b, $a + b = b + a$.

R-3. (*Associative Property of Addition*)   For all a, b, and c, $(a + b) + c = a + (b + c)$.

R-4. (*Identity Element for Addition*)   There exists in R an element, 0, called the **identity element for addition,** or the **additive identity,** with the property that, for all a, $a + 0 = 0 + a = a$.

R-5. (*Additive Inverse Property*)   Corresponding to every real number, a, there is a unique real number, $-a$, called the **opposite** or **additive inverse** of a, such that $a + (-a) = (-a) + a = 0$.

R-6. (*Closure Property of Multiplication*)   For all a and b, their product, denoted by ab, is a unique element of R.

R-7. (*Commutative Property of Multiplication*)   For all a and b, $ab = ba$.

R-8. (*Associative Property of Multiplication*)   For all a, b, and c, $(ab)c = a(bc)$.

R-9. (*Identity Element for Multiplication*)   There exists in R an element, 1, called the **identity element for multiplication,** or the **multiplicative identity,** such that for all a, $a \cdot 1 = 1 \cdot a = a$.

R-10. (*Multiplicative Inverse Property*)   Corresponding to every real number a, other than zero ($a \neq 0$), there is a unique real number denoted by $\dfrac{1}{a}$, called the **reciprocal** or **multiplicative inverse** of a, such that

$$a \cdot \frac{1}{a} = \frac{1}{a} \cdot a = 1.$$

R-11. (*Distributive Property*)   For all a, b, and c, $a(b + c) = ab + ac$ and $(b + c)a = ba + ca$.

Axioms R-1 and R-6, which are called the closure properties, state that the sum and product of two real numbers are always real numbers and that each sum and product is uniquely determined. Since the sum of two real numbers is a real number, we say that the set of real numbers is *closed under the operation of addition* or is *closed with respect to addition.* Similarly, the set of real numbers is closed with respect to multiplication. When the result of a binary operation on the elements of some set is an element of that same set, the set is said to be closed under that operation. For example, the set of integers is closed under the operations of addition and multiplication because the sum and product of any two integers is again an integer. The set of odd integers is closed under the operation of multiplication, but is not closed under the operation of addition because the product of any two odd integers is an odd integer, but the sum of any two odd integers is an even integer, not an odd integer.

Axioms R-2 and R-7 are called the commutative properties. These axioms state that the order in which we add or multiply two real numbers does not affect the sum or product. Thus

$$3 + 6 = 6 + 3 \qquad \text{and} \qquad 3 \cdot 6 = 6 \cdot 3$$

The operations of subtraction and division, which we shall define later, do not have this property.

The associative properties, R-3 and R-8, state that three numbers in a sum or product may be grouped in either of two ways without changing the result. These associative properties may be generalized to apply to the sum and product of four or more numbers.

Axioms R-4 and R-9 tell of the existence of certain special real numbers and of their properties. Axioms R-5 and R-10 tell how certain pairs of real numbers are related to these special numbers 0 and 1.

Axiom R-11 connects the operations of addition and multiplication, and is called the distributive property.

The distributive property may be generalized for the sum of three or more addends. Thus

$$a(b + c + d) = ab + ac + ad$$
$$a(b + c + d + e) = ab + ac + ad + ae, \text{ and so forth.}$$

This is sometimes called the **generalized distributive property.**

If a given set, F, and a pair of binary operations defined on the elements of F, satisfy axioms R-1 through R-11, then F is called a **field** under these operations. The set of real numbers with the operations of addition and multiplication is called the **real number field.** The set of rational numbers with the operations of addition and multiplication is also a field called the **rational number field.**

EXAMPLE 1. What axiom of the real numbers assures us that

$$(-3) + [-(-3)] = 0?$$

SOLUTION. The additive inverse axiom.

EXAMPLE 2. What axiom of the real numbers assures us that for all real numbers $y \neq 0$,

$$y \cdot \frac{1}{y} = 1$$

SOLUTION. The multiplicative inverse axiom.

We now define the operations of subtraction and division of real numbers.

DEFINITION 1.14. If a and b are real numbers, their **difference** denoted by $a - b$ is defined to be $a + (-b)$:

$$a - b = a + (-b)$$

The operation of finding the difference is called **subtraction.**

DEFINITION 1.15. If a and b are real numbers and $b \neq 0$, their **quotient** denoted by $\frac{a}{b}$ is defined to be $a\left(\frac{1}{b}\right)$:

$$\frac{a}{b} = a \cdot \frac{1}{b}$$

The operation of finding the quotient is called **division.**

EXERCISES 1.4

Justify each statement below by citing the appropriate axiom of equality. Each variable in Exercises 1–10 denotes a real number.

1. If $a = 3$, then $a + 2 = 5$.
2. If $7 = x$, then $x = 7$.
3. If $x = 6$ and $y = x - 4$, then $y = 6 - 4$.
4. If $a = 9$, then $2a = 18$.
5. If $7x = 21$, then $x = 3$.
6. If $x - 3 = 9$, then $x = 12$.
7. If $x + 2 = y + 3$ and $y + 3 = 12$, then $x + 2 = 12$.
8. If $y = 3x$ and $3x = 7$, then $y = 7$.
9. If $2 = y + 7$, then $y + 7 = 2$.
10. If $a + b = 9$ and $b = 4$, then $a + 4 = 9$.

Justify each statement by the appropriate real number axiom. Each variable denotes a real number (Exercises 11–25).

11. $x + y$ is a real number.
12. $7 + 3 = 3 + 7$.
13. $(4 \times 2) \times 7 = 4 \times (2 \times 7)$.
14. $2(\frac{1}{2}) = 1$.
15. $(-3) + 0 = -3$.
16. $(-6) + 6 = 0$.
17. $3 + (-6) = (-6) + 3$.
18. $x \cdot \dfrac{1}{x} = 1,\ x \neq 0$.
19. $x(y + z) = xy + xz$.
20. $(a + b) + c = (b + a) + c$.
21. $3(a + b) = 3a + 3b$.
22. $\dfrac{1}{x}(y + z) = \dfrac{1}{x} \cdot y + \dfrac{1}{x} \cdot z,\ x \neq 0$.
23. $(x + y) + [-(x + y)] = 0$.
24. $\dfrac{1}{x + y} \cdot (x + y) = 1,\ x \neq -y$.
25. $xy + x(z + x) = xy + (xz - xw)$.
26. Which of the following sets are closed under addition?
    (a) $\{1, 2, 3, 4, \ldots\}$.
    (b) $\{0, 2, 4, 6, \ldots\}$.
    (c) $\{1, 3, 4, 7, 11, 18, \ldots\}$.
    (d) $\{0\}$.

27. Which of the following sets are closed under multiplication?
    (a) $\{1, 2, 3, 4, \ldots\}$.
    (b) $\{0, 2, 4, 6, \ldots\}$.
    (c) $\{1, 3, 5, 7, \ldots\}$.
    (d) $\{0\}$.
28. Is the set of real numbers closed under the operation of subtraction? Give an example to support your answer.
29. Is the set of real numbers closed under the operation of division except by zero? Give an example to support your answer.
30. Let $*$ be an operation such that for all real numbers a and b, $a * b = (a + b) - ab$. For example, $3 * 4 = (3 + 4) - (3 \times 4) = 7 - 12 = -5$.
    (a) Find $4 * 5$.
    (b) Find $6 * 3$.
    (c) Find $5 * 4$.
    (d) Find $3 * 6$.
    (e) Do you think that the operation $*$ is commutative? (That is, is $a * b = b * a$?)

1.5 ◆ Field Properties

The basic properties of the real numbers are described by the axioms of equality, the axioms of the real numbers, and definitions. A **definition,** in mathematical usage, is simply an agreement to regard one expression as being equivalent to another. Thus, in the definition of division we agree that $\dfrac{a}{b}$ and $a \cdot \dfrac{1}{b}$ are equivalent. From these basic properties we can derive other properties of real numbers such as:

THEOREM 1.1. For every real number a, $-(-a) = a$.

PROOF

$a + (-a) = 0$	additive inverse axiom
$[a + (-a)] + [-(-a)] = 0 + [-(-a)]$	addition property of equality
$[a + (-a)] + [-(-a)] = -(-a)$	additive identity axiom
$a + \{(-a) + [-(-a)]\} = -(-a)$	associative property of addition
$a + 0 = -(-a)$	additive inverse axiom
$a = -(-a)$	additive identity axiom
$-(-a) = a$	symmetric property of equality

This form of logical reasoning from **hypothesis** to **conclusion** is called a **proof.** Assertions proved in this way are called **theorems.** Proving a theorem consists of showing that the conclusion follows as a logical consequence of the hypothesis and the axioms.

Mathematical theorems usually appear in the form "if h then c" where h is the hypothesis and c is the conclusion. Suppose we know that "if h then c_1" where c_1 is a conclusion different from the desired conclusion c. Suppose also that we can prove "if c_1 then c_2," "if c_2 then c_3," and "if c_3 then c." We now have a chain of implications:

$$\text{If } h \text{ then } c_1.$$
$$\text{If } c_1 \text{ then } c_2.$$
$$\text{If } c_2 \text{ then } c_3.$$
$$\text{If } c_3 \text{ then } c.$$

We conclude, using a law of logic called the **chain rule,** that "if h then c," the desired conclusion.

We shall prove some theorems about the real numbers in this section. Our purpose in proving these familiar theorems is not merely to recall known facts, but also to show why these facts are logical consequences of the axioms.

THEOREM 1.2. (*Cancellation Law of Addition*) **For all real numbers a, b, and c, if $a + c = b + c$, then $a = b$.**

PROOF

$a + c = b + c$	hypothesis
$(a + c) + (-c) = (b + c) + (-c)$	addition property of equality
$a + [c + (-c)] = b + [c + (-c)]$	associative property of addition
$a + 0 = b + 0$	additive inverse axiom
$a = b$	additive identity axiom

THEOREM 1.3. (*Cancellation Law of Multiplication*) **For all real numbers, a, b, and c, if $ab = ac$ and $a \neq 0$, then $b = c$.**

PROOF

$ab = ac$	hypothesis
$a \neq 0$	hypothesis
$\dfrac{1}{a}$ is a real number	multiplicative inverse axiom

not equal to zero*

$$\frac{1}{a} \cdot (ab) = \frac{1}{a} \cdot (ac) \qquad \text{multiplication property of equality}$$

$$\left(\frac{1}{a} \cdot a\right)b = \left(\frac{1}{a} \cdot a\right)c \qquad \text{associative property of multiplication}$$

$$1 \cdot b = 1 \cdot c \qquad \text{multiplicative inverse axiom}$$

$$b = c \qquad \text{multiplicative identity axiom}$$

THEOREM 1.4. (*Multiplication Property of Zero*) **For any real number, a,**
   $a \cdot 0 = 0$.

PROOF

$$0 + 0 = 0 \qquad \text{additive identity axiom}$$
$$a(0 + 0) = a \cdot 0 \qquad \text{multiplication property of equality}$$
$$a \cdot 0 + a \cdot 0 = a \cdot 0 \qquad \text{distributive property}$$
$$[a \cdot 0 + a \cdot 0] + [-(a \cdot 0)] \qquad \text{addition property of equality}$$
$$= a \cdot 0 + [-(a \cdot 0)]$$
$$a \cdot 0 + \{a \cdot 0 + [-(a \cdot 0)]\} \qquad \text{associative property of addition}$$
$$= a \cdot 0 + [-(a \cdot 0)]$$
$$a \cdot 0 + 0 = 0 \qquad \text{additive inverse axiom}$$
$$a \cdot 0 = 0 \qquad \text{additive identity axiom}$$

THEOREM 1.5. If a and b are real numbers, then $a(-b) = -(ab)$.

PROOF

$$b + (-b) = 0 \qquad \text{additive inverse axiom}$$
$$a[b + (-b)] = a \cdot 0 \qquad \text{multiplication property of equality}$$

$$a[b + (-b)] = 0 \qquad \text{Theorem 1.4}$$
$$a[(-b) + b] = 0 \qquad \text{commutative property of addition}$$

$$a(-b) + ab = 0 \qquad \text{distributive property}$$
$$[a(-b) + ab] + (-ab) = 0 + (-ab) \qquad \text{addition property of equality}$$
$$a(-b) + [ab + (-ab)] = 0 + (-ab) \qquad \text{associative property of addition}$$

---

* If a is a real number not equal to zero it has a multiplicative inverse $\frac{1}{a}$ by the multiplicative inverse axiom. That this multiplicative inverse is not equal to zero can be proved using the axioms of the real number field. We shall accept this theorem without proof.

$$a(-b) + 0 = 0 + (-ab) \qquad \text{additive inverse axiom}$$
$$a(-b) = -ab \qquad \text{additive identity axiom}$$

A consequence of Theorem 1.5 is that the product of a positive real number and a negative real number is a negative real number.

THEOREM 1.6. If a and b are real numbers, then $(-a)(-b) = ab$.

PROOF

$b + (-b) = 0$	additive inverse axiom
$(-a)[b + (-b)] = (-a) \cdot 0$	multiplication property of equality
$(-a)[b + (-b)] = 0$	Theorem 1.4
$(-a)b + (-a)(-b) = 0$	distributive property
$ab + [(-a)b + (-a)(-b)] = ab + 0$	addition property of equality
$ab + [(-a)b + (-a)(-b)] = ab$	additive identity axiom
$[ab + (-a)b] + (-a)(-b) = ab$	associative property of addition
$[ba + b(-a)] + (-a)(-b) = ab$	commutative property of multiplication
$b[a + (-a)] + (-a)(-b) = ab$	distributive property
$b \cdot 0 + (-a)(-b) = ab$	additive inverse axiom
$0 + (-a)(-b) = ab$	Theorem 1.4
$(-a)(-b) = ab$	additive identity axiom

A consequence of Theorem 1.6 is that the product of two negative numbers is a positive number. For example:

$$(-2)(-3) = 2 \cdot 3 = 6$$
$$(-5)(-1) = 5 \cdot 1 = 5$$

THEOREM 1.7. For real numbers a and b, if $ab = 0$, then $a = 0$ or $b = 0$.

PROOF. If $a = 0$, then the conclusion of the theorem is true. Suppose $a \neq 0$. Then, by the multiplicative inverse axiom, a has a multiplicative inverse, $\frac{1}{a}$, which is not equal to zero. Then

$$\frac{1}{a}(ab) = \frac{1}{a} \cdot 0$$

by the multiplication property of equality. But, by the associative property of multiplication,

$$\frac{1}{a}(ab) = \left(\frac{1}{a} \cdot a\right)b$$

and, by Theorem 1.4, $\frac{1}{a} \cdot 0 = 0$; hence we have

$$\left(\frac{1}{a} \cdot a\right) \cdot b = 0$$

But $\frac{1}{a} \cdot a = 1$ by the multiplicative inverse axiom; hence

$$1 \cdot b = 0$$

and

$$b = 0$$

by the multiplicative identity axiom. Thus we see that if $a \neq 0$, then $b = 0$ and the theorem is proved.

THEOREM 1.8. **For real numbers $a, b, c,$ and $d, b \neq 0$ and $d \neq 0, \frac{a}{b} = \frac{c}{d}$ if and only if $ad = bc$.**

PROOF. This theorem is really two theorems stated at once. The first part says if $ad = bc$, then $\frac{a}{b} = \frac{c}{d}$. The second part says if $\frac{a}{b} = \frac{c}{d}$, then $ad = bc$. Notice that the second part of the theorem is the converse of the first part. We first prove: If $ad = bc$, then $\frac{a}{b} = \frac{c}{d}$.

Since b and d are not 0, their multiplicative inverses, $\frac{1}{b}$ and $\frac{1}{d}$, exist. Now

$ad = bc$	hypothesis
$\left(\frac{1}{b} \cdot \frac{1}{d}\right)(ad) = \left(\frac{1}{b} \cdot \frac{1}{d}\right)(bc)$	multiplication property of equality
$\left(\frac{1}{b} \cdot \frac{1}{d}\right)(da) = \left(\frac{1}{d} \cdot \frac{1}{b}\right)(bc)$	commutative property of multiplication
$\frac{1}{b}\left(\frac{1}{d} \cdot d\right)a = \frac{1}{d}\left(\frac{1}{b} \cdot b\right)c$	associative property of multiplication
$\frac{1}{b} \cdot 1 \cdot a = \frac{1}{d} \cdot 1 \cdot c$	multiplicative inverse axiom
$\frac{1}{b} \cdot a = \frac{1}{d} \cdot c$	multiplicative identity axiom
$a \cdot \frac{1}{b} = c \cdot \frac{1}{d}$	commutative property of multiplication
$\frac{a}{b} = \frac{c}{d}$	definition of division

We now prove: If $\dfrac{a}{b} = \dfrac{c}{d}$, then $ad = bc$.

$\dfrac{a}{b} = \dfrac{c}{d}$	hypothesis
$(bd)\left(\dfrac{a}{b}\right) = (bd)\left(\dfrac{c}{d}\right)$	multiplication property of equality
$(bd)\left(a \cdot \dfrac{1}{b}\right) = (bd)\left(c \cdot \dfrac{1}{d}\right)$	definition of division
$(db)\left(\dfrac{1}{b} \cdot a\right) = (bd)\left(\dfrac{1}{d} \cdot c\right)$	commutative property of multiplication
$d\left(b \cdot \dfrac{1}{b}\right)a = b\left(d \cdot \dfrac{1}{d}\right)c$	associative property of multiplication
$d \cdot 1 \cdot a = b \cdot 1 \cdot c$	multiplicative inverse axiom
$da = bc$	multiplicative identity axiom
$ad = bc$	commutative property of multiplication

THEOREM 1.9. For all real numbers a, b, and c, if $a = b$ and $c = d$, then
$a + c = b + d$.

PROOF

$a = b$	hypothesis
$a + c = b + c$	addition property of equality
$a + c = b + d$	substitution property of equality

Theorem 1.9 is important in that it allows us to add two equalities (equations) member by member. Thus, if $x + y = 9$ and $x - y = 12$, then

$$(x + y) + (x - y) = 9 + 12$$
$$2x = 21$$

THEOREM 1.10. If a and b are real numbers, then $-(a + b) = (-a) + (-b)$.

PROOF

$$a + (-a) = 0$$
$$b + (-b) = 0$$
$$[a + (-a)] + [b + (-b)] = 0 + 0$$
$$[a + (-a)] + [b + (-b)] = 0$$
$$\{[a + (-a)] + b\} + (-b) = 0$$
$$\{a + [(-a) + b]\} + (-b) = 0$$
$$\{a + [b + (-a)]\} + (-b) = 0$$

$[(a + b) + (-a)] + (-b) = 0$

$(a + b) + [(-a) + (-b)] = 0$

$$0 = (a + b) + [-(a + b)]$$

$(a + b) + [(-a) + (-b)] = (a + b) + [-(a + b)]$

$$(-a) + (-b) = -(a + b)$$

The reasons in the proof of Theorem 1.10 have been omitted. The reader should supply a reason for each step in the proof.

EXERCISES 1.5

Give a reason for each step in the proofs given below. (Exercises 1-11)

1. Prove: If a is a real number, then $(-1)a = -a$.
   Proof: (a) $(-1)a = a(-1)$
         (b) $\qquad = -(a \cdot 1)$
         (c) $\qquad = -a$

2. Prove: For real numbers a, b, and c if $c + a = c + b$, then $a = b$.

   Proof: (a) $\qquad c + a = c + b$
         (b) $(-c) + (c + a) = (-c) + (c + b)$
         (c) $[(-c) + c] + a = [(-c) + c] + b$
         (d) $\qquad 0 + a = 0 + b$
         (e) $\qquad a = b$

3. Prove: For real numbers a, b, and c if $ba = ca$, $a \neq 0$, then $b = c$.
   Proof: (a) $ba = ca$

         (b) $\quad a \neq 0$

         (c) $\dfrac{1}{a}$ is a real number not equal to zero

         (d) $\quad (ba)\dfrac{1}{a} = (ca)\dfrac{1}{a}$

         (e) $b\left(a \cdot \dfrac{1}{a}\right) = c\left(a \cdot \dfrac{1}{a}\right)$

         (f) $\qquad b \cdot 1 = c \cdot 1$

         (g) $\qquad b = c$

4. Prove: For real numbers a and b if $a + b = 0$, then $b = -a$ and $a = -b$.
   Proof: Part I: If $a + b = 0$, then $b = -a$.
         (a) $\qquad a + b = 0$
         (b) $(-a) + (a + b) = (-a) + 0$

(c) $[(-a) + a] + b = (-a) + 0$

(d) $\qquad 0 + b = (-a) + 0$

(e) $\qquad b = -a$

Part II: If $a + b = 0$, then $a = -b$.

(f) $\qquad a + b = 0$

(g) $(a + b) + (-b) = 0 + (-b)$

(h) $a + [b + (-b)] = 0 + (-b)$

(i) $a + [b + (-b)] = -b$

(j) $\qquad a + 0 = -b$

(k) $\qquad a = -b$

5. Prove: If a is a real number and $a \neq 0$, then $\dfrac{1}{\frac{1}{a}} = a$.

Proof: (a) $\dfrac{1}{\frac{1}{a}} \cdot \dfrac{1}{a} = 1$

(b) $\qquad 1 = a \cdot \dfrac{1}{a}$

(c) $\dfrac{1}{\frac{1}{a}} \cdot \dfrac{1}{a} = a \cdot \dfrac{1}{a}$

(d) $\dfrac{1}{a} \cdot \dfrac{1}{\frac{1}{a}} = \dfrac{1}{a} \cdot a$

(e) $\qquad \dfrac{1}{\frac{1}{a}} = a$

6. Prove: (Right Distributive Property). For all real numbers a, b, and c,

$$(b + c)a = ba + ca.$$

Proof: (a) $(b + c)a = a(b + c)$

(b) $\qquad = ab + ac$

(c) $\qquad = ba + ca$

7. Prove: If a, b, and c are real numbers, then

$$(a + b) + c = (c + b) + a$$

Proof: (a) $(a + b) + c = a + (b + c)$

(b) $\qquad = a + (c + b)$

(c) $\qquad = (c + b) + a$

8. Prove: For real numbers a, b, and c if $a + b = c$, then $b = (-a) + c$.

    Proof: (a) $\qquad\qquad a + b = c$

          (b) $(-a) + (a + b) = (-a) + c$

          (c) $[(-a) + a] + b = (-a) + c$

          (d) $\qquad\quad 0 + b = (-a) + c$

          (e) $\qquad\qquad b = (-a) + c$

9. Prove: For real numbers a, b, and c if $ab = c$, $b \neq 0$, then $a = c\left(\dfrac{1}{b}\right)$.

    Proof: (a) $\qquad ab = c$

          (b) $(ab)\left(\dfrac{1}{b}\right) = c\left(\dfrac{1}{b}\right)$

          (c) $a\left[b\left(\dfrac{1}{b}\right)\right] = c\left(\dfrac{1}{b}\right)$

          (d) $\qquad a \cdot 1 = c\left(\dfrac{1}{b}\right)$

          (e) $\qquad\quad a = c\left(\dfrac{1}{b}\right)$

10. Prove: For all real numbers a, b, and c, $a(b - c) = ab - ac$.

    Proof: (a) $a(b - c) = a[b + (-c)]$

          (b) $\qquad\quad = ab + a(-c)$

          (c) $\qquad\quad = ab + (-ac)$

          (d) $\qquad\quad = ab - ac$

11. Prove: For all real numbers a and b, if $a \neq 0$ and $b \neq 0$, then $(ab)\left(\dfrac{1}{a} \cdot \dfrac{1}{b}\right) = 1$.

    Proof: (a) $(ab)\left(\dfrac{1}{a} \cdot \dfrac{1}{b}\right) = (ba)\left(\dfrac{1}{a} \cdot \dfrac{1}{b}\right)$

          (b) $\qquad\qquad = \left[(ba)\dfrac{1}{a}\right] \cdot \dfrac{1}{b}$

          (c) $\qquad\qquad = \left[b \cdot \left(a \cdot \dfrac{1}{a}\right)\right]\dfrac{1}{b}$

          (d) $\qquad\qquad = (b \cdot 1) \cdot \dfrac{1}{b}$

          (e) $\qquad\qquad = b \cdot \dfrac{1}{b}$

          (f) $\qquad\qquad = 1$

12. Prove: For all real numbers a, $a \neq 0$, if $ax = a$, then $x = 1$.

13. Prove: For all real numbers a and b if $a = b$, then $-a = -b$.

14. Prove: If a, b, and c are real numbers, $(a + b) - c = a + (b - c)$.

15. Prove: If a, b, and c are real numbers, $(a - b) + c = a - (b - c)$.

16. Prove: For all real numbers a and b if $ab = 1$, $b \neq 0$, then $a = \dfrac{1}{b}$.

17. Prove: The multiplicative inverse of a real number not equal to zero is unique. (*Hint.* Assume that there are two and prove that they are equal.)

1.6 ◆ Geometric Representation of the Real Numbers

The real numbers can be represented by points on a line as shown in Figure 1.2 in the following way. A line is drawn with a sequence of equally spaced points marked on it. We choose a point and label it 0. The number zero is associated with the point labeled 0. With the consecutive equally spaced points to the right of the point labeled 0, the positive integers $1, 2, 3, 4, \ldots$ are associated in order. With the consecutive equally spaced points to the left of the point labeled 0 are associated the negative integers, -1, -2, $-3, \ldots$ in order. The points between those points associated with the integers represent real numbers that are not integers.

It can be proved, although we shall not do it here, that there is a **one-to-one correspondence** between the points on a line and the real numbers. This means that to each point there corresponds exactly one real number and to each real number there corresponds exactly one point on the line. We call the points corresponding to real numbers the **graphs** of the numbers. We call the real numbers corresponding to the points on a line the **coordinates** of the points. The line on which the real numbers are graphed is called the **real number line.**

Figure 1.2

1.7 ◆ Order Properties of the Real Numbers.

We now present the axioms dealing with the order relation "is less than." The relation "is less than" is denoted by the symbol $<$. Just as we did not attempt to formulate a definition of the equality relation but, instead, stated the axioms of equality that completely determine the properties of this relation, we do not give a formal definition of the relation "is less than." Rather, we state the order axioms below, which completely determine the properties of this order relation.

ORDER AXIOMS

O-1. (*Transitive Property*)   For all real numbers a, b, and c, if $a < b$ and $b < c$, then $a < c$.

O-2. (*Trichotomy Property*)   For all real numbers a and b, one and only one of the following holds: $a = b$, $a < b$, or $b < a$.

O-3. (*Addition Property*)   For all real numbers a, b, and c, if $a < b$, then $a + c < b + c$ and $c + a < c + b$.

O-4. (*Multiplication Property for Positive Numbers*)   For all real numbers a, b, and c, if $a < b$ and $0 < c$, then $ac < bc$ and $ca < cb$.

O-5. (*Multiplication Property for Negative Numbers*)   For all real numbers a, b, and c, if $a < b$ and $c < 0$, then $bc < ac$ and $cb < ca$.

O-6. (*Substitution Property*)   For all real numbers a, b, a', and b', if $a = a'$, $b = b'$, and $a < b$, then $a' < b'$.

If we accept the relation "is less than" as undefined, we can define the relation "is greater than" as follows.

DEFINITION 1.16. If a and b are real numbers, then a is greater than b, denoted by $a > b$, if and only if $b < a$.

Any field that satisfies the order axioms is called an **ordered field.**

We now define what we mean when we say a real number is "positive" and a real number is "negative."

DEFINITION 1.17. A real number a is **positive** if and only if $a > 0$. A real number a is **negative** if and only if $a < 0$. The real number 0 is neither positive nor negative.

We now prove some familiar theorems for real numbers using the order axioms.

THEOREM 1.11. For real numbers a, and b, if $a < b$, then $b - a$ is positive.

PROOF. Since $a < b$, we use the addition property of order, adding $-a$, giving

$$a + (-a) < b + (-a)$$
$$0 < b + (-a) \qquad \text{additive inverse axiom}$$
$$0 < b - a \qquad \text{definition of subtraction}$$
$$b - a \text{ is positive} \qquad \text{definition of positive}$$

THEOREM 1.12. For real numbers a and b if $b - a$ is positive, then $a < b$.

PROOF. Since $b - a$ is positive,

$$b - a > 0 \qquad \text{definition of positive}$$
$$(b - a) + a > 0 + a \qquad \text{addition property of order}$$
$$(b - a) + a > a \qquad \text{additive identity axiom}$$
$$[b + (-a)] + a > a \qquad \text{definition of subtraction}$$
$$b + [(-a) + a] > a \qquad \text{associative property of addition}$$
$$b + 0 > a \qquad \text{additive inverse axiom}$$
$$b > a \qquad \text{additive identity axiom}$$
$$a < b \qquad \text{definition of greater than}$$

THEOREM 1.13. The sum of two positive numbers is positive.

PROOF. Let a and b be two positive numbers. Then

$$0 < b \qquad \text{definition of positive}$$
$$a + 0 < a + b \qquad \text{addition property of order}$$
$$a < a + b \qquad \text{additive identity axiom}$$
$$0 < a \qquad \text{definition of positive}$$
$$0 < a + b \qquad \text{transitive property of order } (0 < a \text{ and}$$
$$a < a + b, \text{ hence } 0 < a + b)$$
$$a + b \text{ is positive} \qquad \text{definition of positive}$$

THEOREM 1.14. The product of two positive numbers is positive.

PROOF. Let a and b be two positive numbers. Then

$$0 < a \qquad \text{definition of positive}$$
$$(0)(b) < ab \qquad \text{multiplication property of positive numbers}$$
$$0 < ab \qquad \text{Theorem 1.4}$$
$$ab \text{ is positive} \qquad \text{definition of positive}$$

THEOREM 1.15. If $a \neq 0$, then either a is positive or the opposite of a is positive.

PROOF. By the trichotomy property, there are two cases:

Case 1. $a > 0$. In this case a is positive and the theorem is true.
Case 2. $a < 0$. Since $-1 < 0$, we have

$$a(-1) > 0(-1) \qquad \text{multiplication property of negative numbers}$$
$$a(-1) > 0 \qquad \text{Theorem 1.4}$$
$$-a > 0 \qquad \text{Exercise 1.5, Problem 1}$$
$$-a \text{ is positive} \qquad \text{definition of positive}$$

EXERCISES 1.6

1. Justify each step in the proof below.
   Prove: For real numbers a, b, c, and d, if $a > b$ and $c > d$, then $a + c > b + d$.

   Proof: (a) Since $a > b$, $a - b$ is positive.
        (b) Since $c > d$, then $c - d$ is positive.
        (c) $(a - b) + (c - d)$ is positive.
        (d) $(a - b) + (c - d) = [a + (-b)] + [c + (-d)]$
        (e) $\qquad\qquad\qquad = [(-b) + a] + [c + (-d)]$
        (f) $\qquad\qquad\qquad = \{[(-b) + a] + c\} + (-d)$
        (g) $\qquad\qquad\qquad = [(-b) + (a + c)] + (-d)$
        (h) $\qquad\qquad\qquad = [(a + c) + (-b)] + (-d)$
        (i) $\qquad\qquad\qquad = (a + c) + [(-b) + (-d)]$
        (j) $\qquad\qquad\qquad = (a + c) + [(-1)(b) + (-1)(d)]$
        (k) $\qquad\qquad\qquad = (a + c) + [(-1)(b + d)]$
        (l) $\qquad\qquad\qquad = (a + b) + [-(b + d)]$
        (m) $\qquad\qquad\qquad = (a + c) - (b + d)$
        (n) Therefore $a + c > b + d$.

Prove the following theorems. The variables in each case represent real numbers.

2. If $a < b$ and $c < d$, then $a + c < b + d$.
3. The product of a positive number and a negative number is a negative number.
4. The product of two negative numbers is a positive number.
5. If $a < b$, then $-a > -b$.
6. If $1 < a$, then $a < a^2$.
7. If $a + c < b + c$, then $a < b$.
8. If $ac < bc$, $c > 0$, then $a < b$.
9. If $ac > bc$, $c < 0$, then $a < b$.
10. If $0 < a < 1$, then $0 < a^2 < a$.

11. If $a < b$, then $a < \dfrac{a + b}{2} < b$.

1.8 ✦ Absolute Value

With every real number, a, there is associated a nonnegative number called the **absolute value** of a, denoted by $|a|$. We define the absolute value of a real number as follows.

DEFINITION 1.18. If a is a real number, the **absolute value** of a, denoted by $|a|$, is a if a is nonnegative and the opposite of a if a is negative.

That is,

$$|a| = \begin{cases} a \text{ if } a \text{ is nonnegative} \\ -a \text{ if } a \text{ is negative} \end{cases}$$

We see from Definition 1.18 that the absolute value of a real number is always positive or zero, since $|0| = 0$; if $a > 0$, $|a| = a$ which is positive; if $a < 0$, $|a| = -a$ which is positive, since the opposite of a negative number is positive. For example,

$$|2| = 2 \qquad\qquad |-7| = -(-7) = 7$$
$$|-8| = -(-8) = 8 \qquad\qquad |12| = 12$$

The geometric interpretation of $|a|$ is the (undirected) distance between the point with coordinate a on the real number line and the point with coordinate 0, (Figure 1.3).

Figure 1.3

EXAMPLE 1. Find the value of $|-8| + |7|$.

SOLUTION $\qquad\qquad\qquad |-8| + |7| = 8 + 7 = 15$

EXAMPLE 2. Find the value of $\dfrac{|-36|}{|3|}$.

SOLUTION $\qquad\qquad \dfrac{|-36|}{|3|} = \dfrac{-(-36)}{3} = \dfrac{36}{3} = 12$

EXAMPLE 3. Find the value of $|2| + |-3| + |0|$.

SOLUTION $\qquad |2| + |-3| + |0| = 2 + [-(-3)] + 0$
$$= 2 + 3 + 0 = 5$$

EXERCISES 1.7

1. Give the absolute value of each of the following.
    (a) -6 (d) 7
    (b) 5 (e) $8 - 2$
    (c) -3 (f) $-12 - 5$

2. Find the values of x such that $|x| = 5$.
3. Find the values of x such that $|x| = 12$.
4. Find the values of x such that $|x + 5| = 7$.
5. By considering the various possibilities for negative and positive values of a and b, prove that $|ab| = |a| \cdot |b|$.
6. By considering the various possibilities for negative and positive values of a and b, prove that $|a + b| \le |a| + |b|$.
7. Find the value of each of the following.

   (a) $|8| + |-3|$ (e) $|-9| \cdot |3|$
   (b) $|-4| + |-5|$ (f) $|-4| \cdot |-5|$
   (c) $|-6| - |8|$ (g) $|-8| \cdot |8|$
   (d) $|3| \cdot |-4|$ (h) $|-9| \cdot |-8|$

8. Find the value of each of the following.

   (a) $\dfrac{|-8|}{|4|}$

   (b) $\dfrac{|-24|}{|6|}$

   (c) $\dfrac{|360|}{|-12|}$

   (d) $\dfrac{\left|\dfrac{1}{\sqrt{2}}\right|}{\left|-\dfrac{1}{\sqrt{2}}\right|}$

   (e) $\dfrac{\left|\dfrac{a}{b}\right|}{\left|-\dfrac{b}{a}\right|}$, $a \ne 0, b \ne 0$

9. Show that for all real numbers a and b, $|a - b| = |b - a|$.
10. What conclusion can you draw if you know that $|x| > x$?
11. What conclusion can you draw if you know that $|x + y| = 0$?
12. It is known that $|x + y| = x$ but $y \ne 0$. What conclusion can you draw?
13. It is known that $|x + y| = |x - y|$. What conclusion can you draw?

CHAPTER REVIEW

1. Name the property of the real numbers and their operations illustrated by each of the following statements.

   (a) $\frac{1}{2} \cdot 2 = 1$.
   (b) $2 + (6 + 7) = (2 + 6) + 7$.

(c) $7 \cdot \frac{1}{4} = \frac{1}{4} \cdot 7$.

(d) $3(1.8 + 6.7) = (3)(1.8) + (3)(6.7)$.

(e) $-\sqrt{2} + 0 = -\sqrt{2}$.

(f) $\frac{1}{2} + (-\frac{1}{2}) = 0$.

(g) $4.68 \times 1 = 4.68$.

(h) $7 \cdot (3 \cdot 6) = (7 \cdot 3) \cdot 6$.

(i) $\frac{1}{4} + \frac{1}{2} = \frac{1}{2} + \frac{1}{4}$.

(j) $3(-7)$ is a real number.

2. Let $Q = \{x \mid x$ is a rational number$\}$, $S = \{x \mid x$ is an irrational number$\}$, and $R = \{x \mid x$ is a real number$\}$. Which of the following are true statements?

(a) $Q \cup S = R$ (f) $Q \cap R = \varnothing$

(b) $Q \cap S = R$ (g) $R \cup S = R$

(c) $Q \subseteq R$ (h) $R \cup Q = Q$

(d) $R \subseteq S$ (i) $S \cap R = \varnothing$

(e) $Q \cap R = Q$ (j) $S \subseteq R$

3. Give a reason for each step in the proof below.

Theorem: If $a \neq 0$ and $b \neq 0$ are real numbers and $ab = 1$, then $b = \frac{1}{a}$.

Proof: (a) $ab = 1$.

(b) $\frac{1}{a}$ exists.

(c) $\frac{1}{a}(ab) = \frac{1}{a} \cdot 1$.

(d) $\left(\frac{1}{a} \cdot a\right) \cdot b = \frac{1}{a} \cdot 1$.

(e) $1 \cdot b = \frac{1}{a} \cdot 1$.

(f) $b = \frac{1}{a}$.

4. Give a reason for each step in the proof below.

Theorem: If $a < 0$ and $b < 0$, then $a + b < 0$.

Proof: (a) $a < 0$.

(b) $a + b < 0 + b$.

(c) $a + b < b$.

(d) $b < 0$.

(e) $a + b < 0$.

5. Use set builder notation to denote the set of all real numbers greater than 7.8 and less than 12.7.

6. List the elements in the set $\{x \mid x$ is an integer and $-4 < x < 5\}$.

7. Name the property of equality illustrated by each of the following statements.
   (a) If $a = 7$, then $a + 3 = 10$.
   (b) If $x + y = z + t$, then $z + t = x + y$.
   (c) If $x + y = a$ and $a = 3 - k$, then $x + y = 3 - k$.
   (d) If $a + b = k$ and $b = 3$, then $a + 3 = k$.
   (e) If $x - 5 = 11$, then $x = 16$.

8. Which of the following sets are closed under the operation of addition?
   (a) $\{\ldots, -4, -2, 0, 2, 4, \ldots\}$.
   (b) $\{0, 1, 2, 3\}$.
   (c) $\{1, 3, 5, 7, \ldots\}$.
   (d) $\{x \mid x \text{ is a real number}\}$.

9. Find the value of each of the following:
   (a) $|8| + |-7|$.
   (b) $|3| \cdot |-5|$.
   (c) $\dfrac{|6| \cdot |-9|}{|3|}$.
   (d) $|-9 + 3| - |-6 - 8|$.
   (e) $\dfrac{|-6 - 12|}{|6 + 3|}$.

10. Is the set of integers and the operations of addition and multiplication a field? Why or why not?

In each of the following give a word, phrase, or symbol to make true statements (Exercises 11–20).

11. The natural numbers, their opposites, and _____ make up the set of _____ .

12. The sum of two positive real numbers is a(n) _____ real number.

13. The _____ or _____ of 6 is -6.

14. If the product of two real numbers is the multiplicative identity, then each number is called the _____ of the other.

15. If $a < 0$ then $|a|$ is equal to _____ .

16. If a and b are real numbers $a - b$ is defined to be _____ .

17. If a and b are real numbers $(-a)(-b) =$ _____ .

18. The opposite of $x - y$ is _____ .

19. $\left\{ \dfrac{a}{b} \;\middle|\; a \in I,\ b \in I, \text{ and } b \neq 0 \right\}$ is the set of _____ .

20. If a is a real number, then $a \cdot 0 =$ _____ .

CHAPTER 2

Algebraic Expressions

2.1 ◆ Positive Integer Exponents

Positive integer exponents are associated with a real number to indicate repeated multiplication of that number. We write $x \cdot x \cdot x = x^3$, where the positive integer 3 is called the **exponent** and indicates that the real number x is used as a factor three times. In general, for any positive integer, n, and any real number, x,

$$x^n = \underbrace{x \cdot x \cdot x \cdot \cdots \cdot x}_{n \text{ factors}}$$

The positive integer n is called the **exponent** of x and the real number x is called the **base.** The expression x^n is called a **power** and is read "x to the nth power" or "x to the nth." As illustrations we have

$$3^2 = 3 \cdot 3 = 9$$
$$(-2)^3 = (-2)(-2)(-2) = -8$$
$$(\tfrac{1}{4})^4 = \tfrac{1}{4} \cdot \tfrac{1}{4} \cdot \tfrac{1}{4} \cdot \tfrac{1}{4} = \tfrac{1}{16}$$
$$(2)^5 = 2 \cdot 2 \cdot 2 \cdot 2 \cdot 2 = 32$$

When a number is expressed as a positive integer power of some base, we say that it has been written in **exponential notation.** For example, the exponential notation of 8 is 2^3.

DEFINITION 2.1. For any real number x, n a positive integer greater than 1, $x^1 = x$, and $x^n = x \cdot x \cdot x \cdot \cdots \cdot x$, where there are n factors.

THEOREM 2.1. For all real numbers x, n a positive integer:

1. If $x > 0$, then $x^n > 0$.

2. If $x < 0$, then $x^n > 0$ if n is even.

3. If $x < 0$, then $x^n < 0$ if n is odd.

4. If $x = 0$, then $x^n = 0$.

We do not give a formal proof of Theorem 2.1, but rather establish its validity intuitively. Let us first consider part 1 of the theorem. Since the product of two positive real numbers is a positive real number, the associative property of multiplication assures us that the product of n positive real numbers is a positive real number. Hence

$$\underbrace{x \cdot x \cdot x \cdots x}_{n \text{ factors}} = x^n > 0$$

In part 2, x is negative and n is an even positive integer, hence x^n is the product of an even number of negative factors. The product of any two of these factors is positive since the product of two negative real numbers is a positive real number. We see, then, that x^n is the product of $\dfrac{n}{2}$ positive factors and hence is positive.

Now let us discuss part 3. Since n is negative and n is odd, x^n is the product of an odd number of negative factors. Since n is odd, $(n - 1)$ is even. We have in this case

$$\underbrace{x \cdot x \cdot x \cdot x \cdot x \cdots x}_{(n-1) \text{ factors}} > 0$$

Then

$$\underbrace{(x \cdot x \cdot x \cdot x \cdots)}_{(n-1) \text{ factors}} \cdot x < 0$$

Part 4 of the theorem is obvious, since we have the product of n factors, each of which is zero.

As a consequence of Theorem 2.1 we have

$$2^4 = 2 \cdot 2 \cdot 2 \cdot 2 = 16 > 0$$
$$(-3)^2 = (-3)(-3) = 9 > 0$$
$$(-4)^3 = (-4)(-4)(-4) = -64 < 0$$

The basic **laws of exponents** are stated in the following theorem.

THEOREM 2.2. For x and y real numbers and m and n positive integers:

e-1. $x^n \cdot x^m = x^{m+n}$.

e-2. $[x^m]^n = x^{mn}$.

e-3. $[xy]^n = x^n \cdot y^n$.

e-4. For $y \neq 0$, $\left(\dfrac{x}{y}\right)^n = \dfrac{x^n}{y^n}$.

e-5. For $x \neq 0$, if $m = n$, then $\dfrac{x^m}{x^n} = 1$,

if $m > n$, then $\dfrac{x^m}{x^n} = x^{m-n}$,

if $m < n$, then $\dfrac{x^m}{x^n} = \dfrac{1}{x^{n-m}}$.

Counting the factors in the expressions in these laws of exponents gives an easy argument to establish their validity. For example, in e-1 we write

$$x^m \cdot x^n = \underbrace{[x \cdot x \cdot \,\cdots\, \cdot x]}_{m \text{ factors}} \cdot \underbrace{[x \cdot x \cdot \,\cdots\, \cdot x]}_{n \text{ factors}}$$

Since the total number of factors in the right member is $m + n$, then

$$x^m \cdot x^n = x^{m+n}$$

Similarly, for e-2 we write

$$[x^m]^n = \underbrace{x^m \cdot x^m \cdot \,\cdots\, \cdot x^m}_{n \text{ factors}}$$

and count the number of times x appears as a factor in the right member. Since x^m has x occurring as a factor m times, and since there are n groups of such factors, the total number of factors is mn and we have

$$(x^m)^n = x^{mn}$$

Laws e-3 and e-4 are obtained in a similar fashion. The reader should provide his own proofs for these.

We now prove e-5.

PROOF

If $m = n$, then $x^m = x^n$ and

$$\frac{x^m}{x^n} = \frac{x^n}{x^n} = 1, \text{ provided } x \neq 0$$

If $m \neq n$, then

$$\frac{x^m}{x^n} = \frac{\overbrace{x \cdot x \cdot x \cdots x}^{m \text{ factors}}}{\underbrace{x \cdot x \cdot x \cdots x}_{n \text{ factors}}}$$

If $m > n$, then the numerator has more factors than the denominator and we can regroup the factors of the numerator to give

$$\frac{x^m}{x^n} = \frac{\overbrace{(x \cdot x \cdots x)}^{n \text{ factors}} \cdot \overbrace{(x \cdot x \cdots x)}^{m\text{-}n \text{ factors}}}{\underbrace{x \cdot x \cdot x \cdots x}_{n \text{ factors}}}$$

$$= \frac{\overbrace{x \cdot x \cdots x}^{n \text{ factors}}}{\underbrace{x \cdot x \cdots x}_{n \text{ factors}}} \cdot \overbrace{(x \cdot x \cdots x)}^{m\text{-}n \text{ factors}}$$

$$= x^{m-n}$$

On the other hand, if $m < n$, then the denominator has more factors than the numerator and we can regroup the factors of the denominator to give

$$\frac{x^m}{x^n} = \frac{\overbrace{x \cdot x \cdots x}^{m \text{ factors}}}{\underbrace{(x \cdot x \cdots x)}_{n\text{-}m \text{ factors}} \cdot \underbrace{(x \cdot x \cdots x)}_{m \text{ factors}}}$$

$$= \frac{1}{\underbrace{(x \cdot x \cdots x)}_{n\text{-}m \text{ factors}}} \cdot \frac{\overbrace{x \cdot x \cdots x}^{m \text{ factors}}}{\underbrace{(x \cdot x \cdots x)}_{m \text{ factors}}}$$

$$= \frac{1}{x^{n-m}}$$

These laws of exponents may be extended to rules involving three or more powers. For example:

$$x^m \cdot x^n \cdot x^p = x^{m+n+p} \qquad x^a \cdot x^b \cdot x^c \cdot x^d = x^{a+b+c+d}$$
$$(xyz)^n = x^n \cdot y^n \cdot z^n \qquad (xyzw)^m = x^m \cdot y^m \cdot z^m \cdot w^m$$

and so forth.

EXAMPLE 1. Find the product $(2x^3y^2)(3x^2y^4)$.

SOLUTION

$$(2x^3y^2)(3x^2y^4) = (2)(3)(x^3)(x^2)(y^2)(y^4) \qquad \text{associative and commutative}$$
properties of multiplication
$$= 6x^5y^6 \qquad \text{e-1}$$

EXAMPLE 2. Find the product $(3x^3y^2z^4)^3$.

SOLUTION

$$(3x^3y^2z^4)^3 = (3^3)(x^3)^3(y^2)^3(z^4)^3 \qquad \text{e-3}$$
$$= 27x^9y^6z^{12} \qquad \text{e-2 and e-1}$$

EXAMPLE 3. Find the product $\left(\dfrac{2x^2}{y^3}\right)^2 \cdot \left(\dfrac{y^3}{x^3}\right)^3$.

SOLUTION

$$\left(\frac{2x^2}{y^3}\right)^2 \cdot \left(\frac{y^3}{x^3}\right)^3 = \left(\frac{2^2x^4}{y^6}\right) \cdot \left(\frac{y^9}{x^9}\right) \qquad \text{e-4 and e-3}$$

$$= 2^2\left(\frac{x^4}{x^9}\right)\left(\frac{y^9}{y^6}\right) \qquad \text{Why?}$$

$$= 4 \cdot \frac{1}{x^5} \cdot y^3 \qquad \text{e-5}$$

$$= \frac{4y^3}{x^5}$$

EXERCISES 2.1

Perform the indicated operations.

1. 3^3
2. 2^4
3. $(\tfrac{1}{2})^3$
4. $(-\tfrac{1}{4})^4$
5. $a^4 \cdot a^3$
6. $(3^2)(3^3)$
7. $(10^2)(10^5)$
8. $(2^3)(2^5)(2^2)$
9. $(2^2)^3$
10. $(2a^3b)(3ab^2)$

11. $(3a^2b^3)(5ab^4)$

12. $(a^4b^2)^3(a^2b^4)^3$

13. $(3a^2b^3)^4(-2ab^5)^5$

14. $(6x^2y^3z)^3(\frac{1}{2}x^2y^4)^4$

15. $\dfrac{a^5}{a^3}$

16. $\dfrac{(3^4)(7^3)}{(3^9)(7^2)}$

17. $\dfrac{a^3b^2}{a^5b}$

18. $\dfrac{2x^{12}}{8x^3}$

19. $\dfrac{x^3y^4}{x^5y^6}$

20. $\dfrac{2^3a^4b^3c^7}{2^2a^9b^8c^4}$

21. $\dfrac{p^7q^3r^2s^5}{p^5q^9rs^5}$

22. $\dfrac{36x^3y^4z^7}{-12x^7y^2z^6}$

23. $\dfrac{m^{8k}n^{6k}}{3m^{11k}n^{6k}}$

24. $\dfrac{a^{6n+1}b^{5n-2}}{a^{6n-1}b^{5n+1}}$

25. $\dfrac{(3a^kb^{4k})^3}{(5a^{k+1}b^{2k+1})^2}$

26. $\left(\dfrac{28x^4y^3}{7x^2y^6}\right)^2 \cdot \left(\dfrac{81x^5y}{54x^3y^2}\right)^4$

27. $\dfrac{(2x^2)^3(3y^4)^2}{(5xy^2)^4(x^3y)^5}$

28. $[(2x^2y^3)^4 \cdot (3x^3y)^5]^2$

29. $\dfrac{[(3xy^2)^4 \cdot (2x^3y^4)^3]^2}{(4x^2y^3)^5}$

30. $\dfrac{[(5x^2y)^2 \cdot (2x^3y^2)^3]^5}{[(3x^3y)^4 \cdot (5x^2y^4)^3]^6}$

31. Prove: If x and y are real numbers and n is a positive integer, $(xy)^n = x^ny^n$.

32. Prove: If x and y are real numbers, $y \neq 0$, and n is a positive integer, then
$$\left(\frac{x}{y}\right)^n = \frac{x^n}{y^n}.$$

2.2 ◆ Negative Integer Exponents and Zero Exponents

It is possible to extend our work to include as an exponent the number 0 or a negative integer. This extension must be carried out so that we will have the same laws of exponents as we did for positive integer exponents. Thus, Law e-1 states that

$$x^n \cdot x^m = x^{n+m}$$

where n and m are positive integers. If this same law is to hold whenever m and n are 0, we would have

$$x^0 \cdot x^0 = x^{0+0} = x^0$$

But

$$x^0 \cdot x^0 = x^0$$

may be written [using the addition property of equality and adding $(-x^0)$ to each member]

$$x^0 \cdot x^0 - x^0 = 0$$

and, by the distributive property,

$$x^0 \cdot (x^0 - 1) = 0$$

We must now construct our definition for exponent 0 so that x^0 is either 0 or 1 (since by Theorem 1.7, $x^0 = 0$ or $x^0 - 1 = 0$).

However, we also want

$$x^m \cdot x^0 = x^{m+0} = x^m$$

where m is any positive integer and x is any real number. Therefore we define x^0 to be 1, when x is any real number.

DEFINITION 2.2. For any real number x.

$$x^0 = 1.$$

This definition is consistent with Law e-5, namely,

$$\frac{x^m}{x^m} = 1 = x^0 = x^{m-m}$$

We next construct a definition for negative integer exponents: x^{-1}, x^{-2}, x^{-3}, and so forth. We want to be consistent with Law e-5 for positive integer exponents, which says that

$$\frac{x^m}{x^n} = x^{m-n}$$

whenever $m > n$ and $x \neq 0$; and

$$\frac{x^m}{x^n} = \frac{1}{x^{n-m}}$$

whenever $m < n$ and $x \neq 0$. Since $n - m = -(m - n)$, we now state the following.

DEFINITION 2.3. If x is any real number except zero and p is a positive integer,

$$x^{-p} = \frac{1}{x^p}$$

Thus $x^{-1} = \frac{1}{x^1} = \frac{1}{x}$, $x^{-2} = \frac{1}{x^2}$, $x^{-3} = \frac{1}{x^3}$, and so forth.

Definitions 2.2 and 2.3 then give us

$$\frac{x^m}{x^n} = x^{m-n}$$

whenever $x \neq 0$ and m and n are any integers: positive, negative, or zero. In fact, all five of the laws of exponents given in Theorem 2.2 remain valid for all integer exponents.

EXAMPLE 1. Express $(x^{-3}y^{-2}z^0)^{-2}$ without negative exponents and simplify.

SOLUTION

$$
\begin{aligned}
(x^{-3}y^{-2}z^0)^{-2} &= [(x^{-3}y^{-2})(1)]^{-2} &&\text{Definition 2.2}\\
&= [x^{-3}y^{-2}]^{-2} &&\text{multiplicative}\\
& &&\quad\text{identity axiom}\\
&= x^{(-3)(-2)}y^{(-2)(-2)} &&\text{Laws e-2 and e-3}\\
&= x^6y^4 &&\text{Theorem 1.6}
\end{aligned}
$$

EXAMPLE 2. Express $\dfrac{x^{-1} + y^{-1}}{(xy)^{-1}}$ without negative exponents and simplify.

SOLUTION

$$
\begin{aligned}
\frac{x^{-1} + y^{-1}}{(xy)^{-1}} &= \frac{\dfrac{1}{x} + \dfrac{1}{y}}{\dfrac{1}{xy}} &&\text{Definition 2.3}\\[3em]
&= \frac{\dfrac{y + x}{xy}}{\dfrac{1}{xy}} &&\text{Why?}\\[3em]
&= \frac{y + x}{xy} \cdot \frac{xy}{1} &&\text{Why?}\\[1.5em]
&= y + x
\end{aligned}
$$

EXERCISES 2.2

Write without negative or zero exponents and perform the indicated operations.

1. $(a^{-5}b^7)(a^{-2}b^{-7}c^0)$
2. $(x^2y^3z^{-2})(x^{-1}y^{-3}z^0)$
3. $(2a^0b^{-1})(ab^2)^{-2}$
4. $(c^2d^3y^{-4})^0(2c^3d^{-2}y^3)$

5. $(x + y)^0$

6. $(-2x^{-2})(4x^2)(-\frac{1}{6}x^{-3})$

7. $(x^{-2}y^{-2})^{-2}$

8. $(-\frac{1}{2}x^{-3}y^2)^{-4}$

9. $\left(\dfrac{1}{4a^2b^3}\right)^{-3}$

10. $(x^2y^{-3}z^{-5})(x^{-2}y^3z^8)$

11. $\left(\dfrac{x^{-5}y^4}{x^2y^{-2}}\right)^2 \cdot \left(\dfrac{x^4y^{-5}}{x^3y^{-7}}\right)^{-3}$

12. $\left(\dfrac{a^{-1}b^{-2}}{c^3}\right)^2 \cdot \left(\dfrac{a^{-4}b^2}{c^{-3}}\right)^{-2}$

13. $\dfrac{a^{-3} - b^{-3}}{a^{-1} - b^{-1}}$

14. $\dfrac{4x^3 - x^{-1}}{2 - x^{-2}}$

15. $\dfrac{9 - x^{-4}}{3x^{-1} - x^{-3}}$

16. $\dfrac{4x^{-2} - 4x^{-1} + 1}{2x^{-2} - x^{-1}}$

17. $\dfrac{x^{-2} - y^{-2}}{x^{-1} - y^{-1}}$

18. $\dfrac{a^{-2} - b^{-2}}{a^2 - b^2}$

19. $\dfrac{(x^{-1} - y^{-1})^{-1}}{xy}$

20. $\dfrac{(a^{-1} + b^{-1})^{-1}}{(a + b)^{-1}}$

21. $\dfrac{a^0 - b^0}{a^0 + b^0}$

22. $\dfrac{a^{2n-3}}{a^{3n+1}} \cdot \dfrac{a^{n+5}}{a^{n-3}}$

23. $\dfrac{(3x^{n+1})^2}{x^{2(n+1)}} \cdot \dfrac{x^{-n}}{(x^{-n})^3}$

24. $\dfrac{a^{-3} - b^{-3}}{a^{-3} + b^{-3}}$

2.3 ◆ Radicals

If $a \geq 0$ is a real number and n is a positive integer, then there is a unique real number $b \geq 0$ such that $b^n = a$. This number, b, is called the **principal nth root** of a and is denoted by $\sqrt[n]{a}$. It can also be shown that if $a < 0$ and n is an odd positive integer, then there is a unique real number $b < 0$ such that $b^n = a$. In this case we again write $b = \sqrt[n]{a}$ and also call b the principal nth root of a. For example, $\sqrt[4]{16} = 2$, $\sqrt[3]{27} = 3$, $\sqrt[3]{-1} = -1$. We summarize these remarks by the following definition.

DEFINITION 2.4. For a and b nonnegative real numbers, n a positive integer, or for a and b negative real numbers, n a positive odd integer, if $b^n = a$, then a is the **principal nth root** of b, denoted by $\sqrt[n]{a}$.

In the case $n = 2$, we write $\sqrt{a}$ rather than $\sqrt[2]{a}$ and call $\sqrt{a}$ the **principal square root** of a. The number $\sqrt[3]{a}$ is called the **principal cube root** of a.

Note that when n is even, $\sqrt[n]{a}$ is defined only when a is nonnegative. The reason for this is that for all real numbers, b, when n is even, b^n is nonnegative. We give a method for taking even roots of negative numbers in Chapter 12 on complex numbers.

When $\sqrt[n]{a}$ exists it is a unique real number. However, for n a positive integer, it is possible that $c^n = a$ even though c is *not* the principal nth root of a. In this case, c is called an nth root of a. For example, -2 and 2 are both square roots of 4, since $(-2)^2 = 4$ and $2^2 = 4$, but -2 is not the principal square root of 4. We summarize by saying that the principal square root of a nonnegative real number is that *nonnegative* real number whose square is the given real number.

In the symbol $\sqrt[n]{a}$, called a **radical**, the number a is called the **radicand**, n is called the **index** of the radical, and the symbol $\sqrt{}$ is called a **radical sign**.

THEOREM 2.3. If n is a positive integer and x and y are real numbers such that $\sqrt[n]{x}$ and $\sqrt[n]{y}$ exist, then

    r-1. $(\sqrt[n]{x})^n = x.$

    r-2. $\sqrt[n]{x}\,\sqrt[n]{y} = \sqrt[n]{xy}.$

    r-3. **For** $y \neq 0,$ $\dfrac{\sqrt[n]{x}}{\sqrt[n]{y}} = \sqrt[n]{\dfrac{x}{y}}.$

    r-4. **For** $x \geq 0$ **and** n **any positive integer,** $\sqrt[n]{x^n} = x$; **for** $x < 0$ **and** n **an odd positive integer,** $\sqrt[n]{x^n} = x.$

The formulas in Theorem 2.3 are called the **laws of radicals.** We shall prove Law r-2. The proofs of the other laws of radicals are left to the reader.

PROOF of r-2. Let $a = \sqrt[n]{x}$ and $b = \sqrt[n]{y}.$

    Then $a^n = x$ and $b^n = y$ Definition 2.4

                  $a^n \cdot b^n = xy$ Why?

                  $(ab)^n = xy$ Law e-3

                    $ab = \sqrt[n]{xy}$ Definition 2.4

          $\sqrt[n]{x} \cdot \sqrt[n]{y} = \sqrt[n]{xy}$ substitution property of equality

We now prove an important theorem about square roots.

THEOREM 2.4. For any real number, $x,$

$$\sqrt{x^2} = |x|$$

PROOF. For $x = 0,$ we have $\sqrt{0^2} = \sqrt{0} = 0 = |0|.$

    For $x > 0,$ $\sqrt{x^2}$ is the positive number whose square is $x^2.$

    But of the two square roots of $x^2,$ $-x$ and $x,$ x is the principal square root, so that $\sqrt{x^2} = x.$ But for $x > 0$ we also have $|x| = x,$ so that $\sqrt{x^2} = |x|.$

On the other hand, when $x < 0$, $\sqrt{x^2}$ is still the positive number whose square is x. Of the two square roots of x^2, $-x$ and x, $-x$ is the one which is positive so that $\sqrt{x^2} = -x$. By the definition of absolute value, when $x < 0$, $|x| = -x$ so that we again have $\sqrt{x^2} = |x|$. For example, $\sqrt{(-3)^2} = |-3| = 3$.

We use the laws of radicals to **simplify** expressions involving radicals. The examples below illustrate some of the techniques used to simplify radical expressions.

EXAMPLE 1. Write the radical $\sqrt[5]{32x^6y^{12}}$ with all possible factors removed from the radical sign.

SOLUTION

$$
\begin{aligned}
\sqrt[5]{32x^6y^{12}} &= \sqrt[5]{(2^5x^5y^{10})(xy^2)} \\
&= \sqrt[5]{(2xy^2)^5(xy^2)} \\
&= \sqrt[5]{(2xy^2)^5}\,\sqrt[5]{xy^2} \\
&= 2xy^2\,\sqrt[5]{xy^2}
\end{aligned}
$$

EXAMPLE 2. Write $\sqrt{\dfrac{2x^3}{5yz^5}}$ as an expression with no radicals in the denominator.

SOLUTION

$$
\begin{aligned}
\sqrt{\frac{2x^3}{5yz^5}} &= \sqrt{\frac{2x^3}{5yz^5} \cdot \frac{5yz}{5yz}} \\
&= \sqrt{\frac{(2)(5)x^2 \cdot xyz}{5^2y^2z^6}} \\
&= \sqrt{\left(\frac{x}{5yz^3}\right)^2(10xyz)} \\
&= \left|\frac{x}{5yz^3}\right|\sqrt{10xyz}
\end{aligned}
$$

The technique used in Example 2 is called **rationalizing the denominator.** When this technique is used, no radicals appear in the denominator of the simplified expression. To employ this technique we multiply the numerator and denominator of the radicand by a factor which produces a denominator which is the nth power of some number, where n is the index of the radical. Thus in Example 2 we multiplied numerator and denominator by $5yz$ to give a new denominator of the form $(5yz^3)^2$ under the radical sign.

A radical of index n is said to be in **simplest form** if no perfect nth power or any negative integer exponents appears in the radicand and if no radical appears in the denominator.

EXAMPLE 3. Simplify $\dfrac{1}{\sqrt{x} + \sqrt{y}}$.

SOLUTION. (The reader should supply the reasons for each step.)

$$\frac{1}{\sqrt{x} + \sqrt{y}} = \frac{1}{\sqrt{x} + \sqrt{y}} \cdot \frac{\sqrt{x} - \sqrt{y}}{\sqrt{x} - \sqrt{y}}$$

$$= \frac{\sqrt{x} - \sqrt{y}}{(\sqrt{x} + \sqrt{y})(\sqrt{x} - \sqrt{y})}$$

$$= \frac{\sqrt{x} - \sqrt{y}}{x - y}$$

EXAMPLE 4. Simplify $\sqrt{12} + \sqrt{27} + \sqrt{2x^3} - 8\sqrt{8x^3}$.

SOLUTION

$$\sqrt{12} + \sqrt{27} + \sqrt{2x^3} - 8\sqrt{8x^3}$$
$$= \sqrt{2^2 \cdot 3} + \sqrt{3^2 \cdot 3} + \sqrt{x^2 \cdot 2x} - 8\sqrt{2^2 x^2 \cdot 2 \cdot x}$$
$$= 2\sqrt{3} + 3\sqrt{3} + |x|\sqrt{2x} - 16|x|\sqrt{2x}$$
$$= (2 + 3)\sqrt{3} + (1 - 16)|x|\sqrt{2x}$$
$$= 5\sqrt{3} - 15|x|\sqrt{2x}$$

EXERCISES 2.3

Simplify.

1. $\sqrt{50}$

2. $\sqrt{162}$

3. $\sqrt[3]{-375}$

4. $\sqrt[6]{x^{19}y^8}$

5. $\sqrt[5]{100{,}000}$

6. $\sqrt{\tfrac{1}{2}}$

7. $\sqrt{50p^5r^9}$

8. $\dfrac{\sqrt[3]{16}}{\sqrt[3]{2}}$

9. $\dfrac{\sqrt[3]{243}}{\sqrt[3]{3}}$

10. $\sqrt{\dfrac{2a^5}{3b^4}}$

11. $\sqrt[3]{\dfrac{-27a^7}{b^9}}$

12. $\sqrt[3]{x^{-6}y^{12}}$

13. $\sqrt{\dfrac{x^2y^6}{x^6y^4}}$

14. $\sqrt{\dfrac{27a^8b^3}{c^4d^4}}$

15. $(\sqrt[3]{2x^3y^4})^4$

16. $\sqrt{6x^3y^2}\sqrt{2x^2y^3}$

17. $\sqrt[4]{8x^5y^3}\sqrt[4]{4x^3y^4}$

18. $\dfrac{\sqrt[3]{-250x^{-3}y^2}}{\sqrt[3]{x^2y^{-3}}}$

19. $\sqrt{(a-b)^2}$

20. $\sqrt[3]{(a-b)^3}$

21. $\dfrac{\sqrt{x}-\sqrt{y}}{\sqrt{x}+\sqrt{y}}$

22. $\dfrac{1}{\sqrt{x+y}}$

23. $\dfrac{1}{\sqrt[4]{x^3}}$

24. $2\sqrt{20}+\sqrt{125}-\sqrt{45}$

25. $\sqrt[3]{-32}-2\sqrt[3]{108}$

26. $\sqrt{4x^3}-\sqrt{16x^5}+\sqrt{x^9}$

27. $\dfrac{1}{1+\sqrt{x}}$

28. $\dfrac{x+3}{\sqrt{x+2}+\sqrt{x-3}}$

29. $\dfrac{x}{1+\sqrt{x}}+\dfrac{2}{\sqrt{1+x}}$

30. $\dfrac{1}{\sqrt{x+1}+2}+\dfrac{1}{x}$

31. $\dfrac{3}{4+\sqrt{x+1}}$

32. $\dfrac{2}{\sqrt{x+1}-2}$

33. $\dfrac{1}{\sqrt{x+y}+\sqrt{x-y}}$

34. $\dfrac{\sqrt{2a}-\sqrt{2b}}{\sqrt{2a}+\sqrt{2b}}$

35. $\dfrac{\sqrt{x+2}+\sqrt{x-2}}{\sqrt{x+3}-\sqrt{x-5}}$

36. Prove Theorem 2.3, Part r-1.
37. Prove Theorem 2.3, Part r-3.
38. Prove Theorem 2.3, Part r-4.

2.4 ◆ Rational Number Exponents

We now define rational number exponents. We recall that a rational number is one that can be expressed in the form $\dfrac{m}{n}$ where m and n are integers and $n \neq 0$. Again we wish to construct our definition in such a manner that the laws of exponents that we have established for integer exponents will also be valid for rational number exponents.

DEFINITION 2.5. If x is a real number and n is a positive integer, then

$$x^{1/n} = \sqrt[n]{x}$$

provided $\sqrt[n]{x}$ exists.

We can see how this definition is consistent with our previous work. In particular, for Law e-2,

$$(x^{1/n})^n = x^{(1/n)\cdot n} = x^1 = x$$

Thus, for $x^{1/n}$ to have a meaning consistent with the laws of exponents, it must be taken as the principal nth root of x.

We now need to consider $x^{p/q}$ where p and q are integers and $q > 0$. If

Law e-2 is to hold and since $\dfrac{p}{q} = p \cdot \dfrac{1}{q} = \dfrac{1}{q} \cdot p$, we must have

$$x^{p/q} = (x^p)^{1/q} = \sqrt[q]{x^p}$$
$$= (x^{1/q})^p = (\sqrt[q]{x})^p$$

We must also make sure that $\sqrt[q]{x}$ exists. Thus we have:

DEFINITION 2.6. If $\dfrac{p}{q}$ is a rational number, with p and q integers and $q > 0$, then

$$x^{p/q} = \sqrt[q]{x^p} = (\sqrt[q]{x})^p$$

provided $\sqrt[q]{x}$ exists.

We have proved that the laws of exponents hold for all integer exponents. It is not difficult to prove that these same laws are also true for rational number exponents. We shall assume the truth of these laws of exponents for all rational number exponents.

EXAMPLE 1. Simplify $8^{-2/3}$.

SOLUTION

$$8^{-2/3} = \frac{1}{8^{2/3}} \qquad \text{Definition 2.3}$$

$$= \frac{1}{(\sqrt[3]{8})^2} \qquad \text{Definition 2.6}$$

$$= \frac{1}{(\sqrt[3]{2^3})^2} \qquad \text{Definition 2.1}$$

$$= \frac{1}{2^2} \qquad \text{Law r-1}$$

$$= \frac{1}{4} \qquad \text{Definition 2.1}$$

EXAMPLE 2. Simplify $(27)^{2/3} \cdot (4)^{-3/2}$.

SOLUTION. The reader should supply a reason for each step.

$$(27)^{2/3} \cdot (4)^{-3/2} = (\sqrt[3]{27})^2 (\sqrt{4})^{-3}$$
$$= (\sqrt[3]{3^3})^2 (\sqrt{2^2})^{-3}$$
$$= (3^2)(2^{-3})$$
$$= 9 \cdot \tfrac{1}{8}$$
$$= \tfrac{9}{8}$$

EXAMPLE 3. Simplify. $\left(\dfrac{3x^{2/3}}{y^{1/2}}\right)^2 \cdot \left(\dfrac{x^{-5/6}}{2y^{1/3}}\right)$.

SOLUTION

$$\left(\frac{3x^{2/3}}{y^{1/2}}\right)^2 \cdot \left(\frac{x^{-5/6}}{2y^{1/3}}\right) = \left(\frac{9x^{4/3}}{y}\right) \cdot \left(\frac{x^{-5/6}}{2y^{1/3}}\right)$$

$$= \frac{9x^{3/6}}{2y^{4/3}}$$

$$= \frac{9x^{1/2}}{2y^{4/3}} \cdot \frac{y^{2/3}}{y^{2/3}}$$

$$= \frac{9x^{1/2}y^{2/3}}{2y^2}$$

EXERCISES 2.4

Simplify.

1. $(27)^{2/3}$

2. $(16)^{-1/4}$

3. $(25)^{3/2}$

4. $(9)^{-1/2}$

5. $(-8)^{2/3}$

6. $(0.01)^{1/2}$

7. $\left(-\dfrac{1}{125}\right)^{2/3}$

8. $(64)^{-1/2}$

9. $(125x^4)^{-2/3}$

10. $\left(\dfrac{x^{-4}}{4y^6}\right)^{-1/2}$

11. $\left(\dfrac{-27a^3}{b^6}\right)^{2/3}$

12. $(8x^{-2/3})^{1/2}$

13. $\left(\dfrac{a^{2/3}}{a^{-1/3}}\right)^3$

14. $\left(\dfrac{3x^{2/3}}{5y^{1/6}}\right)^3$

15. $\left(\dfrac{b^{-2}x^3}{b^{-3}x^{-6}}\right)^{1/2}$

16. $\left(\dfrac{x^{1/3}}{y^{3/2}}\right)\left(\dfrac{y^{1/2}}{x^{2/3}}\right)^{-2}$

17. $x^{1/2} \cdot x^{1/3} \cdot x^{-1/6}$

18. $\left(\dfrac{a^{4/3}b^{-1/2}}{a^{2/3}b^{-3/2}}\right)^3$

19. $\left(\dfrac{x^{-2}y^3}{3x^4y^{-5}}\right)^{2/3}$

20. $\left(\dfrac{(x^{1/2}y^{-3}z^2)^6}{(x^5y^3z^{1/4})^{12}}\right)^3$

21. $(x^{1/3} - y^{1/3})(x^{2/3} + x^{1/3}y^{1/3} + y^{2/3})$

22. $(x - y^{-1})^2$

23. $(x - y)^{-2}(x^{-2} + y^{-2})$

24. $(x^{1/2} - y^{1/2})(x^{1/2} + y^{1/2})$

25. $(x^{1/4} + y^{1/4})(x^{1/2} - x^{1/4}y^{1/4} + y^{1/2})$

26. $(x^{1/2} + x^{-1/2})^2$

27. $(a^{2/3} + b^{2/3})(a^{4/3} - a^{2/3}b^{2/3} + b^{4/3})$

28. $(x^{k/2} + y^{-k/2})^2$

29. $(w^{2/3} - u^{2/3})^2$

30. $(c^n - k^{3n})(c^{2n} + c^nk^{3n} + k^{6n})$

2.5 ◆ Special Products

The properties of the real numbers discussed in Chapter 1 can be applied to algebraic expressions in which the variables represent real numbers. In particular, certain multiplication formulas occur so frequently that they warrant special recognition.

THEOREM 2.5. If x and y are real numbers:

1. $(x + y)(x - y) = x^2 - y^2$.
2a. $(x + y)^2 = x^2 + 2xy + y^2$.
2b. $(x - y)^2 = x^2 - 2xy + y^2$.
3. $(ax + by)(cx + dy) = acx^2 + (ad + bc)xy + bdy^2$.
4a. $(x + y)(x^2 - xy + y^2) = x^3 + y^3$.
4b. $(x - y)(x^2 + xy + y^2) = x^3 - y^3$.

The proof of the formulas in this theorem depend on the properties of the real number field, particularly the distributive property. The reader should prove each formula in the theorem.

Because each variable in each of the formulas in Theorem 2.5 may be replaced by any algebraic expression in which the variables represent real numbers, the formulas are very general in nature. We give some examples which illustrate the techniques used.

EXAMPLE 1. Find the product $(x - 3y)(x + 3y)$.

SOLUTION. Formula 1 of Theorem 2.5 and the laws of exponents give

$$
\begin{aligned}
(x - 3y)(x + 3y) &= x^2 - (3y)^2 \\
&= x^2 - 3^2 y^2 \\
&= x^2 - 9y^2
\end{aligned}
$$

EXAMPLE 2. Find the product $(\sqrt{x} - \sqrt{y})^2$.

SOLUTION. Using Formula 2b of Theorem 2.5 and the laws of exponents, we obtain

$$
\begin{aligned}
(\sqrt{x} - \sqrt{y})^2 &= (\sqrt{x})^2 - 2\sqrt{x}\sqrt{y} + (\sqrt{y})^2 \\
&= x - 2\sqrt{xy} + y
\end{aligned}
$$

EXAMPLE 3. Find the product $(a + b - 3)^2$.

SOLUTION. Considering $a + b$ as one number, we have

$$
\begin{aligned}
(a + b - 3)^2 &= [(a + b) - 3]^2 \\
&= (a + b)^2 - 2(3)(a + b) + 3^2 \quad \text{Formula 2b, Theorem 2.5} \\
&= a^2 + 2ab + b^2 - 6(a + b) + 9 \quad \text{Formula 2a, Theorem 2.5} \\
&= a^2 + 2ab + b^2 - 6a - 6b + 9 \quad \text{distributive property}
\end{aligned}
$$

EXAMPLE 4. Find the product $(x - 2y - 4)(x - 2y + 4)$.

SOLUTION

$$
\begin{aligned}
(x - 2y - 4)(x - 2y + 4) &= [(x - 2y) - 4][(x - 2y) + 4] \quad \text{Why?} \\
&= (x - 2y)^2 - 4^2 \quad \text{Why?} \\
&= x^2 - 4xy + 4y^2 - 16
\end{aligned}
$$

EXERCISES 2.5

Find the products (Exercises 1–40).

1. $(a + b)(a - b)$
2. $(a + b)^2$
3. $(a - b)^2$
4. $(x + 3)(x + 2)$
5. $(y - 3)(y + 4)$
6. $(x + 7)(x - 9)$
7. $(y - 8)(y - 2)$
8. $(3x + y)(3x - y)$
9. $(2x + 3y)(2x + y)$
10. $(y + 7)(3y - 2)$
11. $(5x - 2y)(3x + 2y)$
12. $(2a - 3b)^2$
13. $(3x + 2y)^2$
14. $(\frac{1}{2}x - y)^2$
15. $(x^2 + y^2)^2$
16. $\left(3x^2 + \dfrac{1}{x}\right)\left(2x^2 - \dfrac{3}{x}\right)$
17. $(2a + 4b)(6a - 7b)$
18. $(a - b)(a^2 + ab + b^2)$
19. $(a + b)(a^2 - ab + b^2)$
20. $(3x - y)(9x^2 + 3xy + y^2)$

21. $(x + 2y)(x^2 - 2xy + 4y^2)$
22. $(x + 4y)(x^2 - 4xy + 16y^2)$
23. $(\frac{1}{2}x^2 + 2y^3)^2$
24. $(\sqrt{x} + \sqrt{y})(\sqrt{x} - \sqrt{y})$
25. $(\sqrt{x} - \sqrt{y})^2$
26. $(\sqrt[3]{x} - \sqrt[3]{y})(\sqrt[3]{x^2} + \sqrt[3]{xy} + \sqrt[3]{y^2})$
27. $(\sqrt[4]{a} - \sqrt[4]{b})(\sqrt{a} + \sqrt[4]{ab} + \sqrt{b})$
28. $(2\sqrt{x} - \sqrt{y})(2\sqrt{x} + \sqrt{y})$
29. $(-3a + 4\sqrt{b})^2$
30. $(x^{1/3} + y^{1/3})(x^{2/3} - x^{1/3}y^{1/3} + y^{2/3})$
31. $(a^{1/6} - b^{1/6})(a^{1/3} + a^{1/6}b^{1/6} + b^{1/3})$
32. $(x + 3 - y)^2$
33. $(2x + y - 3z)^2$
34. $(a^2 + b^2 + c^2)^2$
35. $(a^2 - b^2 + c^2)^2$
36. $[4 - (x + 3y)][3 + 2(x + 3y)]$
37. $(2a^2b - 2c^2 + 3)^2$
38. $(8a - 3b - 3c)(2a + 5b + 5c)$
39. $(4 - x^2 - 2y^3)(4 + x^2 + 2y^3)$
40. $(5k^2 - 3m^2 + 2)(5k^2 - 3m^2 - 6)$

41. If x and y are real numbers, prove: $(x + y)(x - y) = x^2 - y^2$.
42. If x and y are real numbers, prove: $(x + y)^2 = x^2 + 2xy + y^2$.
43. If x and y are real numbers, prove: $(x - y)^2 = x^2 - 2xy + y^2$.

44. If x, y, a, b, c, and d are real numbers, prove:
    $(ax + by)(cx + dy) = acx^2 + (ad + bc)xy + bdy^2$.
45. If x and y are real numbers, prove: $(x + y)(x^2 - xy + y^2) = x^3 + y^3$.
46. If x and y are real numbers, prove: $(x - y)(x^2 + xy + y^2) = x^3 - y^3$.

2.6 ◆ Factoring

When the formulas in Theorem 2.5 are read from right to left, they may be considered as factorization formulas, that is, formulas which express various polynomials in the form of a product.

Before carrying out factoring of polynomial expressions, it is necessary to specify the system from which the coefficients of the factors are to be chosen. In general, we shall agree that if an expression with integer coefficients is given, then the factors should also be expressions with integer coefficients. When we say that a polynomial expression is **factored completely** we mean that no factor in the product may again be factored into factors which are in the same system.

EXAMPLE 1. Factor $4x^2 - 9y^2$.

SOLUTION. We recognize the difference of two squares. Hence by Formula 1, Theorem 2.5, we obtain

$$4x^2 - 9y^2 = (2x)^2 - (3y)^2$$
$$= (2x + 3y)(2x - 3y).$$

EXAMPLE 2. Factor $a^6 - 64$.

SOLUTION. We see that Theorem 2.5, Formula 1 applies with $x = a^3$ and $y = 8$, and also that Formula 4b applies with $x = a^2$ and $y = 4$. When an expression can be factored as the difference of two squares (Formula 1) or as the difference of two cubes (Formula 4b), it is desirable to factor it initially as the difference of two squares. Thus

$$a^6 - 64 = (a^3)^2 - 8^2$$
$$= (a^3 + 8)(a^3 - 8)$$

The first of these factors is the sum of two cubes, (Formula 4a) and the second is the difference of two cubes (Formula 4b). Hence

$$a^6 - 64 = (a^3 + 8)(a^3 - 8)$$
$$= (a^3 + 2^3)(a^3 - 2^3)$$
$$= (a + 2)(a^2 - 2a + 4)(a - 2)(a^2 + 2a + 4)$$

EXAMPLE 3. Factor $3x^2 - 7x - 6$.

SOLUTION. Writing

$$3x^2 - 7x - 6 = (ax + b)(cx + d)$$
$$= acx^2 + (bc + ad)x + bd$$

we see that $ac = 3$ and $bd = -6$.

    Since the coefficients of the given polynomial expression are integers, we need to find integer values of a, b, c, and d for which $ac = 3$, $bd = -6$, and $bc + ad = -7$. We can try all the various combinations by making a table:

a	b	c	d	$bc + ad$
1	1	3	-6	-3
1	6	3	-1	17
1	2	3	-3	4
1	-3	3	2	-7, and so forth

We see that

$$3x^2 - 7x - 6 = (x - 3)(3x + 2)$$

EXERCISES 2.6

Factor completely.

1. $9x^2 - 25$
2. $25a^2 - 144$
3. $16z^6 - 64t^4$
4. $x^2 - 6x + 9$
5. $y^2 - 2y + 1$
6. $x^2 + 2x - 8$
7. $y^2 + 3y - 10$
8. $x^2 + 4x + 4$
9. $a^2 + 10a + 25$
10. $a^2 + 2a - 15$
11. $x^2 - 6x - 27$
12. $x^4 - 10x^2 + 9$
13. $2a^2 - ab - b^2$
14. $3x^2 + x - 4$
15. $25a^2 - 10a + 1$
16. $81a^2 - 16b^2$
17. $16a^2 + 24ab + 9b^2$

18. $5a^2 + 6a + 1$
19. $5a^2 + 20ab + 15b^2$
20. $49a^2 - 14a + 1$
21. $1 + 6a + 9a^2$
22. $6a^2 - 11a - 10$
23. $x^3 - y^3$
24. $8a^3 - b^3$
25. $z^3 + 125b^3$
26. $64x^3 - 8y^3$
27. $6x^2 + 11xy - 10y^2$
28. $8x^2y^2 - 18y^2$
29. $x^4 - y^4$
30. $x^6 - y^6$
31. $64x^6 - y^6$
32. $(x^2 - y^2)^2 - z^4$
33. $8a^3 - (c + d)^3$
34. $x^9 - (a^3 - 1)^3$

35. $6(x - 3)^2 - (x - 3) - 15$
36. $8(x - 1)^2 + 2(x - 1)(y + 2) - 21(y + 2)^2$
37. $(3x - 4)^3 - (2y + 7)^3$
38. $(2x + 3)^3 + (2x - 5)^3$
39. $(2x - 1)^2 - (2x - 1)(2x - 3) - 12(2x - 3)^2$
40. $(x - 2)^3 - 8$

2.7 ◆ Rational Expressions

Expressions that indicate the division of polynomials such as

$$\frac{x^2 - 4x + 4}{x + 1}, \quad \frac{x^3 - 8}{x^2 + 2x + 1}, \quad \frac{1}{4xy^2},$$

are called **rational expressions.** Many problems of mathematics involve combining rational expressions and then simplifying the results. Since rational expressions are quotients, the basic tool used in simplifying them is the following property of real numbers:

If a, b, and c are real numbers, $b \neq 0$, $c \neq 0$, $\dfrac{ac}{bc} = \dfrac{a}{b}$.

The examples below illustrate techniques used in simplifying rational expressions.

EXAMPLE 1. Simplify $\dfrac{x^3 - 8}{x^2 - 4x + 4}$, $x \neq 2$.

SOLUTION

$$\frac{x^3 - 8}{x^2 - 4x + 4} = \frac{(x - 2)(x^2 + 2x + 4)}{(x - 2)^2}$$

$$= \frac{x^2 + 2x + 4}{x - 2}$$

EXAMPLE 2. Simplify: $\dfrac{9 - x^4}{x^4 - x^2 - 6}$, $x^4 - x^2 - 6 \neq 0$.

SOLUTION

$$\frac{9 - x^4}{x^4 - x^2 - 6} = \frac{(3 - x^2)(3 + x^2)}{(x^2 - 3)(x^2 + 2)}$$

$$= \frac{-(x^2 - 3)(x^2 + 3)}{(x^2 - 3)(x^2 + 2)}$$

$$= -\frac{x^2 + 3}{x^2 + 2}$$

$9x^2 - 25 = (3x+5)(3x^2 +18x+36)(3x-5)(3x+ \times)$

EXAMPLE 3. Perform the indicated operations and simplify. Assume no zero divisors.

$$\frac{(x-y)^2}{x^4-y^4} \cdot \frac{x^2+y^2}{x-y}$$

SOLUTION

$$\frac{(x-y)^2}{x^4-y^4} \cdot \frac{x^2+y^2}{x-y} = \frac{(x-y)^2}{(x-y)(x+y)(x^2+y^2)} \cdot \frac{x^2+y^2}{x-y}$$

$$= \frac{(x-y)^2(x^2+y^2)}{(x-y)^2(x+y)(x^2+y^2)}$$

$$= \frac{1}{x+y}$$

EXAMPLE 4. Perform the indicated operations and simplify. Assume no zero divisors.

$$\frac{x^2-9}{x^2-2x-15} \div \frac{x^2-6x+9}{12-4x}$$

SOLUTION

$$\frac{x^2-9}{x^2-2x-15} \div \frac{x^2-6x+9}{12-4x} = \frac{x^2-9}{x^2-2x-15} \cdot \frac{12-4x}{x^2-6x+9}$$

$$= \frac{(x-3)(x+3)}{(x-5)(x+3)} \cdot \frac{(-4)(x-3)}{(x-3)^2}$$

$$= \frac{-4}{x-5}$$

EXAMPLE 5. Simplify $\dfrac{1}{x+2} + \dfrac{2x+9}{x^2-x-6} - \dfrac{2x}{x^2-2x-3}$. Assume no zero divisors.

SOLUTION. The factored denominators are $(x+2)$, $(x-3)(x+2)$, and $(x-3)(x+1)$. Hence the least common denominator (L.C.D.) is $(x+1)(x+2)(x-3)$. We may write:

$$\left[\frac{1}{x+2}\cdot\frac{(x+1)(x-3)}{(x+1)(x-3)}\right]+\left[\frac{2x+9}{(x+2)(x-3)}\cdot\frac{x+1}{x+1}\right]-\left[\frac{2x}{(x-3)(x+1)}\cdot\frac{x+2}{x+2}\right]$$

$$=\frac{(x+1)(x-3)+(2x+9)(x+1)-(2x)(x+2)}{(x+1)(x+2)(x-3)}$$

$$=\frac{(x^2-2x-3)+(2x^2+11x+9)-(4x^2+4x)}{(x+1)(x+2)(x-3)}$$

$$=\frac{-x^2+5x+6}{(x+1)(x+2)(x-3)}$$

$$=\frac{-(x^2-5x-6)}{(x+1)(x+2)(x-3)}$$

$$=\frac{-(x-6)(x+1)}{(x+1)(x+2)(x-3)}$$

$$=-\frac{x-6}{(x+2)(x-3)}$$

EXAMPLE 6. Simplify $\dfrac{1+\dfrac{2}{x-1}}{\dfrac{x^2+x}{x^2+x-2}}$. Assume no zero divisors.

SOLUTION. We first simplify the numerator of the expression:

$$1+\frac{2}{x-1}=\frac{x-1}{x-1}+\frac{2}{x-1}$$

$$=\frac{x+1}{x-1}$$

We now have

$$\frac{1+\dfrac{2}{x-1}}{\dfrac{x^2+x}{x^2+x-2}}=\frac{\dfrac{x+1}{x-1}}{\dfrac{x^2+x}{x^2+x-2}}$$

$$=\frac{\dfrac{x+1}{x-1}}{\dfrac{x(x+1)}{(x+2)(x-1)}}$$

$$=\frac{x+1}{x-1}\cdot\frac{(x+2)(x-1)}{x(x+1)}$$

$$=\frac{(x+1)(x+2)(x-1)}{x(x+1)(x-1)}$$

$$=\frac{x+2}{x}$$

ALTERNATE SOLUTION. We first multiply numerator and denominator of the given expression by $(x - 1)(x + 2)$ to obtain

$$\frac{1 + \dfrac{2}{x - 1}}{\dfrac{x^2 + x}{x^2 + x - 2}} = \frac{1 + \dfrac{2}{x - 1}}{\dfrac{x(x + 1)}{(x + 2)(x - 1)}} \cdot \frac{(x - 1)(x + 2)}{(x - 1)(x + 2)}$$

$$= \frac{1 \cdot (x - 1)(x + 2) + 2(x + 2)}{x(x + 1)}$$

$$= \frac{x^2 + x - 2 + 2x + 4}{x(x + 1)}$$

$$= \frac{x^2 + 3x + 2}{x(x + 1)}$$

$$= \frac{(x + 1)(x + 2)}{x(x + 1)}$$

$$= \frac{x + 2}{x}$$

Expressions like the one in Example 6, in which the numerator or the denominator or both the numerator and denominator of a rational expression are rational expressions, are called **complex fractions.**

EXERCISES 2.7

Simplify. Assume no zero divisors. (Exercises 1–16.)

1. $\dfrac{125}{360}$

2. $\dfrac{27a^3b^2}{36a^5b^7}$

3. $\dfrac{a^2 + ab}{a^3 + 3ab^2}$

4. $\dfrac{a^2b^2 - a^2c^2}{a^2b - a^2c}$

5. $\dfrac{y^2 - 1}{y^3 - 1}$

6. $\dfrac{x^2 - 1}{x^4 - x^2}$

7. $\dfrac{x^2 - 9}{x^2 - 6x + 9}$

8. $\dfrac{6x^2 - 9x}{24x^3}$

9. $\dfrac{a^2 - 4a + 4}{a^2 - 4}$

10. $\dfrac{4x^2 - 4x - 3}{2x^2 + 5x - 12}$

11. $\dfrac{9a^2 - 4b^2}{3a^2 + 13ab - 10b^2}$

12. $\dfrac{125x^3 - 64y^3}{25x^2 - 16y^2}$

13. $\dfrac{x^6 - y^6}{x^4 - y^4}$

15. $\dfrac{6x^2 - 17xy + 12y^2}{16x^2 - 4xy - 30y^2}$

14. $\dfrac{a^2 + a - 20}{7a - 12 - a^2}$

16. $\dfrac{8a^2 - 52a + 60}{72 - 30a - 12a^2}$

Perform the indicated operations and simplify. Assume no zero divisors.

17. $\dfrac{3x^3}{5y^5} \cdot \dfrac{25y^2}{27x^7}$

24. $\dfrac{2x^2 + 3x - 2}{x^2 + 5x + 6} \cdot \dfrac{3x^2 - 10x + 3}{6x^2 - 5x + 1}$

18. $\dfrac{8a^2}{15b^3} \cdot \dfrac{5b^6}{36a^7}$

25. $\dfrac{x^2 - 4x + 3}{x^2 - 4x + 4} \cdot \dfrac{x^2 - x - 2}{x^2 - 3x + 2}$

19. $\dfrac{40x^2y^2}{24xy^3} \cdot \dfrac{36x^2y^2}{120x^3y^4}$

26. $\dfrac{a^2 + a - 12}{a^2 - 2a - 3} \cdot \dfrac{a^2 + 5a + 4}{a^2 - 9}$

20. $\dfrac{125a^3b^2}{8ab^6} \cdot \dfrac{364a^3b^4}{25ab^4}$

27. $\dfrac{x^2 - 16}{2x + 5} \div \dfrac{x^2 - 8x + 16}{4x^2 - 25}$

21. $\dfrac{36x^2y^3}{5xy^4} \div \dfrac{32x^3y^7}{25x^4y^6}$

28. $\dfrac{x^2 + 3xy + 2y^2}{x^2 - 2xy - 3y^2} \div \dfrac{x^2 - y^2}{x^2 - 9y^2}$

22. $\dfrac{156a^3b}{35a^4b^3} \div \dfrac{48a^4b^6}{55ab^3}$

29. $\dfrac{a^2 - 2ab + b^2}{a^3 - b^3} \div \dfrac{a - b}{a^2 + ab + b^2}$

23. $\dfrac{x^2 - y^2}{x^3 - y^3} \cdot \dfrac{x^2 + xy + y^2}{x^2 + 2xy + y^2}$

30. $\dfrac{a^4 - b^4}{a^2 + b^2} \cdot \dfrac{a^2}{(a - b)^2} \div \dfrac{ab + b^2}{a - b}$

Simplify. Assume no zero divisors (Exercises 31–46).

31. $\dfrac{3x + 2}{3} + \dfrac{4x - 3}{6}$

36. $\dfrac{a}{b} + \dfrac{a}{a - b} - \dfrac{2ab - a^2}{a^2 - ab}$

32. $\dfrac{a - 3b}{a} + \dfrac{b - 2a}{b}$

37. $\dfrac{4}{a} - \dfrac{6}{3a + 2} - \dfrac{4}{a(3a - 2)}$

33. $\dfrac{x + 1}{x} - \dfrac{x + 3}{x + 1}$

38. $\dfrac{2a - b}{5a - 3} + \dfrac{3a + b}{3 - 5a}$

34. $\dfrac{3a}{5bc} + \dfrac{2b}{3ac} - \dfrac{5c}{7ab}$

39. $\dfrac{a + b}{4 - a} + \dfrac{2a - 1}{3a - 12}$

35. $\dfrac{x}{3y^2z} + \dfrac{5y}{18x^2z} - \dfrac{2z}{9xy^2}$

40. $\dfrac{2a - 3b}{5a - 3} + \dfrac{2a - b}{3 - 5a}$

41. $\dfrac{2xy}{x^3 + y^3} - \dfrac{x}{x^2 - xy + y^2}$

42. $\dfrac{1}{(a - b)(b - c)} + \dfrac{1}{(b - c)(c - a)} + \dfrac{1}{(c - a)(a - b)}$

43. $\dfrac{x - y}{x + y} - \dfrac{x + y}{x - y}$

44. $\dfrac{2x}{x^2 - y^2} - \dfrac{x - y}{x^2 + 2xy + y^2} - \dfrac{4xy}{(x + y)^2(x - y)}$

45. $\dfrac{x^2 + x - 1}{x^2 - x + 1} - \dfrac{x^2 - x - 1}{x^2 + x + 1} + \dfrac{4}{x^4 + x^2 + 1}$

46. $\dfrac{2a - 1}{2a^2 - a - 6} - \dfrac{2a - 3}{3a^2 - 10a + 8} + \dfrac{a + 3}{6a^2 + a - 12}$

Simplify. Assume no zero divisors (Exercises 47–60).

47. $\dfrac{1 + \frac{1}{2}}{2 - \frac{1}{2}}$

48. $\dfrac{3 + \frac{1}{3}}{2 - \frac{1}{3}}$

49. $\dfrac{5 + \frac{1}{8}}{3 - \frac{1}{8}}$

50. $\dfrac{5 - \frac{2}{9}}{2 + \frac{2}{9}}$

51. $\dfrac{x - \frac{1}{x}}{x + \frac{1}{x}}$

52. $\dfrac{1 - \frac{1}{x}}{1 + \frac{1}{x}}$

53. $\dfrac{1 - \frac{1}{x^2}}{1 - \frac{1}{x}}$

54. $\dfrac{1 + \frac{1}{y} - \frac{2}{y^2}}{1 + \frac{5}{y} + \frac{6}{y^2}}$

55. $\dfrac{1 - \frac{1}{x} - \frac{2}{x^2}}{\frac{1}{x} + \frac{1}{x^2}}$

56. $\dfrac{\frac{1}{x} + \frac{1}{y}}{\frac{1}{x} - \frac{1}{y}}$

57. $\dfrac{\frac{x^2}{y^2} + \frac{y}{x}}{\frac{1}{x} - \frac{1}{y} + \frac{x}{y^2}}$

58. $\dfrac{\frac{x}{y} - 1}{\frac{x}{y} - 2 + \frac{2}{1 + \frac{y}{x}}}$

59. $\dfrac{\frac{x + y}{x - y} - \frac{x - y}{x + y}}{\frac{x + y}{x - y} + \frac{x - y}{x + y}}$

60. $\dfrac{\frac{1}{x^2 + y^2}}{1 - \frac{x}{x + \frac{y^2}{x - y}}}$

CHAPTER REVIEW

Simplify. Assume no zero divisors (Exercises 1–9).

1. (a) $x^4 \cdot x^7$ (b) $\dfrac{x^8}{x^5}$

2. (a) $(x^3 y^2)^4$ (b) $\left(\dfrac{x^2}{y^3}\right)^3$

3. (a) $x^{-2} x^5$ (b) $\dfrac{x^{-3}}{x^{-5}}$

4. (a) $\left(\dfrac{x^{-1}}{y^{-3}}\right)^2$ (b) $\left(\dfrac{x^2}{y^{-3}}\right)^{-1}$

5. (a) $\sqrt{125}$ (b) $\sqrt{x^4 y^2}$

6. (a) $\dfrac{1}{\sqrt{x^4y}}$ (b) $\sqrt[3]{\dfrac{8x^8}{3y}}$

7. (a) $a^{1/3} \cdot a^{2/5}$ (b) $\dfrac{a^{1/4}}{a^{1/3}}$

8. (a) $\left(\dfrac{a^{-1/2}}{b^{-2/3}c^{1/4}}\right)^3$ (b) $\left(\dfrac{x^3}{y^{-4}z^5}\right)^{1/2}$

9. (a) $\left(\dfrac{a^5b^0}{c^{-3}}\right)^{1/4}$ (b) $\left(\dfrac{\sqrt[3]{x}y^{1/2}}{z}\right)^{1/4}$

10. Perform the indicated multiplications.

(a) $(2x - 3)(x + 1)$ (c) $(2x - y)(4x^2 + 2xy + y^2)$
(b) $(x - 3y)(x + 3y)$ (d) $(2x + 1)^2$

Factor completely (Exercises 11–14).

11. $32 - 2x^2$
12. $6x^5 - 13x^4 - 5x^3$
13. $6x^4 - 67x^2 - 60$
14. $x^4 - 16x^2 + 63$

Perform the indicated operation and simplify. Assume no zero divisors (Exercises 15–20).

15. $\dfrac{3ab}{4a^2} \cdot \dfrac{8a^4b^2}{9a^2b^3}$

16. $\dfrac{x^2 + x - 12}{x^2 - 1} \cdot \dfrac{4x^3 + 4x^2}{x + 4}$

17. $\dfrac{2x^2 - x - 28}{3x^2 - x - 2} \div \dfrac{4x^2 + 16x + 7}{3x^2 + 11x + 6}$

18. $\dfrac{7 - 11x - 6x^2}{(2x - 1)^2} \div \dfrac{14 - x - 3x^2}{x^3 - 4x^5}$

19. $\dfrac{3x^2 - 15}{20x^3 - 20} - \dfrac{x + 2}{4x^2 + 4x + 4} - \dfrac{1}{5 - 5x}$

20. $\dfrac{\dfrac{1}{x} + \dfrac{1}{3y}}{\dfrac{1}{x^2} - \dfrac{1}{9y^2}} + \dfrac{\dfrac{1}{3x} - \dfrac{1}{y}}{\dfrac{1}{9x^2} - \dfrac{1}{y^2}}$

CHAPTER 3

Equations and Inequalities over the Real Numbers

3.1 ✦ Definitions

An **equation** is a statement that two given expressions name the same number. To write an equation we set one of the given expressions equal to the other. Thus

$$6x + 3 = 5$$

is an equation; it says that $6x + 3$ and 5 name the same number. The expression on the left of the equal sign is called the **left member** of the equation; the expression on the right of the equal sign is called the **right member** of the equation.

Statements in which one expression names a number greater than, greater than or equal to, less than, or less than or equal to another expression are called **inequalities.** Thus

$$3x + 2 < 5 \qquad x - 3 > 5$$
$$2x + 7 \le 9 \qquad 5x - 2 \ge 8$$

are inequalities.

Equations and inequalities which involve variables are called **open sentences** or simply **sentences.** The equation and the inequalities listed above are open sentences.

A set of numbers which are either implicitly or explicitly stated as permissible values of a variable is called the **replacement set** or the **domain** of that variable. That subset of the replacement set whose elements make an open sentence a true statement is called the **solution set** of the open sentence. Every element of the solution set is called a **solution** of the open sentence and is said to **satisfy** that sentence. Open sentences that have the same solution set are called **equivalent sentences.** In this book, unless stated otherwise, the domain is the set R of real numbers.

Open sentences are either conditional sentences or identities. An **identity** is an open sentence that is true for every number in the replacement set of any variable or variables involved. A **conditional sentence** is an open sentence which is false for at least one number in the replacement set. For example, the open sentences on the left below are conditional sentences, and those on the right are identities (with the replacement set being the set of real numbers).

Conditional sentences	*Identities*
$3x + 2 = 8$	$5x - 3x + 2 = 2x + 2$
$x^2 + 3x + 4 > 0$	$x + 1 < x + 2$

To find the solution set of an open sentence we either determine the members of the solution set by inspection or trial and error or we generate a sequence of equivalent sentences until we reach one with an obvious solution set.

Before stating ways to generate equivalent equations, we recall that an **algebraic expression** is any collection of variables and real numbers obtained by applying to this collection a finite number of additions, subtractions, multiplications, and divisions together with the process of taking roots. Some examples of algebraic expressions are:

$$3x + 2 \qquad \frac{x^2 + 4}{x - 2}$$

$$\frac{1}{x^2 + 1} \qquad \frac{5x + 4}{x - 7}$$

If the variables in an algebraic expression are replaced by specific real numbers, the resulting real number is called the **value** of the expression. When working with algebraic expressions, it is assumed that domains are chosen so that variables do not represent numbers that result in divisions by zero or even roots of negative numbers.

We now state the following theorems for generating equivalent equations. We shall accept these theorems without proof.

THEOREM 3.1. Let p, q, and r be algebraic expressions in the variable x. Then (a) for all values of x for which p, q, and r name real numbers, the open sentences

$$p = r$$
$$p + q = r + q$$

and

$$q + p = q + r$$

are equivalent sentences; (b) for all values of x for which p, q, and r name real numbers, and for which $q \neq 0$, the open sentences

$$p = r$$
$$pq = rq$$

and

$$qp = qr$$

are equivalent.

THEOREM 3.2. Let p, q, and r be expressions in the variable x. Then (a) for all values for x for which p, q, and x name real numbers, the open sentences

$$p < r$$
$$p + q < r + q$$

and

$$q + p < q + r$$

are equivalent sentences; (b) for all values of x for which p, q, and r name real numbers and for which $q > 0$, the open sentences

$$p < r$$
$$pq < rq$$

and

$$qp < qr$$

are equivalent; (c) for all values of x for which p, q, and r are real numbers, and for which $q < 0$, the open sentences

$$p < r$$
$$pq > rq$$

and

$$qp > qr$$

are equivalent sentences.

When Theorems 3.1 and 3.2 are applied to equations, the resulting equations are called **transformed equations** and an application of these theorems is called an **elementary transformation.** For example, the equation

$$3x + 5 = 11$$

in which $p = 3x + 5$ and $r = 11$, is equivalent to the equation

$$3x = 6$$

by using Theorem 3.1a in which we take $q = -5$. In this case the elementary transformation consists of adding -5 to both members of the given equation.

We now apply Theorem 3.1b to get the equivalent equation

$$x = 2$$

where we take $p = 3x$, $r = 6$ and $q = \frac{1}{3}$. Since the equations

$$3x + 5 = 11$$
$$3x = 6$$
$$x = 2$$

are all equivalent, we see that $\{2\}$ is the solution set of each of them.

EXAMPLE 1. Given $3x + 7 = 9$, find an equivalent equation whose solution set is obvious.

SOLUTION

$3x + 7 = 9$	
$(3x + 7) + (-7) = 9 + (-7)$	Theorem 3.1a, $q = -7$
$3x = 2$	Why?
$(\frac{1}{3})(3x) = (\frac{1}{3})(2)$	Theorem 3.1b, $q = \frac{1}{3}$
$x = \frac{2}{3}$	

The solution set of the given equation is $\{\frac{2}{3}\}$.

EXAMPLE 2. Given $3x < 9$. Find an equivalent sentence whose solution set is obvious.

SOLUTION

$3x < 9$	
$(\frac{1}{3})(3x) < (\frac{1}{3})(9)$	Theorem 3.2b, $q = \frac{1}{3}$
$x < 3$	

The solution set of $x < 3$ is the set of all real numbers that are less than 3. Since the sentence $x < 3$ is equivalent to the given sentence, the solution set of the given sentence is the set of all real numbers less than 3. Using set-builder notation this solution set is denoted by

$$\{x \mid x < 3 \text{ and } x \in R\}$$

EXAMPLE 3. Given $8 - 5x > 3$. Find an equivalent sentence whose solution set is obvious.

SOLUTION

$8 - 5x > 3$	
$(8 - 5x) + (-8) > 3 + (-8)$	Theorem 3.2a, $q = -8$

$$-5x > -5 \qquad \text{Why?}$$
$$(-\tfrac{1}{5})(-5x) < (-\tfrac{1}{5})(-5) \qquad \text{Theorem 3.2c, } q = -\tfrac{1}{5}$$
$$x < 1$$

The solution set of $x < 1$ and of the given sentence is

$$\{x \mid x < 1 \text{ and } x \in R\}.$$

3.2 ◆ Linear Equations

The equation

$$ax + b = 0$$

where a and b are real numbers and $a \neq 0$ is called a **first degree** or **linear** equation over the real numbers. An equation which is equivalent to an equation of this form will also be called a **linear equation.**

We now show that the equation $ax + b = 0$ has one and only one real number as a solution. By Theorem 3.1a,

(1) $$ax + b = 0$$

is equivalent to

(2) $$ax = -b$$

since we need only perform the elementary transformation that consists of adding $-b$ to both members of equation 1. Since $a \neq 0$, $\dfrac{1}{a} \neq 0$ exists and equation (2) is equivalent to

(3) $$x = -\frac{b}{a}$$

by Theorem 1.3b $\left(q = \dfrac{1}{a} \right)$.

Since equations 1, 2, and 3 are equivalent, they have the same solution set. But the solution set of equation 3 is $\left\{ -\dfrac{b}{a} \right\}$. Thus the solution set of equation 1 is also $\left\{ -\dfrac{b}{a} \right\}$ and $-\dfrac{b}{a}$ is the one and only real number that satisfies equation 1.

EXAMPLE 1. Find the solution set of $3x + 9 = 7x + 1$.

SOLUTION

$$3x + 9 = 7x + 1$$
$$-4x = -8 \qquad \text{Theorem 3.1a (adding } -9 - 7x)$$
$$x = 2 \qquad \text{Theorem 3.1b (multiplying by } -\tfrac{1}{4})$$

The solution set is $\{2\}$.

EXAMPLE 2. Find the solution set of $x^2 - 3x + 2 = (x - 4)(x + 3) - 1$.

SOLUTION

$$x^2 - 3x + 2 = (x - 4)(x + 3) - 1$$
$$x^2 - 3x + 2 = x^2 - x - 12 - 1$$
$$-2x = -15$$
$$x = \tfrac{15}{2}$$

The solution set is $\{\tfrac{15}{2}\}$.

EXAMPLE 3. Richard is twice as old as Jay. In two years the sum of their ages will be five times as much as Jay's age was four years ago. How old are each of the boys now?

SOLUTION. Let x represent Jay's age now. Then $2x$ represents Richards' present age. In two years Jay will be $(x + 2)$ years old and Richard will be $(2x + 2)$ years old. Four years ago Jays' age was $x - 4$. Then the conditions of the problem tell us

$$(x + 2) + (2x + 2) = 5(x - 4)$$

This equation is equivalent to

$$3x + 4 = 5x - 20$$
$$-2x = -24$$
$$x = 12$$

Jay is 12 years old and Richard is $(2)(12) = 24$ years old.

EXERCISES 3.1

Find the solution set of each of the following first degree equations. The domain is the set R of real numbers. (Exercises 1–20).

1. $2x + 4 = 5x - 5$. $2(x + 2) + 5(x - 1)$
2. $2x + 7(x + 6) = 3x + 2(x + 7) - 8$.
3. $6x + 7 = 5x + 6$.
4. $10x + 4 = 9x - 3$.
5. $8 - (2x - 5) = 12 - 3x$.
6. $10x + 4 - 2x = 3x + 7 + 6x$.
7. $2 - (4x - 3) = 7 - (5x + 4)$.
8. $3(4 + 3x) - 2(4 - 3x) = 0$.
9. $4 + x(6x - 11) = 4 + x(6x - 10)$.
10. $5x(x + 2) - 3x(2x + 4) = 5 - x^2 - x$.
11. $8(x - 2) = 9(x - 4) + 13$.

12. $7x + 2(x - 4) + 3(x - 2) = 8x$.
13. $6 + 4(2x - 1) = 9 + 3(2x + 5)$.
14. $(x - 4)(x - 3) = (x - 6)(x - 2)$.
15. $(5 - 2x)(4 + x) = (10 + 2x)(3 - x)$.
16. $x^2 - (x - 2)(x - 5) = 8 - 2x$.
17. $(x + 4)(x + 3) = (x + 1)(x + 2)$.
18. $(3 - 4x)(2 - 3x) + (1 + x)(19 - 12x) = 0$.
19. $(5x + 2)(x - 4) - (x - 3)(7x + 4) = 7x - 2x^2 - 6$.
20. $(2x - 5)(4x + 3) - (3x + 4)(x - 6) = 5x^2 + 3x$.
21. Mr. Moser is three times as old as his daughter. In 13 years he will be twice as old as his daughter. What is Mr. Moser's present age?
22. Joyce has $1.44 in pennies and nickels. She has 6 more pennies than nickels. How many nickels does Joyce have?
23. Three consecutive integers have a sum of 15. Find the integers.
24. Glenn leaves Boston for Montreal at 9:00 A.M. and drives at an average speed of 40 miles per hour. Lyman leaves Boston at 10:00 A.M. and takes the same route as Glenn at an average speed of 50 miles per hour. At what time will Lyman overtake Glenn?
25. Leo has $10.60 in nickels, dimes, and quarters. He has 7 more dimes than nickels, and 3 times as many quarters as nickels. How many of each coin does Leo have?
26. Find three consecutive odd integers such that twice the sum of the first and third is equal to four times the second.
27. Barry leaves St. Louis for New Orleans at 7:00 A.M. and drives at an average speed of 30 miles per hour. Richard leaves St. Louis at 10:00 A.M. and follows Barry at an average speed of 40 miles per hour. At what time will Richard overtake Barry?
28. Ruth leaves Las Vegas for Los Angeles at noon and travels at an average speed of 40 miles per hour. Two hours later Joan leaves Las Vegas and travels in the opposite direction at an average speed of 50 miles per hour. At what time are they 170 miles apart?
29. Find three consecutive odd integers whose sum is 63.
30. Find three consecutive even integers whose sum is 366.

3.3 ✦ Equations Involving Absolute Value

Solution sets of equations involving absolute value can be found by using the fact that the absolute value of a real number is the (undirected) distance between the point whose coordinate is 0 and the point with that real number as its coordinate.

EXAMPLE 1. Find the solution set of $|x - 3| = 7$.

SOLUTION. The equation $|x - 3| = 7$ is equivalent to: the distance between $x - 3$ and 0 is equal to 7. If $y = x - 3$, we see that $|y| = 7$ means that the set of all values of y which satisfy the equation have graphs on the number line which are 7 units from the point whose graph is 0 (Figure 3.1). Thus we see that

$$x - 3 = 7 \quad \text{or} \quad x - 3 = -7$$
and
$$x = 10 \quad \text{or} \quad x = -4$$

The solution set is $\{-4, 10\}$.

CHECK

$$|-4 - 3| = |-7| = 7$$
$$|10 - 3| = |7| = 7$$

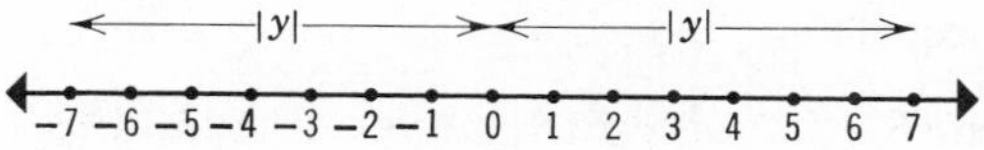

Figure 3.1

EXAMPLE 2. Find the solution set of $|3x - 5| = 10$.

SOLUTION. The equation $|3x - 5| = 10$ is equivalent to: the distance between $3x - 5$ and 0 is equal to 10. If $y = 3x - 5$, we see that $|y| = 10$ means that the set of all values of y which satisfy the equation have graphs on the number line which are 10 units from the point which represents 0 (Figure 3.2). Thus we see that

$$3x - 5 = 10 \quad \text{or} \quad 3x - 5 = -10$$
and
$$3x = 15 \quad \text{or} \quad 3x = -5$$
$$x = 5 \quad \text{or} \quad x = -\tfrac{5}{3}$$

The solution set is $\{-\tfrac{5}{3}, 5\}$.

CHECK

$$|3(-\tfrac{5}{3}) - 5| = |-5 - 5| = |-10| = 10$$
$$|3 \cdot 5 - 5| = |15 - 5| = |10| = 10$$

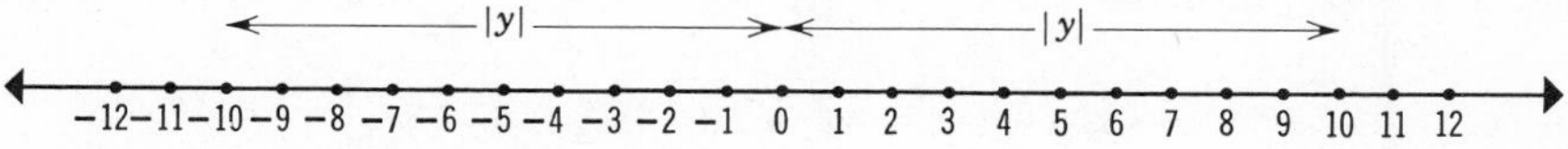

Figure 3.2

EXERCISES 3.2

Find the solution set of each of the following. The domain is the set R of real numbers.

1. $|x| = 3$
2. $|x| = 12$
3. $|x + 3| = 5$
4. $|x - 7| = 8$
5. $|x + 3| = 11$
6. $|2x - 5| = 7$
7. $|3x - 4| = 11$
8. $|5x + 1| = 9$
9. $|2x + 7| = 1$
10. $|3x + 6| = 0$

11. $|5x - 7| = 3$
12. $|2x + 9| = 21$
13. $|5x - 1| = 1$
14. $|4x - 6| = 6$
15. $|4x - 6| = 18$
16. $|3x + 6| = 9$
17. $|3x + 6| = 1$
18. $|x - 2| = |x - 4|$
19. $|2x - 6| = |3x - 12|$
20. $|5x + 1| = 3$

3.4 ◆ First Degree Inequalities

We use Theorem 3.2 to solve first degree inequalities in a manner similar to the use of Theorem 3.1 to solve first degree equations. As an example, let us find the solution set of

$$3x - 6 < 0$$

Theorem 3.2a tells us that adding 6 to each member gives the equivalent inequality

$$3x < 6$$

Since $\frac{1}{3}$ is positive, Theorem 3.2b tells us that multiplying each member by $\frac{1}{3}$ gives the equivalent inequality

$$x < 2$$

Since all three inequalities are equivalent to each other, they have the solution set $\{x \mid x < 2 \text{ and } x \in R\}$.

We can graph the solution set of this inequality on the number line as shown in Figure 3.3. The heavy black trace indicates the points whose coordinates are elements of the solution set. The open circle around the point with coordinate 2 tells us that the number 2 is not included in the solution set. The arrow at the end of the heavy trace tells us that all numbers which are coordinates of points to the left of the heavy trace are also in the solution set.

Figure 3.3

EXAMPLE 1. Find the solution set of $2x - 4 < 8$.

SOLUTION. Using Theorem 3.2a and adding 4 to each member of the given inequality,

$$2x - 4 < 8$$

gives the equivalent inequality

$$2x < 12$$

Since $\frac{1}{2}$ is positive, Theorem 3.2b tells us that multiplying each member of the above inequality by $\frac{1}{2}$ gives the equivalent inequality

$$x < 6$$

The solution set is $\{x \mid x < 6 \text{ and } x \in R\}$ or $\{x \mid x < 6\}$.

EXAMPLE 2. Find the solution set of $2x + 3 > 9$.

SOLUTION. Using Theorem 3.2a and adding -3 to each member of the given inequality, we have the equivalent inequality

$$2x > 6$$

Using Theorem 3.2b, multiplying each member by $\frac{1}{2}$, we have the equivalent inequality

$$x > 3$$

The solution set is $\{x \mid x > 3 \text{ and } x \in R\}$ or $\{x \mid x > 3\}$.

EXAMPLE 3. Find the solution set of $3x + 4 \geq 5x - 8$.

SOLUTION. The sentence $3x + 4 \geq 5x - 8$ is a compound sentence. It says

$$3x + 4 > 5x - 8 \text{ or } 3x + 4 = 5x - 8.$$

The solution set of this sentence is the union of the solution sets of $3x + 4 > 5x - 8$ and $3x + 4 = 5x - 8$. We first solve

$$\begin{aligned}
3x + 4 &> 5x - 8 \\
-2x &> -12 & \text{Theorem 3.2a} \\
x &< 6 & \text{Theorem 3.2c}
\end{aligned}$$

The solution set is $\{x \mid x < 6 \text{ and } x \in R\}$ or $\{x \mid x < 6\}$. We now solve

$$\begin{aligned}
3x + 4 &= 5x - 8 \\
-2x &= -12 & \text{Theorem 3.1a} \\
x &= 6 & \text{Theorem 3.1b}
\end{aligned}$$

The solution set is $\{6\}$. The solution set of the given sentence is

$$\{x \mid x < 6\} \cup \{x \mid x = 6\}$$

which may be written $\{x \mid x \leq 6\}$. Theorem 3.2 remains true if the symbol $<$ is replaced by $\leq$. Hence we may solve the given sentence as shown below.

$$
\begin{aligned}
3x + 4 &\geq 5x - 8 \\
-2x + 4 &\geq -8 &&\text{Theorem 3.2a} \\
-2x &\geq -12 &&\text{Theorem 3.2a} \\
x &\leq 6 &&\text{Theorem 3.2c}
\end{aligned}
$$

The solution set is $\{x \mid x \leq 6\}$.

EXAMPLE 4. Said Jerry to Ann: "I earn just \$25 less than twice as much as you do each week." If this couple needed a joint weekly income of at least \$170 to make ends meet, what is the least each of them must earn to avoid going into debt?

SOLUTION. Let x represent the number of dollars Ann earns each week. Then $2x - 25$ represents the number of dollars Jerry earns each week. Since their joint incomes must be greater than or equal to \$170 for them to make ends meet, we see that

$$
\begin{aligned}
x + (2x - 25) &\geq 170 \\
3x - 25 &\geq 170 \\
3x &\geq 195 \\
x &\geq 65
\end{aligned}
$$

Thus Ann must earn \$65 or more each week.

$$
\begin{aligned}
\text{Since } 2(65) - 25 &= 130 - 25 \\
&= 105
\end{aligned}
$$

Jerry must earn at least \$105 each week.

EXERCISES 3.3

Find the solution sets and graph. The domain is the set R of real numbers.

1. $3x + 9 < 0$ 5. $6x - 7 < -13$
2. $4x + 8 > 0$ 6. $7x + 1 \geq 15$
3. $5x - 15 > 0$ 7. $2x - 7 \leq 7$
4. $4x + 2 < 12$ 8. $3x + 5 > 20$

9. $5x - 7 < 18$

10. $4x + 1 \geq 10$

11. $2x - 7 < 10$

12. $4x + 7 < 2x + 1$

13. $5x + 11 > 8x + 26$

14. $2x - 1 < x + 4$

15. $3x + 5 > 5x - 4$

16. $(x - 1)(x + 2) > (x + 3)(x - 2)$

17. $x^2 - x + 1 \geq (x - 2)(x - 3)$

18. $(3x + 1)(2x - 2) > (6x - 7)(x + 1)$

19. $2x(x + 1) > (x - 2)(2x + 3)$

20. $x^3 < x(x - 1)(x + 3) - 2x^2$

21. Phil's age is 5 more than 3 times Terry's age and Phil is more than 31 years old. What is the youngest Terry can be?

22. On a three-day hike Marguerite, Frances, and Mary walked twice as far on the second day as the first and four miles on the third day. If the total distance they walked was not more than 22 miles, what is the longest distance they could have walked on the first day?

23. The Bryant's want to build a house that will have a floor area between 2400 square feet and 3200 square feet. The type of house that they want will cost $12 per square foot. Find the price range of their new house.

24. Jim saves stamps. On his birthday he received twice as many stamps as he had. After he gave one stamp away he had more than 59 stamps left. What is the smallest number of stamps Jim could have had before his birthday?

25. Henry's father is three times as old as Henry. In four years the sum of their ages will be less than 84 years. What is Henry's maximum present age?

3.5 ◆ Inequalities Involving Absolute Values

In finding solution sets of inequalities involving absolute values, we use the same method as we used to find solution sets of equations involving absolute values.

EXAMPLE 1. Find the solution set of $|2x + 3| < 5$. The domain is the set R.

SOLUTION. The given inequality means that on the number line the distance between the point labeled $2x + 3$ and the point labeled 0 is less than 5. If $y = 2x + 3$, we see that $|y| < 5$ means that the set of all values of y that satisfy the inequality have graphs on the number line between the point whose graph is -5 and the point whose graph is 5. These points are indicated by the heavy trace in Figure 3.4.

Figure 3.4

We see, then, that

$$-5 < 2x + 3 \quad \text{and} \quad 2x + 3 < 5$$
$$-8 < 2x \quad \text{and} \quad 2x < 2$$
$$-4 < x \quad \text{and} \quad x < 1$$

The solution set of $-4 < x$ is $\{x \mid -4 < x\}$. The solution set of $x < 1$ is $\{x \mid x < 1\}$. The solution set of the compound sentence $-5 < 2x + 3$ and $2x + 3 < 5$ is

$$\{x \mid -4 < x\} \cap \{x \mid x < 1\}$$

This set is usually written

$$\{x \mid -4 < x < 1\}$$

EXAMPLE 2. Find the solution set of $|x + 1| > 9$. The domain is the set R.

SOLUTION. The given inequality says that on the number line the distance between the point labeled $x + 1$ and the point labeled 0 is greater than 9. If $y = x + 1$, we see that $|y| > 9$ means that the set of all values that satisfy the given inequality have graphs on the number line to the right of the point labeled 9 or to the left of the point labeled -9 (Figure 3.5).

Then

$$
\begin{array}{ccc}
x + 1 < -9 & \text{or} & x + 1 > 9 \\
x < -10 & \text{or} & x > 8
\end{array}
$$

The solution set of $x < -10$ is $\{x \mid x < -10\}$. The solution set of $x > 8$ is $\{x \mid x > 8\}$. The solution set of the compound sentence $x + 1 < -9$ or $x + 1 > 9$ is

$$\{x \mid x < -10\} \cup \{x \mid x > 8\}$$

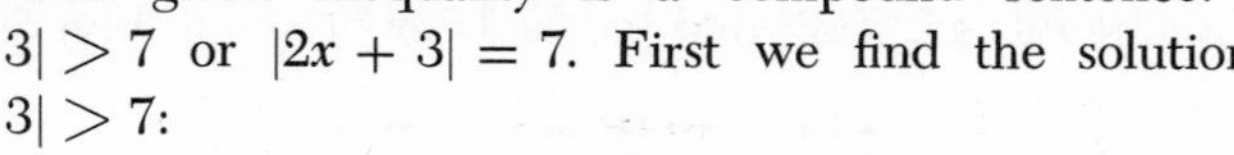

Figure 3.5

EXAMPLE 3. Find the solution set of $|2x + 3| \geq 7$. The domain is the set R.

SOLUTION. The given inequality is a compound sentence. It says $|2x + 3| > 7$ or $|2x + 3| = 7$. First we find the solution set of $|2x + 3| > 7$:

$$
\begin{array}{ccc}
2x + 3 < -7 & \text{or} & 2x + 3 > 7 \\
2x < -10 & \text{or} & 2x > 4 \\
x < -5 & \text{or} & x > 2
\end{array}
$$

The solution set is

$$\{x \mid x < -5\} \cup \{x \mid x > 2\}.$$

Now we find the solution set of $|2x + 3| = 7$.

$$2x + 3 = 7 \quad \text{or} \quad 2x + 3 = -7$$
$$2x = 4 \quad \text{or} \quad 2x = -10$$
$$x = 2 \quad \text{or} \quad x = -5$$

The solution set is

$$\{x \mid x = 2\} \cup \{x \mid x = -5\}$$

The solution set of $|2x + 3| \geq 7$ is

$$\{x \mid x < -5\} \cup \{x \mid x > 2\} \cup \{x \mid x = 2\} \cup \{x \mid x = -5\}$$

This set is usually written

$$\{x \mid x \leq -5\} \cup \{x \mid x \geq 2\}$$

EXERCISES 3.4

Find the solution sets. The domain is the set R of real numbers.

1. $|x - 3| > 2$
2. $|x - 3| < 2$
3. $|x + 1| < 5$
4. $|x + 1| > 5$
5. $|2x + 7| < 9$
6. $|2x + 7| \geq 9$
7. $|3x - 6| \geq 18$
8. $|2x - 3| < 3$
9. $|5x + 7| < 3$
10. $|6x - 5| > 7$

11. $|2x + 1| \leq 3$
12. $|3x + 2| \geq 5$
13. $|2x - 3| < 4$
14. $|4x + 3| < 1$
15. $|3x - 4| > 3$
16. $|7x - 8| < 9$
17. $|8x + 7| > 6$
18. $|12x + 5| \leq 13$
19. $|5x - 7| > 1$
20. $|7x + 4| < 2$

3.6 ◆ Quadratic Equations

An equation of the form

$$ax^2 + bx + c = 0$$

in which a, b, and c are real numbers, and $a \neq 0$, is called a **second degree** or **quadratic equation** over the real numbers. Any equation which is equivalent to an equation of this form will also be called a quadratic equation. We will use the following theorem to find the solution sets of quadratic equations.

THEOREM 3.3. If p and q are algebraic expressions in the variable x, then

for all values of x for which p and q name real numbers, a real number, r, is a solution of the equation

$$pq = 0$$

if and only if the value of p is equal to 0 when x is replaced by r or the value of q is equal to 0 when x is replaced by r or both.

Theorem 3.3 is a direct consequence of Theorem 1.7 which says that for real numbers a and b, $ab = 0$ if and only if $a = 0$, or $b = 0$, or both.

The solution set of a quadratic equation over R the set of real numbers, may have one member, or two members, or no members. For example, the quadratic equation

$$x^2 = 9$$

has the solution set $\{-3, 3\}$. The quadratic equation

$$x^2 = 0$$

has the solution set $\{0\}$. The quadratic equation

$$x^2 = -1$$

has no real number as its solution. The solution set in R is thus the empty set, $\varnothing$.

A quadratic equation of the form

$$(x - a)^2 = b^2, \text{ where } b > 0$$

has the solution set

$$\{a + b, a - b\}$$

We now find the solution set of the general quadratic equation:

(1) $ax^2 + bx + c = 0 \ (a \neq 0)$

We begin by using Theorem 3.1a, adding $-c$ to each member of (1). This produces the equivalent equation

(2) $ax^2 + bx = -c$

Since $a \neq 0$, $\dfrac{1}{a} \neq 0$ exists and we may use Theorem 3.1b, multiplying both members by $\dfrac{1}{a}$. This produces the equivalent equation

(3) $x^2 + \dfrac{b}{a}x = -\dfrac{c}{a}$

We now use the method called **completing the square** to arrive at a new equation whose left member will be the square of a binomial. Theorem 3.1a allows us to add $\left(\dfrac{b}{2a}\right)^2$ to both members of (3), obtaining

(4) $$x^2 + \frac{b}{a}x + \left(\frac{b}{2a}\right)^2 = \left(\frac{b}{2a}\right)^2 - \frac{c}{a}$$

The left member of (4) is the perfect square $\left(x + \dfrac{b}{2a}\right)^2$, giving

(5) $$\left(x + \frac{b}{2a}\right)^2 = \frac{b^2}{4a^2} - \frac{c}{a}$$

Equation 5 is equivalent to Equation 1 and we can find solutions in R, the set of real numbers, provided that

$$\frac{b^2}{4a^2} - \frac{c}{a} = \frac{b^2 - 4ac}{4a^2} \geq 0$$

If this condition is satisfied, then Equation 5 is equivalent to the **quadratic formula**

$$x = \frac{-b \pm \sqrt{b^2 - 4ac}}{2a}$$

where the symbol "$\pm$" is read "plus or minus." The solution set of the quadratic equation $ax^2 + bx + c = 0$ is

$$\left\{\frac{-b - \sqrt{b^2 - 4ac}}{2a}, \frac{-b + \sqrt{b^2 - 4ac}}{2a}\right\}$$

Examination of the quadratic formula thus shows that if $b^2 - 4ac < 0$, the solution set of $ax^2 + bx + c = 0$ over the set of real numbers is the empty set $\varnothing$. If $b^2 - 4ac = 0$, the solution set has one number in it. If $b^2 - 4ac > 0$, the solution set consists of two real numbers. The number represented by $b^2 - 4ac$ is called the **discriminant** of the quadratic equation.

EXAMPLE 1. Find the solution set of $2x^2 - 3x - 4 = 0$ over R, the set of real numbers.

SOLUTION. We see that $a = 2$, $b = -3$, and $c = -4$. Replacing a, b, and c by these values in the quadratic formula, we obtain the equivalent equations

$$x = \frac{-(-3) \pm \sqrt{(-3)^2 - 4(2)(-4)}}{(2)(2)}$$

$$x = \frac{3 \pm \sqrt{9 + 32}}{4}$$

$$x = \frac{3 \pm \sqrt{41}}{4}$$

The solution set is $\left\{ \dfrac{3 + \sqrt{41}}{4}, \dfrac{3 - \sqrt{41}}{4} \right\}$.

EXAMPLE 2. Determine $k \neq 0$ in $kx^2 + 3x - 4 = 0$ such that there (a) is one real solution; (b) are two real solutions.

SOLUTION. Here $a = k$, $b = 3$, and $c = -4$. Then

$$b^2 - 4ac = 9 + 16k$$

(a) If there is one real solution we have

$$9 + 16k = 0$$

and
$$k = -\tfrac{9}{16}$$

(b) If there are two real solutions we have

$$9 + 16k > 0$$

and
$$k > -\tfrac{9}{16}$$

EXAMPLE 3. The product of two consecutive even integers is 960. Find the two integers.

SOLUTION. Let x represent one of the integers. Since even integers differ by 2, the other integer can be represented by $x + 2$. Then

$$x(x + 2) = 960$$
$$x^2 + 2x = 960$$
$$x^2 + 2x - 960 = 0$$

We have a quadratic equation in which a is 1, b is 2, and c is -960. Replacing a, b, and c by these three values in the quadratic formula, we have

$$x = \frac{-2 + \sqrt{2^2 - 4(1)(-960)}}{2(1)}$$

$$= \frac{-2 \pm \sqrt{4 + 3840}}{2}$$

$$= \frac{-2 \pm \sqrt{3844}}{2}$$

$$= \frac{-2 \pm 62}{2}$$

And $x = -32$ or $x = 30$. When $x = -32$, $x + 2 = -30$. When $x = 30$, $x + 2 = 32$. Thus two consecutive even integers whose product is 960 are either -32 and -30 or 30 and 32.

EXERCISES 3.5

Find the solution set in R, the set of real numbers (Exercises 1–20).

1. $x^2 - 6x + 5 = 0$ 11. $3x^2 + x - 5 = 0$
2. $x^2 - 4x + 4 = 0$ 12. $2x^2 - 3x + 1 = 0$
3. $x^2 + x - 2 = 0$ 13. $x^2 - 5x - 7 = 0$
4. $9x^2 + 1 = 6x$ 14. $5x^2 + x - 3 = 0$
5. $x^2 + 8x + 4 = 0$ 15. $2x^2 + 2x - 1 = 0$
6. $4x^2 + 4x = 3$ 16. $4x^2 - 5x = 0$
7. $x^2 - 2x = 1$ 17. $8x^2 + 3x = 3$
8. $x^2 + 3x - 18 = 0$ 18. $3x^2 + 3x = 4$
9. $4x^2 - 6x - 1 = 0$ 19. $5x^2 - 2x - 4 = 0$
10. $7x^2 + x - 3 = 0$ 20. $19x^2 + 11x - 7 = 0$

For each of the following determine the values of k that give one real solution (Exercises 21–30).

21. $x^2 + kx + 25 = 0$ 26. $k^2x^2 + 1 + (4 - k)x = 0$
22. $3x^2 - kx + 3 = 0$ 27. $kx^2 + 5x + k = 0$
23. $x^2 + kx + k = -2 - 3x$ 28. $(2k - 5)x^2 + 2kx = 4$
24. $x^2 + kx + k + 3 = 0$ 29. $kx^2 + 12x + 9k = 0$
25. $2x^2 + 2kx + k^2 = 2$ 30. $(2k + 1)x^2 + 3kx + 4 = 0$

31. Jeanne has 20 more nickels than quarters. The product of the number of nickels and the number of quarters is 800. How many of each coin does she have?

32. On Wednesday Fred worked 10 more problems than on Tuesday. The product of the number of problems he worked on Tuesday and the number he worked on Wednesday is 600. How many problems did he work on Wednesday?

33. The tens digit of a number exceeds the units digit by 5 and the square of the units digit exceeds the tens digit by 7. What is the number?

34. Mrs. Boyle stapled 10 more sets of papers on Monday than on Friday. The square of the number she stapled on Monday exceeds 166 times the number she stapled on Friday by 700. How many sets of papers did she staple on Monday?

35. A rectangular field has an area of 180,000 square feet. The perimeter of the field is 2200 feet. What are the dimensions of the field?

3.7 ◆ Quadratic Inequalities

In solving quadratic inequalities, we first find an equivalent inequality written in **standard form:**

$$ax^2 + bx + c > 0$$

or

$$ax^2 + bx + c < 0$$

where a, b, and c are real numbers and $a \neq 0$. For example, the inequality

$$3x^2 - 2 < 5x$$

is equivalent to

$$3x^2 - 5x - 2 < 0$$

In order to find the solution set of a quadratic inequality, we first determine an equivalent inequality of the form $(x + p)^2 > k$ or of the form $(x + p)^2 < k$, p and k real numbers. We do this by completing the square.

EXAMPLE 1. Find the solution set of $3x^2 - 5x - 2 < 0$.

SOLUTION

$$
\begin{array}{ll}
3x^2 - 5x - 2 < 0 & \\
x^2 - \frac{5}{3}x - \frac{2}{3} < 0 & \text{Theorem 3.2b} \\
x^2 - \frac{5}{3}x < \frac{2}{3} & \text{Theorem 3.2a} \\
x^2 - \frac{5}{3}x + \frac{25}{36} < \frac{2}{3} + \frac{25}{36} & \text{Theorem 3.2a} \\
(x - \frac{5}{6})^2 < \frac{49}{36} & x^2 - 2ax + a^2 = (x - a)^2 \\
-\frac{7}{6} < x - \frac{5}{6} < \frac{7}{6} & \text{definition of square root} \\
-\frac{1}{3} < x < 2 & \text{Theorem 3.2a}
\end{array}
$$

The solution set is $\{x \mid -\frac{1}{3} < x < 2\}$.

EXAMPLE 2. Find the solution set of $x^2 - 6x + 9 > 0$.

SOLUTION

$$x^2 - 6x + 9 > 0$$
$$(x - 3)^2 > 0$$

Each real number except 3 is a solution of $(x - 3)^2 > 0$, since for all real numbers $a \neq 0$, $a^2 > 0$. Hence the solution set is $\{x \mid x \neq 3\}$.

EXAMPLE 3. Find the solution set of $x^2 + 2x + 1 < 0$.

SOLUTION

$$x^2 + 2x + 1 < 0$$
$$(x + 1)^2 < 0$$

There are no real number solutions of $(x + 1)^2 < 0$, since the squares of all real numbers are greater than or equal to 0. Thus the solution set is $\varnothing$.

EXAMPLE 4. Find the solution set of $3x^2 - 2x - 4 > 0$.

SOLUTION

$$3x^2 - 2x - 4 > 0$$
$$3x^2 - 2x > 4$$
$$x^2 - \tfrac{2}{3}x > \tfrac{4}{3}$$
$$x^2 - \tfrac{2}{3}x + \tfrac{1}{9} > \tfrac{4}{3} + \tfrac{1}{9}$$
$$(x - \tfrac{1}{3})^2 > \tfrac{13}{9}$$

$$x - \frac{1}{3} > \frac{\sqrt{13}}{3} \qquad \text{or} \qquad x - \frac{1}{3} < -\frac{\sqrt{13}}{3}$$

$$x > \frac{1}{3} + \frac{\sqrt{13}}{3} \qquad \text{or} \qquad x < \frac{1}{3} - \frac{\sqrt{13}}{3}$$

The solution set is $\left\{ x \mid x > \dfrac{1}{3} + \dfrac{\sqrt{13}}{3} \right\} \cup \left\{ x < \dfrac{1}{3} - \dfrac{\sqrt{13}}{3} \right\}$

EXAMPLE 5. Find the solution set of $-2x^2 + 4x < 1$.

SOLUTION

$$-2x^2 + 4x < 1$$
$$2x^2 - 4x > -1$$
$$x^2 - 2x > -\tfrac{1}{2}$$
$$x^2 - 2x + 1 > \tfrac{1}{2}$$
$$(x - 1)^2 > \tfrac{1}{2}$$
$$x - 1 > \tfrac{1}{2}\sqrt{2} \qquad \text{or} \qquad x - 1 < -\tfrac{1}{2}\sqrt{2}$$
$$x > 1 + \tfrac{1}{2}\sqrt{2} \qquad \text{or} \qquad x < 1 - \tfrac{1}{2}\sqrt{2}$$

The solution set is $\{ x \mid x > 1 + \tfrac{1}{2}\sqrt{2} \} \cup \{ x \mid x < 1 - \tfrac{1}{2}\sqrt{2} \}$.

3.8 ◆ Another Method for Solving Quadratic Inequalities

In solving quadratic inequalities we first find an equivalent inequality written in standard form:

$$ax^2 + bx + c < 0$$
or
$$ax^2 + bx + c > 0$$

where a, b, and c are real numbers and $a \neq 0$. We may approach the quadratic inequality by first considering the solution set of the quadratic equation we would have if we replaced the inequality symbol by an equal sign. Thus, for the example $3x^2 - 5x - 2 < 0$, we need to consider the equation

$$3x^2 - 5x - 2 = 0$$

Since the left member of this equation can be factored, we have

$$(3x + 1)(x - 2) = 0$$

whose solution set is $\{-\frac{1}{3}, 2\}$. (If the left member of the equation cannot be factored readily we use the quadratic formula to determine the solution set.)

We now see that the numbers $-\frac{1}{3}$ and 2 are definitely not solutions of the *inequality.* (Why?) Examining the number line (Figure 3.6), we see that excluding these two numbers leaves us with three sets of numbers which are possible candidates as part of the solution set of the given quadratic inequality:

$$L = \{x \mid x < -\frac{1}{3}\}$$
$$M = \{x \mid -\frac{1}{3} < x < 2\}$$
and
$$P = \{x \mid x > 2\}$$

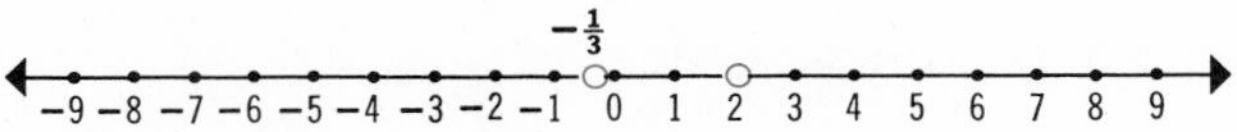

Figure 3.6

We examine each of these sets in turn.

1. If $x \in L$, then examining whether x is or is not a solution, we find

$$x < -\frac{1}{3}$$

$$
\begin{array}{lll}
3x < -1 & \text{and} & x < -\frac{1}{3} \\
3x + 1 < 0 & \text{and} & x - 2 < -\frac{7}{3} \\
3x + 1 < 0 & \text{and} & x - 2 < 0
\end{array}
$$

$$(3x + 1)(x - 2) > 0$$

But we want $(3x + 1)(x - 2) < 2$. Therefore, no member of L is a solution of the given inequality.

2. If $x \in M$, then examining whether x is or is not a solution, we find

$$
\begin{array}{lll}
x > -\frac{1}{3} & \text{and} & x < 2 \\
3x > -1 & \text{and} & x < 2 \\
3x + 1 > 0 & \text{and} & x - 2 < 0
\end{array}
$$

$$(3x + 1)(x - 2) < 0$$

Therefore, every member of M is a solution of the given inequality.

3. If $x \in p$, then examining whether x is or is not a solution, we find

$$x > 2$$
$$3x > 6 \qquad \text{and} \qquad x > 2$$
$$3x + 1 > 7 \qquad \text{and} \qquad x - 2 > 0$$
$$3x + 1 > 0 \qquad \text{and} \qquad x - 2 > 0$$
$$(3x + 1)(x - 2) > 0$$

We want $(3x + 1)(x - 2) < 0$. Therefore, no member of p is a solution of the given inequality. The solution set of $3x^2 - 5x - 2 < 0$ is $\{x \mid -\frac{1}{3} < x < 2\}$. The graph of this solution set is shown in Figure 3.7.

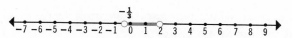

Figure 3.7

The above procedure may be followed whenever the quadratic equation derived from a given quadratic inequality has a solution set with two distinct elements. Thus to find the solution set of

$$ax^2 + bx + c < 0$$

or

$$ax^2 + bx + c > 0$$

we find the solution set of

$$ax^2 + bx + c = 0$$

If this solution set is $\{r_1, r_2\}$, where $r_1 < r_2$, we need only examine the positive or negative character of the quadratic polynomial

$$ax^2 + bx + c$$

on each of the three sets obtained from removing r_1 and r_2 from the set of real numbers.

It may happen that the quadratic equation we obtain from a given quadratic inequality has only one number in its solution set. For example, in seeking the solution set of

$$x^2 - 6x + 9 > 0$$

we find that the quadratic equation

$$x^2 - 6x + 9 = 0$$

is equivalent to

$$(x - 3)^2 = 0$$

whose solution set is $\{3\}$. Following a similar procedure, removing 3 from the set of real numbers leaves us two sets:

$$L = \{x \mid x < 3\}$$
and
$$P = \{x \mid x > 3\}$$

Examining each of these, we have

(i) If $x < 3$, then $x - 3 < 0$ and $(x - 3)^2 > 0$

(ii) If $x > 3$, then $x - 3 > 0$ and $(x - 3)^2 > 0$

We conclude that the solution set of

$$x^2 - 6x + 9 > 0$$

is $\{x \mid x < 3 \text{ or } x > 3\}$ which can also be written $\{x \mid x \neq 3\}$.

In general, if the quadratic equation

$$ax^2 + bx + c = 0$$

has a solution set $\{r\}$ which contains only one number, then the solution set of either of the quadratic inequalities possible is either $\{x \mid x \neq r\}$ or is the empty set, ϕ.

The final case occurs when the quadratic equation has an empty solution set. In this case the solution set of either of the quadratic inequalities is the set of all real numbers or is empty.

EXERCISES 3.6

Find the solution sets over R, the set of real numbers.

1. $x^2 - 2x + 1 < 0$
2. $x^2 - 2x + 1 > 0$
3. $x^2 - 3x - 10 > 0$
4. $x^2 - 3x - 10 < 0$
5. $x^2 + 8x + 15 \leq 0$
6. $x^2 + 7x + 12 > 0$
7. $x^2 - 5x + 6 \leq 0$
8. $x^2 + 10x + 24 > 0$
9. $x^2 - 4x - 21 > 0$
10. $2x^2 + 7x - 15 \geq 0$

11. $2x^2 + 5x - 12 \leq 0$
12. $x^2 - 5x \geq 8$
13. $x^2 - 6x \leq 5$
14. $2x^2 - x - 1 < 0$
15. $6 - x - x^2 \geq 0$
16. $5 - 4x < x^2$
17. $x^2 + 2x - 5 > 0$
18. $x^2 - 3x + 10 < 0$
19. $3x^2 - 5x - 1 > 0$
20. $2x^2 + x - 5 < 0$

3.9 ◆ Equations Containing Rational Expressions

Some equations involving rational expressions may be solved by solving equivalent quadratic equations. In using this method for finding solutions sets of fractional equations, we must be certain that the new equations are truly equivalent to the given equations.

EXAMPLE 1. Find the solution set of $\dfrac{2x + 5}{x} + \dfrac{x - 1}{2} = 5$.

SOLUTION. In examining the given equation we see that 0 is not a permissible replacement for x, since in this case the denominator of $\dfrac{2x + 5}{x}$ vanishes (that is, it is equal to 0). We find an equivalent equation by multiplying each member of the given equation by $2x$ to obtain a new equation that contains no quotients. This procedure produces an equivalent equation if and only if $x \neq 0$, that is, x cannot be 0. Then

$$\frac{2x + 5}{x} + \frac{x - 1}{2} = 5$$

$$2(2x + 5) + x(x - 1) = 5(2x) \qquad \text{and} \qquad x \neq 0$$
$$4x + 10 + x^2 - x = 10x \qquad \text{and} \qquad x \neq 0$$
$$x^2 - 7x + 10 = 0 \qquad \text{and} \qquad x \neq 0$$
$$(x - 5)(x - 2) = 0$$

The solution set is $\{2, 5\}$.

CHECK

$$\frac{2(2) + 5}{2} + \frac{2 - 1}{2} = \frac{4 + 5}{2} + \frac{1}{2}$$

$$= \frac{9}{2} + \frac{1}{2}$$

$$= \frac{10}{2}$$

$$= 5$$

$$\frac{2(5) + 5}{5} + \frac{5 - 1}{2} = \frac{10 + 5}{5} + \frac{4}{2}$$

$$= \frac{15}{5} + \frac{4}{2}$$

$$= 3 + 2$$

$$= 5$$

EXAMPLE 2. Find the solution set of $1 + \dfrac{12}{x^2 - 4} = \dfrac{3}{x - 2}$.

SOLUTION. In examining the given equation, we see that 2 and -2 are not permissible replacements for x, since if x is 2 the denominators of

$\dfrac{12}{x^2 - 4}$ and $\dfrac{3}{x - 2}$ vanish; and if x is -2 the denominator of $\dfrac{12}{x^2 - 4}$ vanishes. We find an equivalent equation by multiplying each member of the given equation by $x^2 - 4$ to obtain a new equation which contains no quotients. This procedure produces an equivalent equation if and only if we also have $x \neq 2$ and $x \neq -2$. Then

$$(x^2 - 4)(1) + (x^2 - 4) \cdot \dfrac{12}{x^2 - 4} = (x^2 - 4) \cdot \dfrac{3}{x - 2} \text{ and } x \neq \pm 2$$

$$x^2 - 4 + 12 = (x + 2)(3) \text{ and } x \neq \pm 2$$
$$x^2 - 4 + 12 = 3x + 6 \text{ and } x \neq \pm 2$$
$$x^2 - 3x + 2 = 0 \text{ and } x \neq \pm 2$$
$$(x - 2)(x - 1) = 0 \text{ and } x \neq \pm 2$$
$$x = 2 \text{ or } x = 1 \text{ and } x \neq \pm 2$$

Since 2 is not a permissible value of x, the solution set is $\{1\}$.

CHECK

$$1 + \dfrac{12}{(1)^2 - 4} = \dfrac{3}{1 - 2}$$

$$1 + \dfrac{12}{-3} = \dfrac{3}{-1}$$

$$1 - 4 = -3$$

$$-3 = -3$$

EXERCISES 3.7

Find the solution sets over R, the set of real numbers.

1. $\dfrac{1}{2x} + \dfrac{1}{x} = \dfrac{1}{6}$

2. $\dfrac{14}{3x} + \dfrac{6}{x} = 2$

3. $\dfrac{x + 5}{x} - \dfrac{x - 1}{4} = \dfrac{5}{4}$

4. $\dfrac{7}{x - 3} + \dfrac{3}{x - 4} = \dfrac{1}{2}$

5. $\dfrac{x - 1}{x - 2} + \dfrac{x - 3}{x - 4} = 2$

6. $\dfrac{2x}{x + 1} = \dfrac{2x - 1}{x - 1}$

7. $\dfrac{x - 4}{x + 4} + \dfrac{x + 4}{x - 4} = 3$

8. $\dfrac{2}{x - 8} - \dfrac{1}{x - 2} = 0$

9. $\dfrac{1}{x^2} = \dfrac{1}{4} + \dfrac{1}{5}$

10. $\dfrac{2}{x + 2} = \dfrac{x^2 + 4}{x^2 - 4} + \dfrac{x}{2 - x}$

11. $\dfrac{x}{x - 1} - \dfrac{8}{x + 1} = \dfrac{2}{1 - x^2}$

12. $\dfrac{3}{3x - 4} = \dfrac{4}{3x + 4} - \dfrac{5}{9x^2 - 16}$

13. $\dfrac{x - 3}{3 - x} + \dfrac{3x + 1}{x^2 - 9} = \dfrac{1 - 5x}{x + 3}$

14. $\dfrac{1}{10 + x} + \dfrac{1}{10 - x} = \dfrac{5}{24}$

15. $\dfrac{3}{x} = \dfrac{x + 13}{x^2 - x} - \dfrac{6}{x - 1}$

16. $\dfrac{4}{x + 1} - \dfrac{x}{3 - x} = \dfrac{x^2 - 2x + 2}{x^2 - 2x - 3}$

17. $\dfrac{2x - 1}{x + 3} + \dfrac{x - 4}{x - 2} = -\dfrac{1}{x^2 + x - 6}$

18. $\dfrac{2x}{x - 3} + \dfrac{3x + 9}{2x^2 - 3x - 9} + \dfrac{1}{2x + 3} = 0$

19. $\dfrac{\dfrac{x}{x + 1} + 1}{\dfrac{x}{x - 1} - 1} = \dfrac{\dfrac{x}{x - 1} + 1}{\dfrac{x}{x + 1} - 1}$

20. $\dfrac{x + 4}{3x^2 - 5x - 2} + \dfrac{2x + 5}{2x^2 - x - 6} + \dfrac{4x + 3}{6x^2 + 11x + 3} = 0$

3.10 ◆ Equations Containing Radicals

Some equations involving radicals may be reduced to <u>linear</u> or <u>quadratic</u> equations by squaring both members until no radicals appear in the resulting equation. Squaring both members of a given equation does *not* produce an equation equivalent to the given equation. It can be shown, although we shall not do it here, that the solution set of an equation is a subset of the solution set of the equation produced by squaring both its members. Extreme care must be taken, therefore, to check in the original equation all solutions found in the manner illustrated in the examples below.

EXAMPLE 1. Find the solution set of $\sqrt{8x - 7} - 3 = 0$.

SOLUTION. Adding 3 to both members of the given equation gives the equivalent equation

$$\sqrt{8x - 7} = 3$$

Squaring both members of the above equation, we have

$$8x - 7 = 9$$

and

$$8x = 16$$
$$x = 2$$

The solution set is $\{2\}$. Checking in the original equation we find that 2 is a solution and the solution set of the given equation is $\{2\}$.

EXAMPLE 2. Find the solution set of $\sqrt{3 - x} = \sqrt{2 + x} + 3$.

SOLUTION. Squaring both members of the given equation, we obtain

$$3 - x = 2 + x + 6\sqrt{2 + x} + 9$$
$$-2x - 8 = 6\sqrt{2 + x}$$
$$x + 4 = -3\sqrt{2 + x}$$

Squaring both members of the preceding equation, we have

$$x^2 + 8x + 16 = 9(2 + x)$$
$$x^2 - x - 2 = 0$$
$$(x - 2)(x + 1) = 0$$

and $x = 2$ or $x = -1$. The solution set is $\{2, -1\}$. Checking in the original equation, we find that neither 2 nor -1 is a solution and the solution set of the given equation is $\varnothing$.

EXERCISES 3.8

Find the solution set over R, the set of real numbers.

1. $\sqrt{x - 1} = 2$
2. $\sqrt{x - 12} = 4$
3. $\sqrt{x - 1} = 4$
4. $\sqrt{25 - x} = 3$
5. $x - 3 = \sqrt{x - 1}$
6. $\sqrt{x + 2} = \sqrt{x} + 3$
7. $\sqrt{x} = 3 - \sqrt{x + 1}$
8. $\sqrt{3x - 11} = 1 + \sqrt{2x - 9}$
9. $\sqrt{3x - 2} = 1 - \sqrt{2x - 2}$
10. $x - 4 = \sqrt{2x - 5}$
11. $\sqrt{5x - 1} + x = 5$

12. $\sqrt{2 + x} = \sqrt{3 - x} - 3$
13. $\sqrt{2x - 5} - 2 = \sqrt{x - 2}$
14. $\sqrt{x^2 + 3x + 7} - \sqrt{x^2 - 3x + 9} + 2 = 0$
15. $\sqrt{3x - \frac{1}{2}} - \sqrt{x + \frac{1}{2}} = \sqrt{6x + 1}$
16. $\sqrt{9x + 10} = \sqrt{4x + 5} + \sqrt{x + 1}$
17. $\sqrt{x - 2} + \sqrt{x + 3} = \sqrt{4x - 1}$
18. $2\sqrt{x + 4} = \sqrt{9 - x} - 2\sqrt{x + 4}$
19. $\dfrac{\sqrt{x + 3}}{\sqrt{x - 3}} - 2 = \dfrac{\sqrt{x - 3}}{\sqrt{x + 3}}$
20. $\dfrac{\sqrt{3x^2 + 4} + \sqrt{x^2 + 5}}{\sqrt{3x^2 + 4} - \sqrt{x^2 + 5}} - 7 = 0$

3.11 ◆ Equations with Quadratic Form

There are many equations which are not themselves quadratic, but which are equivalent to equations having the form

$$aw^2 + bw + c = 0 \qquad a \neq 0$$

where w represents an expression in another variable. Some examples of such equations are:

    (a) $x - 10\sqrt{x} + 9 = 0$
    (b) $x^4 - 5x^2 + 6 = 0$
    (c) $8x^{1/2} + 7x^{1/4} - 1 = 0$

These equations are said to have quadratic form, since:

(a) $x - 10\sqrt{x} + 9 = 0$ is equivalent to $(\sqrt{x})^2 - 10\sqrt{x} + 9 = 0$.
(b) $x^4 - 5x^2 + 6 = 0$ is equivalent to $(x^2)^2 - 5x^2 + 6 = 0$.
(c) $8x^{1/2} + 7x^{1/4} - 1 = 0$ is equivalent to $8(x^{1/4})^2 + 7x^{1/4} - 1 = 0$.

We can solve an equation that has quadratic form by first finding the solutions of the equivalent quadratic equation in the new variable and then solving the resulting equations.

EXAMPLE 1. Find the solution set of $x - 10\sqrt{x} + 9 = 0$.

SOLUTION. If we let $w = \sqrt{x}$, we have the equivalent quadratic equations

$$w^2 - 10w + 9 = 0$$
$$(w - 9)(w - 1) = 0$$

which has solution set $\{9, 1\}$. The given equation is equivalent to

$$\sqrt{x} = 1 \qquad \text{or} \qquad \sqrt{x} = 9$$
$$x = 1 \qquad \text{or} \qquad x = 81$$

from which we obtain the solution set $\{1, 81\}$.

CHECK

$$1 - 10\sqrt{1} + 9 = 1 - 10 + 9 = 0$$
$$81 - 10\sqrt{81} + 9 = 81 - 10 \cdot 9 + 9 = 81 - 90 + 9 = 0$$

EXAMPLE 2. Find the solution set of $x^{-2} - x^{-1} - 12 = 0$.

SOLUTION. Let $w = x^{-1}$. Then the given equation is equivalent to

$$w^2 - w - 12 = 0$$
$$(w - 4)(w + 3)$$

which has solution set $\{4, -3\}$. The given equation is equivalent to

$$x^{-1} = 4 \qquad \text{or} \qquad x^{-1} = -3$$
$$x = \tfrac{1}{4} \qquad \text{or} \qquad x = -\tfrac{1}{3}$$

The solution set is $\{-\tfrac{1}{3}, \tfrac{1}{4}\}$.

CHECK

$$(\tfrac{1}{4})^{-2} - (\tfrac{1}{4})^{-1} - 12 = 4^2 - 4 - 12 = 16 - 4 - 12 = 0$$
$$(-\tfrac{1}{3})^{-2} - (-\tfrac{1}{3})^{-1} - 12 = (-3)^2 - (-3) - 12 = 9 + 3 - 12 = 0$$

EXAMPLE 3. Find the solution set of $x + 2\sqrt{x} - 3 = 0$.

SOLUTION. If we let $w = \sqrt{x}$, we have the equivalent equations

$$w^2 + 2w - 3 = 0$$
$$(w + 3)(w - 1) = 0$$

which has solution set $\{-3, 1\}$. Then the given equation is equivalent to

$$\sqrt{x} = -3 \quad \text{or} \quad \sqrt{x} = 1$$

Since $\sqrt{x}$ must be greater than or equal to 0 when x is a real number, we reject $\sqrt{x} = -3$. Then

$$\sqrt{x} = 1 \quad \text{and} \quad x = 1$$

The solution set is $\{1\}$.

CHECK

$$1^2 + 2\sqrt{1} - 3 = 1 + 2 - 3 = 0$$

EXAMPLE 4. Find the solution set of $x^{-2} + x^{-1} = 0$.

SOLUTION. Let $w = x^{-1}$. Then we have

$$w^2 + w = 0$$
$$w(w + 1) = 0$$
$$w = 0 \quad \text{or} \quad w = -1$$

But the given equation may be written

$$\frac{1}{x^2} + \frac{1}{x} = 0$$

We see that 0 is not a permissible replacement for x. When x is replaced by -1 we have

$$\frac{1}{1} + \frac{1}{-1} = 0$$

and the solution set is $\{-1\}$.

EXERCISES 3.9

Find the solution set over R, the set of real numbers.

1. $x^4 + 8x^2 - 9 = 0$
2. $x^4 + 21x^2 - 100 = 0$
3. $x^4 - 3x^2 - 108 = 0$
4. $4x^4 + 13x^2 - 75 = 0$
5. $4x^4 - 35x^2 - 9 = 0$
6. $x^4 - 10x^2 + 16 = 0$

7. $x^{-4} - 14x^{-2} + 45 = 0$
8. $x^{-6} + 26x^{-3} - 27 = 0$
9. $x^{3/2} - 7x^{3/4} - 8 = 0$
10. $x^{1/2} - 3x^{1/4} - 4 = 0$
11. $x^{1/4} - 6x^{-1/4} - 1 = 0$
12. $x^{1/3} + 6x^{-1/3} - 5 = 0$

13. $x^{-2} - 3x^{-1} + 2 = 0$

14. $8x^{1/2} - 8x^{-1/2} - 63 = 0$

15. $x^3 + x\sqrt{x} - 72 = 0$

16. $(x - 5) + 2(x - 5)^{1/2} - 8 = 0$

17. $(x^2 + 1)^2 + 4(x^2 + 1) - 45 = 0$

18. $(x^2 - x)^{-2} - \frac{2}{3}(x^2 - x)^{-1} + \frac{1}{12} = 0$

19. $(x^2 - x)^2 + 24 = 14(x^2 - x)$

20. $\dfrac{x^2}{x + 1} + 2\left(\dfrac{x + 1}{x^2}\right) = 3$

CHAPTER REVIEW

Find the solution sets over R, the set of real numbers (Exercises 1 and 2).

1. $3x - 6(5 - 2x) = 5(4x - 3)$.

2. $4 + x^2 = 10 + (x + 2)(x - 3)$.

3. The numerator of a quotient is 5 less than the denominator. If the numerator is increased by 1 and the denominator is increased by 6, the quotient is unchanged. Find this quotient.

Find the solution sets (Exercises 4 and 5).

4. $|2x - 4| = 3$

5. $|x - 5| = |x + 3|$

Find the solution sets and graph (Exercises 6 and 7).

6. $3x + 2 < 11$

7. $2x - 5 \geq 13$

8. George received grades of 83 and 74 on the first two tests in a course. His instructor announced that the course grade would be determined by the average of three test grades, with A being assigned for averages 90–100, B for 80–89, C for 70–79, D for 60–69, and F for 0–59. George received a grade of B. What is the lowest grade he could have obtained on the third test?

Find the solution sets and graph (Exercises 9 and 10).

9. $|x - 3| < 2$

10. $|2x + 5| \geq 1$

Find the solution sets (Exercises 11–13).

11. $x^2 - 3x - 4 = 0$

12. $2x^2 + x - 6 = 0$

13. $3x^2 + x - 1 = 0$

14. A rectangular piece of cardboard is twice as long as it is wide. Squares which measure 2 inches on a side are cut from the corners and the ends folded up to form a box with a volume of 896 cubic inches. What were the original dimensions of the cardboard?

Find the solution sets and graph (Exercises 15 and 16).

15. $x^2 - x - 6 > 0$

16. $2x^2 - 3x - 7 < 0$

Find the solution sets (Exercises 17–20).

17. $x - 4 + \dfrac{1}{x} = 0$

18. $\dfrac{1 + x}{1 - x} = \dfrac{x - 1}{x + 1}$

19. $\sqrt{x + 2} + \sqrt{3 - x} = 3$

20. $x^4 + 4x^2 = 12$

CHAPTER **4**

Relations and Functions

4.1 ✦ Cartesian Products

When we form a pair of objects or numbers, in some cases the order in which we present these objects is immaterial, but in other cases it is important that the order of pairings be specified. For example, a list of pairs of names of colleges that appears in Saturday morning's paper is a list of those colleges which will compete in football games that afternoon. If it is agreed that on Sunday the winners will be named first, then the order of each pair in the list in Sunday morning's paper is important. Thus, Yale, Harvard is a perfectly good listing in Saturday morning's paper, but if Harvard wins the Saturday game, it is important that the listing in Sunday's paper be Harvard, Yale.

A pair of objects in which we distinguish the order in which the objects of the pair are given is called an **ordered pair.** If we consider a and b, in that order, we use the symbol (a, b) to denote the ordered pair. In the ordered pair (a, b), a is called the **first component** and b the **second component.** Two ordered pairs (a, b) and (c, d) are said to be **equal,** denoted by $(a, b) = (c, d)$, if and only if $a = c$ and $b = d$. In an ordered pair the first and second components may be the same as in the ordered pair $(1, 1)$.

We use the definition of an ordered pair to define another operation on sets, the Cartesian product.

DEFINITION 4.1. The **Cartesian product** of two sets A and B, denoted by $A \times B$ and read "A cross B," is the set of all ordered pairs of the form (a, b) where a is an element of set A and b is an element of set B.

Using set-builder notation, we write

$$A \times B = \{(a, b) \mid a \in A \text{ and } b \in B\}$$

For example, if $A = \{1, 2, 3\}$ and $B = \{7, 9\}$, then

$$A \times B = \{(1, 7), (2, 7), (3, 7), (1, 9), (2, 9), (3, 9)\}$$

and

$$B \times A = \{(7, 1), (7, 2), (7, 3), (9, 1), (9, 2), (9, 3)\}$$

We can also form the Cartesian product of a set with itself. Thus

$$A \times A = \{(1, 1), (1, 2), (1, 3), (2, 1), (2, 2), (2, 3), (3, 1), (3, 2), (3, 3)\}$$

and

$$B \times B = \{(7, 7), (7, 9), (9, 7), (9, 9)\}$$

We will be particularly concerned with the Cartesian product $R \times R$, where R is the set of real numbers. We call $R \times R$ the **Cartesian plane**. The set $R \times R$ is the set of all possible ordered pairs whose components are real numbers.

On the number line, real numbers are associated with points on the line. We will associate ordered pairs of real numbers with points on a plane and use this association to picture sets of ordered pairs geometrically. We begin by drawing two lines, one placed horizontally and the other vertically, that are perpendicular to each other and intersect at the point on each line with label 0. We agree that the positive direction is upward on the vertical line and to the right on the horizontal line (Figure 4.1). These lines are called the **coordinate axes**. The horizontal line is called the **axis of abscissas** or the **x-axis**

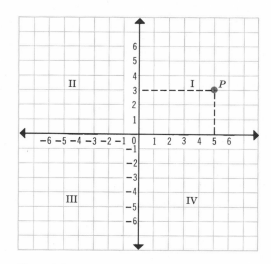

Figure 4.1

and the vertical line is called the **axis of ordinates** or the **y-axis**. The point of intersection of the two axes is called the **origin**.

Let P be any point on the plane. We now construct lines through P perpendicular to the two axes. If x is the number line coordinate of the point where the vertical line through P intersects the horizontal axis and y is the number line coordinate of the point where the horizontal line through P intersects the vertical axis, then the ordered pair of numbers (x, y) will be the label of the point P. The numbers x and y are called the **coordinates** of P. The number x is called the **x-coordinate** or **abscissa** of P. The number y is called the **y-coordinate** or **ordinate** of P. This one-to-one correspondence between points in the plane and ordered pairs of real numbers is called a **Cartesian** or **rectangular coordinate system.**

The two axes separate the plane into four regions called **quadrants**. These quadrants are numbered I, II, III, and IV as shown in Figure 4.1. The coordinate axes are *not* included in any of the quadrants. The first quadrant is, therefore,

$$Q\text{-}I = \{(x, y) \mid x > 0 \text{ and } y > 0\}$$

The other three quadrants are given by:

$$Q\text{-}II = \{(x, y) \mid x < 0 \text{ and } y > 0\}$$
$$Q\text{-}III = \{(x, y) \mid x < 0 \text{ and } y < 0\}$$
$$Q\text{-}IV = \{(x, y) \mid x > 0 \text{ and } y < 0\}$$

The horizontal or x-axis is given by

$$x\text{-axis} = \{(x, y) \mid y = 0\}$$

The vertical or y-axis is given by

$$y\text{-axis} = \{(x, y) \mid x = 0\}$$

EXAMPLE 1. Draw a coordinate system and graph the points whose coordinates are $(0, 4)$, $(3, 0)$, $(1, 2)$, $(2, -1)$, $(-1, -2)$, and $(-3, 1)$.

SOLUTION. See Figure 4.2 on page 94.

Open sentences such as

$$3x + 2y = 6$$
$$2x - y > 4$$

are called **open sentences in two variables** and have ordered pairs of real numbers as **solutions**. If we have an open sentence in the two variables x

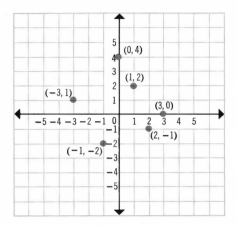

Figure 4.2

and y, we agree that x is a variable representing the first components of the ordered pairs that are solutions and y is the variable representing the second components. For example, the ordered pair $(0, 3)$ is a solution of the open sentence $3x + 2y = 6$ because if the components of $(0, 3)$ are substituted for the variables x and y, in that order, the result is

$$3 \cdot 0 + 2 \cdot 3 = 6$$

which is a true statement. On the other hand, $(3, 0)$ is not a solution of this sentence because replacing x by 3 and y by 0 gives
$$3 \cdot 3 + 2 \cdot 0 = 6$$

which is a false statement.

The solution set over $A \times A$ of an open sentence in two variables is the set of all ordered pairs of $A \times A$ that are solutions of the sentence.

DEFINITION 4.2. For a given set A, the **solution set** over $A \times A$ of an open sentence in two variables is the subset of all ordered pairs of $A \times A$ that are solutions of that open sentence.

We find ordered pairs that are solutions of a given open sentence by assigning values to one variable of the open sentence and determining the paired values for the other variable. Thus, for the open sentence

$$y - x = 1, x \in A, y \in A$$

where

$$A = \{1, 2, 3, 4, 5\}$$

we can obtain all solutions in $A \times A$ by replacing x with 1, 2, 3, 4, and 5 in turn. This gives $(1, 2)$, $(2, 3)$, $(3, 4)$, $(4, 5)$, and $(5, 6)$ as possible solutions. However, since $6 \notin A$, it follows that $(5, 6) \notin A \times A$, and thus is not a solution of the given sentence in $A \times A$. The solution set over $A \times A$ is

$$S = \{(1, 2), (2, 3), (3, 4), (4, 5)\}$$

We may use set-builder notation to denote the solution set S above:

$$S = \{(x, y) \mid y - x = 1, x \in A, y \in A\}$$

In this book we shall be finding solutions sets of open sentences over $R \times R$, where R is the set of real numbers. Since many open sentences have an infinite number of ordered pairs in their solution sets, it will be impossible to list all of the members of the solution sets in these cases. In such cases, we use set builder notation to represent the solution set. For example, the solution set over $R \times R$ of the sentence above, $y - x = 1$, is an infinite set of ordered pairs. We represent the solution set over $R \times R$ by

$$\{(x, y) \mid y - x = 1\}$$

EXAMPLE 2. Let $A = \{1, 2, 3, 4, 5, 6\}$. List the members of $S = \{(x, y) \mid x + y > 8\}$, a subset of $A \times A$.

SOLUTION. The set of ordered pairs in $A \times A$ the sum of whose components is greater than 8 $(x + y > 8)$ is

$$\{(3, 6), (4, 5), (4, 6), (5, 4), (5, 5), (5, 6), (6, 3), (6, 4), (6, 5), (6, 6)\}$$

EXAMPLE 3. Draw a coordinate system and shade the region in which the point (x, y) can be found if $x > 2$ and $y < 3$.

SOLUTION. See Figure 4.3 on page 96. All the points with x-coordinates greater than 2 lie to the right of the vertical line 2 units to the right of the y-axis, that is, the line whose equation is $x = 2$. A portion of this region is shown by \\\\\\. All points whose y-coordinates are less than 3 lie below the horizontal line 3 units above the x-axis, that is, the line whose equation is $y = 3$. A portion of this region is shown by /////. The region in which the point (x, y) has x-coordinate greater than 2 and y-coordinate less than 3 is the region which is cross-hatched XXXXX.

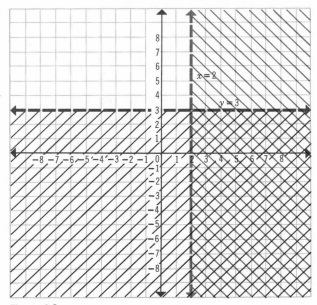

Figure 4.3

EXAMPLE 4. Draw a coordinate system and shade the region in which the point (x, y) can be found such that $|x| \geq 1$ and $|y| \geq 4$.

SOLUTION. See Figure 4.4 on page 97. Any point (x, y) such that $|x| \geq 1$ will lie on the line which is the graph of $x = 1$, or to the right of this line, or on the line which is the graph of $x = -1$, or to the left of this line. A portion of this region is shown by \\\\. Any point (x, y) such that $|y| \geq 4$ will lie on the line that is the graph of $y = 4$, or above this line, or on the line that is the graph of $y = -4$, or below this line. A portion of this region is shown by ////. The region in which the point (x, y) has x-coordinate such that $|x| \geq 1$ and y-coordinate such that $|y| \geq 4$ lies in the region that is crosshatched, XXXX.

EXAMPLE 5. Let $A = \{1, 2, 3, 4, 5\}$. List the members of the set $\{(x, y) \mid y = 5\} \cup \{(x, y) \mid x = 2\}$ which is a subset of $A \times A$.

SOLUTION. The set of ordered pairs of $A \times A$ with y-coordinates equal to 5 is

$$\{(1, 5), (2, 5), (3, 5), (4, 5), (5, 5)\}.$$

The set of ordered pairs of $A \times A$ with x-coordinates equal to 2 is

$$\{(2, 1), (2, 2), (2, 3), (2, 4), (2, 5)\}.$$

The union of these two sets is

$$\{(1, 5), (2, 5), (3, 5), (4, 5), (5, 5), (2, 1), (2, 2), (2, 3), (2, 4)\}.$$

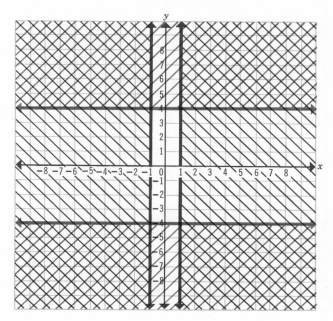

Figure 4.4

EXERCISES 4.1

Determine the Cartesian products $A \times B$ of the following (Exercises 1–6).

1. $A = \{1, 2\}, B = \{3, 6, 7\}$.
2. $A = \{-1, 0, 1\}, B = \{0, 1, 2\}$.
3. $A = \{a, b, c\}, B = \{e, f\}$.
4. $A = \{-3, -2, -1\}, B = \{1, 2, 3\}$.
5. $A = \{c, a, t\}, B = \{d, o, g\}$.
6. $A = \{s, u, z, i, e\}, B = \{g, a, i\}$.
7. If $A = \{1, 2, 3\}$ and $B = \{r, u, t, h\}$, how many elements are in $A \times B$?
8. If there are n elements in the set A and m elements in the set B, how many elements are in $A \times B$? in $B \times A$? in $A \times A$? in $B \times B$?
9. What is the x-coordinate of all points in the y-axis?
10. What is the y-coordinate of all points on the x-axis?

Draw a coordinate system and shade the region(s) in which the point (x, y) can be found if the following is known (Exercises 11–21).

11. $x > 0$ and $y > 0$.
12. $x > 0$ and $y < 0$.
13. $0 < x < 1$ and $1 < y < 2$.
14. $|x| = 2$ and $|y| \geq 1$.

15. $-1 \leq x \leq 1$ and $y < 1$.
16. $1 < x < 3$ and $1 < y < 2$.
17. $|x| \leq 2$ and $|y| \geq 1$.
18. $-1 < x < 1$ and $|y| \geq 1$.
19. $-3 < x < 3$ and $y > 2$.
20. $1 < x < 5$ and $-1 < y < 1$.
21. $|x| > 2$ or $|y| > 1$.

Let $A = \{1, 2, 3, 4, 5, 6, 7, 8, 9, 10\}$. List the members of the following sets which are subsets of $A \times A$ (Exercises 21–30).

22. $S = \{(x, y) \mid x + y = 10\}$.
23. $S = \{(x, y) \mid y = x - 5\}$.
24. $S = \{(x, y) \mid y = 3x + 1\}$.
25. $S = \{(x, y) \mid y = 1\}$.
26. $S = \{(x, y) \mid y = 2x\}$.
27. $S = \{(x, y) \mid y < 3\} \cap \{(x, y) \mid x + y = 3\}$.
28. $S = \{(x, y) \mid y = 0\} \cup \{(x, y) \mid y = 10\}$.
29. $S = \{(x, y) \mid y = 0\} \cap \{(x, y) \mid y = \frac{1}{2}x\}$.
30. $S = \{(x, y) \mid y = 1\} \cap \{(x, y) \mid x = 3\}$.

4.2 ◆ Relations

For a given set, A, the set $A \times A$ is the set of all possible ordered pairs with both components from A. For two given sets, A and B, the set $A \times B$ is the set of all possible ordered pairs with first component from A and second component from B. In this book we shall study ordered pairs of real numbers. Any set of ordered pairs is called a **relation.**

DEFINITION 4.3. Any subset of the cartesian product $A \times A$ is called a **relation** in A.

Relations may be given in one of three ways. If the relation has a finite number of ordered pairs, we can simply list them. This is called the **roster method.** Two relations denoted by the roster method are given below.

(1) $\{(0, 0), (0, 1), (0, 2), (0, 3)\}$
(2) $\{(-2, 3), (-1, 3), (0, 0), (1, 4), (2, 4)\}$

A second way is to display the graph of the relation; that is, the set of points in the Cartesian plane whose labels are the ordered pairs in the given relation. The graphs of relations (1) and (2) above are shown in Figure 4.5.

The third method of giving a relation is as the solution set of an open sen-

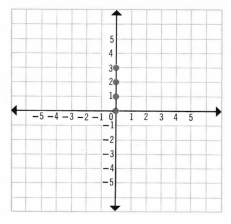 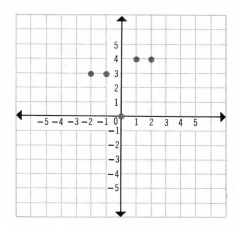

Figure 4.5

tence in two variables x and y. For example, consider the open sentence $y = x^2$. The solution set over $R \times R$ of this equation, namely,

$$\{(x, y) \mid y = x^2\}$$

is the set of ordered pairs and therefore is a relation in R. The solution set of every open sentence in two variables x and y is a relation.

The **domain** of a relation is the set of all the first components of the ordered pair of the relation. The **range** of a relation is the set of all the second components of the ordered pairs in the relation.

EXAMPLE 1. Given the relation

$$S = \{(0, 2), (1, 3), (2, 4), (3, 5), (4, 6)\}$$

what is the domain and range of S?

SOLUTION. The set of all first components of the ordered pairs of this relation is $\{0, 1, 2, 3, 4\}$ and hence is its domain. The set of all second components of the ordered pairs in this relation is $\{2, 3, 4, 5, 6\}$ and hence is its range.

EXAMPLE 2. Let $A = \{-3, -2, -1, 0, 1, 2, 3\}$. The relation R in A is $\{(x, y) \mid y = x^2, x \in A \text{ and } y \in A\}$. What is the domain and the range of R?

SOLUTION. The solution set in A of the open sentence $y = x^2$ is $\{(-1, 1), (0, 0), (1, 1)\}$. (The replacements of $-3, -2, 2,$ and 3 for x give values for y that are not elements of A.) The domain of R is thus $\{-1, 0, 1\}$ and its range is $\{0, 1\}$.

EXAMPLE 3. The relation S in the set R of real numbers is $\left\{(x, y) \mid y = \dfrac{1}{x}\right\}$. What is the domain and the range of S?

SOLUTION. Since for every real number replacement for x, except 0, $\dfrac{1}{x}$ names a real number, the domain of S is $\{x \mid x \in R \text{ and } x \neq 0\}$. To find the range of S we solve the equation $y = \dfrac{1}{x}$ explicitly for x and obtain $x = \dfrac{1}{y}$. Now, since for every real number replacement for y, except 0, $\dfrac{1}{y}$ names a real number, the range of S is $\{y \mid y \in R \text{ and } y \neq 0\}$.

In general, we can determine the domain and range of a relation that is specified by an equation by solving the equation explicitly for each of the variables in turn and looking for excluded values. A possible value of a variable will be excluded if it either results in a division by zero or in taking the square root or any even root of a negative number.

EXAMPLE 4. S is the relation $\{(x, y) \mid x^2 + y^2 = 4\}$. What is the domain and the range of S?

SOLUTION. Solving the equation $x^2 + y^2 = 4$ explicitly for y, we have

$$y = \sqrt{4 - x^2} \qquad \text{or} \qquad y = -\sqrt{4 - x^2}$$

We now seek values of x for which y is a real number. Thus we want the solution set of the inequality

$$4 - x^2 \geq 0$$

This inequality is equivalent to

$$x^2 \leq 4$$
$$|x| \leq 2$$

Hence the domain of S is

$$\{x \mid |x| \leq 2\}$$

Solving $x^2 + y^2 = 4$ explicitly for x, we have

$$x = \sqrt{4 - y^2} \qquad \text{or} \qquad x = -\sqrt{4 - y^2}$$

Now we seek values of y for which x is real. As above we see that the range of S is

$$\{y \mid y \leq |2|\}$$

EXAMPLE 5. Graph the relation

$$\{(-1, 2), (0, 3), (1, -1), (2, -2), (3, 1)\}$$

SOLUTION. See Figure 4.6.

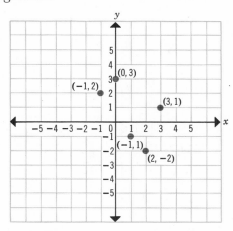

Figure 4.6

EXAMPLE 6. Graph the relation $\{(x, y) \mid y = x\}$.

SOLUTION. See Figure 4.7.

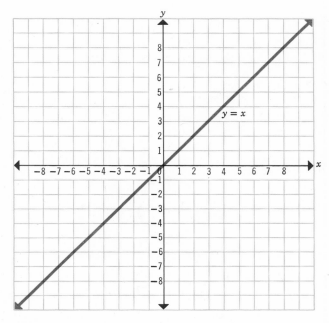

Figure 4.7

EXAMPLE 7. Graph the relation $\{(x,y) \mid 3x + y = 6\}$.

SOLUTION. See Figure 4.8.

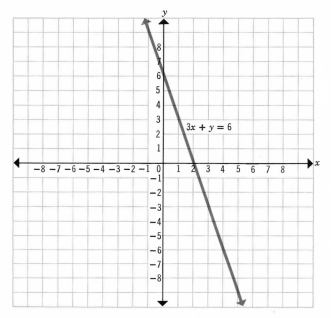

Figure 4.8

EXAMPLE 8. Graph the relation $\{(x, y) \mid x^2 + y^2 = 4\}$.

SOLUTION. See Figure 4.9.

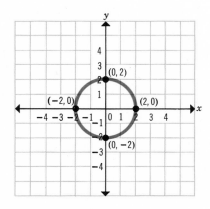

Figure 4.9

EXERCISES 4.2

Specify the domain and range of each of the relations in R the set of real numbers (Exercises 1–20).

1. $\{(-3, 9), (-2, 4), (-1, 1), (0, 0), (1, 1), (2, 4), (3, 9)\}$.
2. $\{(-2, \sqrt{2}), (-1, 1), (0, 0), (1, 1), (2, \sqrt{2})\}$.
3. $\{(x, y) \mid y = 2x\}$.
4. $\left\{(x, y) \mid y = \dfrac{1}{x - 2}\right\}$.
5. $\left\{(x, y) \mid y = \dfrac{3x}{x - 5}\right\}$.
6. $\{(x, y) \mid x^2 + y^2 = 6\}$.
7. $\left\{(x, y) \mid y = \dfrac{1}{x^2 + 1}\right\}$.
8. $\{(x, y) \mid y = \sqrt{9 - x^2}\}$.
9. $\{(x, y) \mid y = \sqrt{x}\}$.
10. $\left\{(x, y) \mid y = \dfrac{1}{(x - 1)(x - 4)}\right\}$.
11. $\{(x, y) \mid x^2 + y^2 = 9\}$.
12. $\{(x, y) \mid x^2 - y^2 = 9\}$.
13. $\{(x, y) \mid 4x^2 + 9y^2 = 36\}$.
14. $\{(x, y) \mid 16x^2 - 25y^2 = 400\}$.
15. $\{(x, y) \mid x = \sqrt{y}\}$.
16. $\{(x, y) \mid x = \sqrt{y - 1}\}$.
17. $\{(x, y) \mid x - 2 = \sqrt{y - 1}\}$.
18. $\{(x, y) \mid (x - 2)(y - 1) = 1\}$.
19. $\{(x, y) \mid (x^2 - x - 2)(y + 1) = 1\}$.
20. $\{(x, y) \mid (x - 2)(y^2 - y - 6) = 2\}$.

Graph the following relations.

21. $\{(-2, 2), (-1, 1), (0, 0), (1, 1), (2, 2)\}$.
22. $\{(-3, -6), (-2, -4), (-1, -2), (0, 0), (1, 2), (2, 4), (3, 6)\}$.
23. $\{(x, y) \mid y = 2x + 3\}$.
24. $\{(x, y) \mid y = 5x - 2\}$.
25. $\{(x, y) \mid x^2 + y^2 = 9\}$.

4.3 ◆ Functions

We may regard a relation as a "matching" or "pairing" operation in which each element of the domain is matched or paired with one or more elements in the range. When relations are interpreted in this way, they are often called **mappings.** Since a relation consists of ordered pairs, we think of the pairing as starting in the domain and proceeding to the range. For example, consider the relation

$$\{(0, 1), (1, 2), (2, 3), (1, 1), (0, 4)\}$$

This relation pairs 0 with 1 and 4, 1 with 1 and 2, and 2 with 3. We may sketch this matching as shown in Figure 4.10.

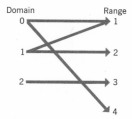

Figure 4.10

Now let us consider a second relation,

$$\{(0, 3), (1, 1), (2, 1), (3, 4)\}$$

The sketch of the matching in this relation is shown in Figure 4.11. The second relation, unlike the first, matches each element of its domain with one and only one element of the range. Relations with this property are called **functions.**

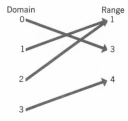

Figure 4.11

DEFINITION 4.4. A **function** is a relation in which no two ordered pairs have the same first component and different second components.

We see that *a function is a special kind of relation.*

It is important to determine whether or not a relation is a function. If a relation is given by the roster method this is easy to do. For example,

$$\{(0, 1), (1, 0), (0, 3), (4, 2), (5, 1)\}$$

is not a function, since the two ordered pairs $(0, 1)$ and $(0, 3)$ have the same first component and different second components. On the other hand,

$$\{(0, 1), (1, 2), (2, 3), (3, 4)\}$$

is a function since no two ordered pairs have the same first component.

It is also easy to determine whether a relation is a function if the relation is given by a graph. If any vertical line intersects the graph of a relation in more than one point, then that relation contains more than one ordered pair with the same first component and is thus not a function. Figure 4.12a shows an example of this. If any vertical line intersects the graph of a relation in at most

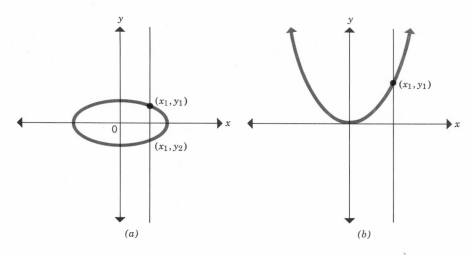

(a)   (b)

Figure 4.12

one point (the vertical line may not intersect the graph at all) then the relation must be a function. Figure 4.12b shows an example of this situation. When we use a vertical line in this manner—to determine if a relation is a function—we are using the **vertical line test.**

For the most part we shall discuss relations and functions which are specified by open sentences in two variables, x and y, and whose domains are subsets of R, the set of real numbers. When a relation is specified by an equation in x and y, one way in which we can determine whether the relation is a function is to solve the equation explicitly for the variable y, which represents an element of the range, and see whether more than one value of y is paired with a single value of x.

EXAMPLE 1. Is the relation $\{(x, y) \mid y^2 = 4 - x^2\}$ a function?

SOLUTION. Since $y^2 = 4 - x^2$ is equivalent to

$$y = \sqrt{4 - x^2} \quad \text{or} \quad y = -\sqrt{4 - x^2}$$

we see that assigning any real number as a value of x, with

$$-2 < x < 2$$

yields two different real values for y. Thus this relation is not a function.

EXAMPLE 2. Determine whether the relation $\{(x, y) \mid x^2 y = 4\}$ is a function.

SOLUTION. We note that the equation $x^2 y = 4$ is equivalent to the equation

$$y = \frac{4}{x^2} \qquad x \neq 0$$

If we assign x any real value, other than 0, we obtain one and only one corresponding value for y so that this relation is a function.

EXAMPLE 3. Graph the function $\{(x, y) \mid y = 2x + 3\}$.

SOLUTION. See Figure 4.13.

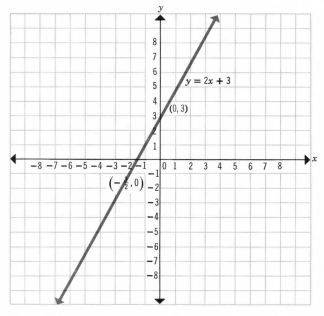

Figure 4.13

EXAMPLE 4. Graph the function $\{(x, y) \mid y = x^2\}$.

SOLUTION. See Figure 4.14.

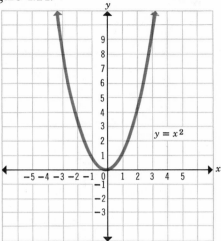

Figure 4.14

EXAMPLE 5. Graph the function $\{(x, y) \mid x^2y = 9\}$.

SOLUTION. See Figure 4.15.

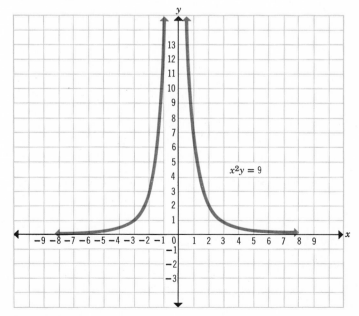

Figure 4.15

State whether or not the given equations specify functions (Exercises 1–10).

1. $y = x + 3$
2. $y^2 = \sqrt{x^2 + 1}$
3. $y^2 + x^2 = 25$
4. $y^2 = x^3$
5. $y = \dfrac{1}{x}$

6. $y^2 = \pm \sqrt{1 - x^2}$
7. $y = \sqrt[3]{x}$
8. $y = x^n$ (n a positive integer)
9. $y^2 = 9 - 4x^2$
10. $x^2 - y^2 = 16$

Find the range of each function with the domain given (Exercises 11–15).

11. $\{(x, y) \mid y = 3x + 2\}$; $\{0, 1, 2\}$.
12. $\{(x, y) \mid y = x^2 - 4\}$; $\{1, 2, 3, 4\}$.
13. $\{(x, y) \mid y = x^2\}$; $\{3, 4, 5\}$.
14. $\{(x, y) \mid y = x^3\}$; $\{-1, 0, 1\}$.
15. $\{(x, y) \mid y + 3x = 10\}$; $\{-2, -1, 0\}$.

Use the vertical line test to determine whether the following graphs specify functions (Exercises 16–25).

16.

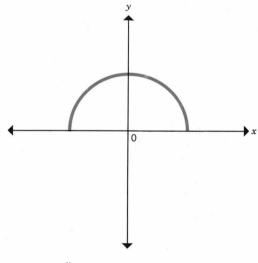

17.

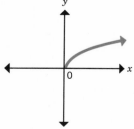

18.

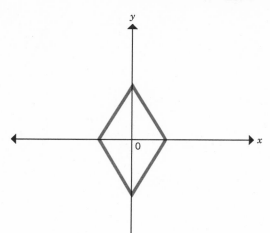

21.

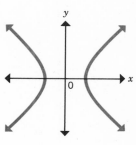

19.

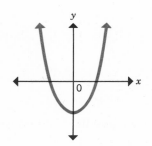

22.

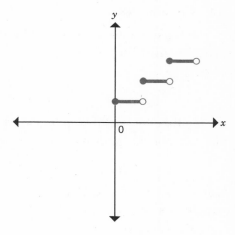

20.

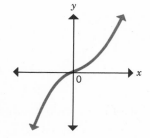

23.

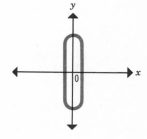

24. 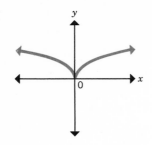   25.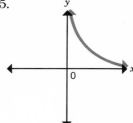

Graph the following functions where $x \in R$ and $y \in R$.

26. $\{(x, y) \mid y = 2x\}$.
27. $\{(x, y) \mid y = x - 3\}$.
28. $\{(x, y) \mid y = 5x + 8\}$.
29. $\{(x, y) \mid y = 2x^2\}$.
30. $\{(x, y) \mid y = |x|\}$.

4.4 ◆ Functional Notation

Functions are usually denoted by single letters. Typically, f, g, h, or E, G, H might be used. For example:

$$f = \{(x, y) \mid y = 3x\}$$
$$g = \{(x, y) \mid y = x^2\}$$

If f is a function and if (x, y) is an ordered pair of f, we usually denote the second component, y, of the ordered pair by the symbol $f(x)$. Thus

$$y = f(x)$$

means that y is the second component of the ordered pair whose first component is x. We read the symbol $f(x)$ "f at x" or "the value of f at x."

With this agreement we can write the ordered pair (x, y) of a function as

$$(x, f(x))$$

Thus for

$$f = \{(x, y) \mid y = 2x^2\}$$

we can use the alternate notation

$$f = \{(x, f(x)) \mid f(x) = 2x^2\}$$

We call the second component of an ordered pair of a function a **value** of that function. For example, in the function

$$f = \{(x, f(x)) \mid f(x) = 2x^2\}$$

the value of f at 2 is $f(2)$ which is 8; the value of f at -1 is $f(-1)$ which is 2.

EXAMPLE 1. Given $g = \{(x, g(x)) \mid g(x) = \sqrt{4 - x^2}\}$, find (a) $g(0)$; (b) $g(1)$;
(c) $g(-1)$; and (c) $g\left(\dfrac{1}{t}\right)$.

SOLUTION. (a) $g(0) = \sqrt{4 - 0} = 2$
          (b) $g(1) = \sqrt{4 - 1} = \sqrt{3}$
          (c) $g(-1) = \sqrt{4 - (-1)^2} = \sqrt{3}$
          (d) $g\left(\dfrac{1}{t}\right) = \sqrt{4 - \left(\dfrac{1}{t}\right)^2} = \dfrac{\sqrt{4t^2 - 1}}{|t|}, \ t \neq 0$

EXAMPLE 2. If $f = \{(x, f(x)) \mid f(x) = x^2\}$ and $g = \{(x, g(x)) \mid g(x) = x + 1\}$,
find (a) $f(g(2))$; (b) $g(g(3))$; (c) $f(g(x))$; and (d) $g(f(x))$.

SOLUTION. (a) $g(2) = 2 + 1 = 3$
           $f(g(2)) = f(3) = 3^2 = 9$
         (b) $g(3) = 3 + 1 = 4$
           $g(g(3)) = g(4) = 4 + 1 = 5$
         (c) $g(x) = x + 1$
           $f(g(x)) = (g(x))^2 = (x + 1)^2 = x^2 + 2x + 1$
         (d) $f(x) = x^2$
           $g(f(x)) = f(x) + 1 = x^2 + 1$

EXERCISES 4.4

1. Given $f = \{(x, f(x)) \mid f(x) = 2x\}$. Find:
    (a) $f(2)$ (d) $f(a + b)$
    (b) $f(-1)$ (e) $f(x + h)$
    (c) $f(-3)$ (f) $f(x + h) - f(x)$

2. Given $f = \{(x, f(x)) \mid f(x) = x + 1\}$. Find:
    (a) $f(0)$ (d) $f(-1)$
    (b) $f(3)$ (e) $f(-a)$
    (c) $f(8)$ (f) $f(x + h)$

3. Given $f = \{(x, f(x)) \mid f(x) = x^2\}$ and $g = \left\{(x, g(x)) \mid g(x) = \dfrac{1}{x}\right\}$. Find:

    (a) $f(2)$ (c) $f(4 + h)$
    (b) $f(1) + g(3)$ (d) $g(-3 + h)$

(e) $f(3 + h) + g(3 + h)$ (g) $f(g(3))$

(f) $\dfrac{f(2)}{g(-1)}$ (h) $f(g(x))$

4. If $f = \{(x, f(x)) \mid f(x) = x^2 + x - 1\}$ and $g = \{(x, g(x)) \mid g(x) = x + 1\}$, find:
   (a) $f(0)$ (d) $f(g(3))$
   (b) $g(3)$ (e) $f(g(-2))$
   (c) $f(3) - g(2)$ (f) $\dfrac{f(g(1))}{g(g(2))}$

5. Suppose $f = \{(x, f(x)) \mid f(x) = x^2 - 2x + 3\}$. Find:
   (a) $f(x + h) - f(x)$ (b) $\dfrac{f(x + h) - f(x)}{h}$

6. Let $f = \{(x, f(x)) \mid f(x) = x^2 + 1\}$. Find:
   (a) $f(s)$ (d) $f(a + b)$
   (b) $f(r)$ (e) $f(a + b) - [f(a) + f(b)]$
   (c) $f(a) + f(b)$ (f) $\dfrac{f(x + h) - f(x)}{h}$

7. Let $F = \{(x, F(x)) \mid F(x) = 3x + 1\}$. Find:
   (a) $F(2)$ (e) $\sqrt{F(x) - 1}$
   (b) $F(\sqrt{y})$ (f) $\sqrt{F(x^2)}$
   (c) $F(z^2)$ (g) $F(x^2) - \sqrt{F(x)}$
   (d) $F(|a|)$ (h) $F(-x)$

8. Let $g = \{(x, g(x)) \mid g(x) = |x| + x\}$. Find:
   (a) $g(2)$ (d) $g(-\frac{3}{4})$
   (b) $g(-2)$ (e) $2g(-1)$
   (c) $g(\frac{1}{2})$ (f) $g(\sqrt{2})$

9. Any function satisfying the condition that $f(x) = f(-x)$ for all x in the domain is called an *even* function. Which of the following functions are even functions?
   (a) $\{(x, f(x)) \mid f(x) = x^2\}$.
   (b) $\{(x, f(x)) \mid f(x) = x^2 - x\}$.
   (c) $\{(x, f(x)) \mid f(x) = x^4 + x^2\}$.
   (d) $\{(x, f(x)) \mid f(x) = -x^3\}$.

10. Any function satisfying the condition that $f(-x) = -f(x)$ for all x in the domain is called an *odd* function. Which of the following functions are odd functions?
    (a) $\{(x, f(x)) \mid f(x) = x^3\}$.
    (b) $\{(x, f(x)) \mid f(x) = x^3 - x\}$.
    (c) $\{(x, f(x)) \mid f(x) = x + x^5\}$.
    (d) $\{(x, f(x)) \mid f(x) = -x^2\}$.

11. What element(s) in the domain of each of the functions below is (are) associated with the element 16 in the range?
    (a) $f = \{(x, f(x)) \mid f(x) = x + 9\}$.
    (b) $f = \{(x, f(x)) \mid f(x) = x^2\}$.

(c) $f = \{(x, f(x)) \mid f(x) = 4x\}$.

(d) $f = \{(x, f(x)) \mid f(x) = x^2 + x + 4\}$.

12. What element(s) in the domain of each of the functions below is (are) associated with the element 0 in the range?

(a) $f = \{(x, f(x) \mid f(x) = \sqrt{1 - x^2}\}$.

(b) $f = \left\{(x, f(x)) \mid f(x) = \dfrac{5x - 7}{x^2 + 2}\right\}$.

(c) $f = \left\{(x, f(x)) \mid f(x) = \dfrac{1}{(x - 1)^2}\right\}$.

4.5 ◆ Linear Functions

A **first degree equation** or **linear equation** in the variables x and y is one of the form

$$Ax + By + C = 0$$

where A and B are not both 0. It can be proved that the graph of any linear equation is a straight line. Functions that are specified by linear equations are called **linear functions.** If $B \neq 0$ the equation $Ax + By + C = 0$ specifies a function whose domain is the set of all real numbers. Since two points determine a straight line, to draw the graph of a linear equation we need to find just two elements of its solution set. The two solutions easiest to find are those ordered pairs with first or second component zero, that is, ordered pairs of the form $(x, 0)$ or $(0, y)$. These are the labels of the points at which the graph intersects the x- and y-axes, respectively. The x-coordinate of the point at which the graph intersects the x-axis is called the **x-intercept** of the line. The y-coordinate of the point at which the graph intersects the y-axis is called the **y-intercept** of the line.

Let us consider the function

$$f = \{(x, y) \mid 2x + 3y = 6\}$$

When 0 is substituted for y we have the equation $2x = 6$ for which 3 is the only solution. Thus $(3, 0)$ is a solution of the equation $2x + 3y = 6$ and the x-intercept of the graph of this equation is 3. When 0 is substituted for x we obtain the equation $3y = 6$ for which 2 is the only solution. Thus $(0, 2)$ is a solution of the equation $2x + 3y = 6$ and the y-intercept of the graph of this equation is 2. The graph of this function is shown in Figure 4.16.

If the graph of a linear function intersects both axes at the origin, the intercepts do not give us separate points so that it is necessary to find at least one other point in order to draw the graph. If the graph intersects the axes at

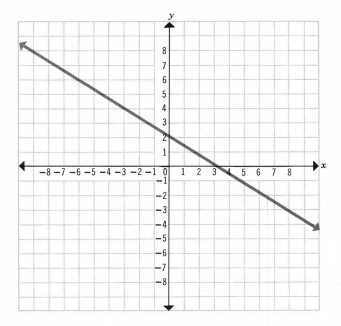

Figure 4.16

points that turn out to appear too close together to be of much help in drawing the line, it is helpful to plot another point in order to draw the graph.

There are two special cases of the linear equation. First, a linear equation of the form

$$By + C = 0, \qquad \text{where } B \neq 0$$

specifies a function in which the second components of the ordered pairs are all the same. Such a function is called a **constant function.**

DEFINITION 4.5. A function, K, is called a **constant function** if and only if all ordered pairs of K have the same second component.

For example, let us consider the function specified by the equation

$$y - 3 = 0$$

Each ordered pair of this function is of the form $(x, 3)$, where x can be any real number. For example, $(1, 3)$, $(3, 3)$, $(0, 3)$, and $(-1, 3)$ are all ordered pairs of this function. The graph of this function is shown in Figure 4.17.

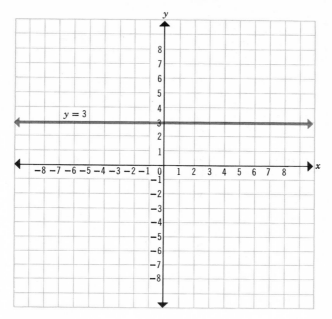

y = 3

Figure 4.17

The second special case of the linear equation is one of the form

$$Ax + C = 0, \qquad \text{where } A \neq 0$$

Here all the ordered pairs that are solutions of this equation have the same first component. It is clear that a linear equation of this type does *not* specify a function. Why? The graph of

$$x + 2 = 0$$

which is an example of such a linear equation, is shown in Figure 4.18.

Any two distinct points in the plane determine a line and are end points of a line segment. The distance between two points is so important that we shall develop a formula for it. Figure 4.19 shows two points labeled P_1 and P_2 with their coordinates, the line segment joining them, and the axes. The position of the two points is immaterial in the development of the formula. If a horizontal line—one parallel to the x-axis—is drawn through P_1 and a vertical line—one parallel to the y-axis—is drawn through P_2, these lines will intersect in a point which we label P_3. The coordinates of P_3 must be (x_2, y_1), since it must have the same x-coordinate as P_2 and the same y-coordinate as P_1. The distance between P_1 and P_3 is $|x_2 - x_1|$ and the

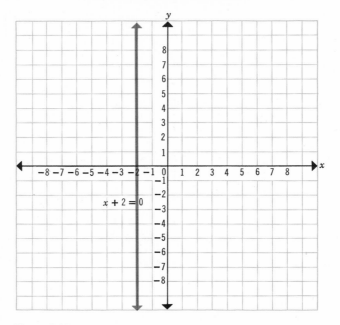

Figure 4.18

distance between P_2 and P_3 is $|y_2 - y_1|$. Since triangle $P_1P_2P_3$ is a right triangle, the Pythagorean theorem assures us that the distance, d, between P_1 and P_2 is

$$d = \sqrt{|x_2 - x_1|^2 + |y_2 - y_1|^2}$$

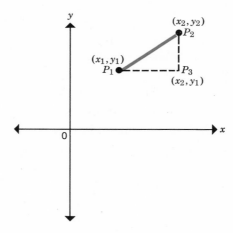

Figure 4.19

Since, for any real number a, $a^2 = (-a)^2$, and since $|a|$ is either a or $-a$, we see that

$$|x_2 - x_1|^2 = (x_2 - x_1)^2 \qquad \text{and} \qquad |y_2 - y_1|^2 = (y_2 - y_1)^2$$

Thus we may write

$$d = \sqrt{(x_2 - x_1)^2 + (y_2 - y_1)^2}$$

This formula is called the **distance formula.**

EXAMPLE 1. Use the distance formula to find the distance between the points whose coordinates are $(4, 2)$ and $(-5, 3)$.

SOLUTION

$$\begin{aligned} d &= \sqrt{[4 - (-5)]^2 + (2 - 3)^2} \\ &= \sqrt{9^2 + (-1)^2} \\ &= \sqrt{81 + 1} \\ &= \sqrt{82} \end{aligned}$$

A measure of the manner in which a line is inclined with respect to the x-axis (horizontal) is called the **slope** of the line.

DEFINITION 4.6. For any two points P_1 and P_2 with coordinates (x_1, y_1) and (x_2, y_2) respectively, and for which $x_1 \neq x_2$, the number

$$m = \frac{y_2 - y_1}{x_2 - x_1}$$

is the **slope** of the line through P_1 and P_2.

If a line is parallel to the x-axis, all of the y-coordinates of its points are the same and its slope is 0. If a line is parallel to the y-axis, all of its points have the same x-coordinates and the slope does not exist for such a line. The slope of a nonvertical line may be determined by using the coordinates of any two points on the line.

Using the concept of slope, we can write linear equations in other forms than $Ax + By + C = 0$, which is called the **standard form** of the equation.

Consider a line having slope m and passing through point P_1 whose coordinates are (x_1, y_1) as shown in Figure 4.20. If we choose another point, P, on that line and assign it coordinates (x, y), the slope of this line is given by

$$m = \frac{y - y_1}{x - x_1}$$

From this we obtain $\qquad y - y_1 = m(x - x_1)$

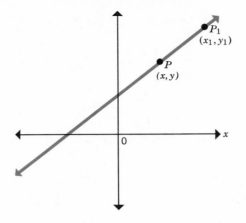

Figure 4.20

Since (x, y) represents any point on the line, this equation is an equation of a line with slope m and passing through (x_1, y_1). An equation in this form is called the **point-slope** form of the linear equation.

Now consider the point-slope form of an equation for the line with slope m and passing through the point on the y-axis having coordinates $(0, b)$. Thus the y-intercept of the line is b. Substituting 0 for x_1 and b for y_1 in the point-slope form, we have

$$y - b = m(x - 0)$$

which is equivalent to $$y = mx + b$$

This equation is called the **slope-intercept** form of a linear equation. Any linear equation in standard form may be written in slope-intercept form by solving it explicitly for y, provided $B \neq 0$. For example,

$$3x + 4y - 12 = 0$$
may be written as $$y = -\tfrac{3}{4}x + 3$$

The graph of these equations is the line with slope $-\tfrac{3}{4}$ and y-intercept 3.

EXAMPLE 2. Find the standard equation of the line through $(3, -1)$ and having slope 2.

SOLUTION. Using the point-slope form of the linear equation, we have

$$y - y_1 = m(x - x_1)$$
$$y - (-1) = 2(x - 3)$$
$$y + 1 = 2x - 6$$
$$2x - y - 7 = 0$$

EXAMPLE 3. Write $3x - 2y + 12 = 0$ in slope-intercept form. What is the y-intercept? What is the slope of the line?

SOLUTION. Solving $3x - 2y + 12 = 0$ explicitly for the variable y, we have

$$y = \frac{3x + 12}{2}$$

$$y = \tfrac{3}{2}x + 6$$

The y-intercept of the graph of this equation is 6; the slope is $\tfrac{3}{2}$.

EXAMPLE 4. Graph the linear function $\{(x, y) \mid y = x + 2\}$.

SOLUTION. If we replace y by 0 in $y = x + 2$, we obtain

$$0 = x + 2$$
$$x = -2$$

Hence the x-intercept of the graph of the function is -2 and the point whose coordinate is $(-2, 0)$ is on the graph of the function. If we replace x by 0 in $y = x + 2$, we obtain $y = 2$. Hence the y-intercept is 2 and the point whose coordinates are $(0, 2)$ is on the graph of the function. The graph of the function is shown in Figure 4.21.

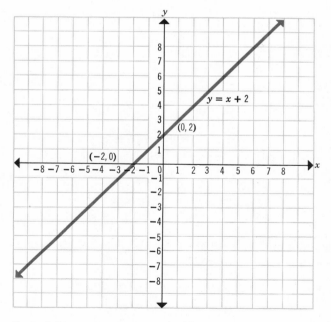

Figure 4.21

EXAMPLE 5. Graph the function $\{(x, y) \mid y = 6\}$.

SOLUTION. This function is a constant function. Each ordered pair of this
  function is of the form $(x, 6)$ where x can be any real number. For
  example, $(0, 6)$, $(1, 6)$, $(-3, 6)$, and $(-6, 6)$ are all ordered pairs of
  this function. The graph of the function is shown in Figure 4.22.

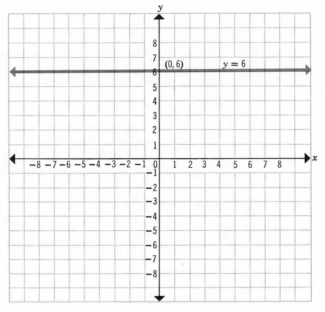

Figure 4.22

EXERCISES 4.5

Find the distance between each of the given pairs of points, and find the slope of
the line determined by them (Exercises 1–9).

1. $(3, 2), (0, -3)$ 4. $(2, -4), (4, 0)$ 7. $(3, -4), (-5, 2)$
2. $(2, 3), (7, -9)$ 5. $(-7, -2), (2, 1)$ 8. $(3, 2), (4, 6)$
3. $(-6, 18), (2, 3)$ 6. $(4, 4), (5, 3)$ 9. $(0, 0), (5, 12)$

Find an equation of the line through each given point and having the given slope.
Write each equation in standard form (Exercises 10–17).

10. $(-1, 0), m = \frac{1}{2}$ 14. $(3, 1), m = -2$
11. $(2, 3), m = 1$ 15. $(-1, -2), m = 0$
12. $(-3, -4), m = 1$ 16. $(-4, 2), m = 3$
13. $(0, 0), m = -1$ 17. $(-2, -2), m = -1$

Write each of the equations in slope-intercept form. Give the slope and the y-intercept of each line (Exercises 18–25).

18. $3x + 6y + 8 = 0$ 22. $x - 2y + 6 = 0$
19. $2x - 3y - 4 = 0$ 23. $3x - 2y = 18$
20. $5x + 3y + 4 = 0$ 24. $5x = 2y + 7$
21. $x - y = 7$ 25. $-x = -y + 2$

26. Show that $A(2, -2)$, $B(4, 0)$, $C(2, 2)$, and $D(0, 0)$ are the vertices of a square.
27. Show that the points $A(2, -1)$, $B(3, 4)$, and $C(-7, 6)$ are the vertices of a right triangle.
28. The lines determined by $A(3, -4)$, $B(1, -2)$, and $C(2, 3)$, $D(1, 2)$ are perpendicular. Find the slope of each line. How do the slopes compare?

Graph the following functions.

29. $\{(x, y) \mid y = x\}$ 32. $\{(x, y) \mid y = 2\}$
30. $\{(x, y) \mid y = 2x - 6\}$ 33. $\{(x, y) \mid y = 5x - 3\}$
31. $\{(x, y) \mid y = x + \frac{1}{2}\}$ 34. $\{(x, y) \mid y = 2 - x\}$

4.6 ◆ The Absolute Value and Bracket Functions

The **absolute value function** is specified by the equation

$$y = |x|$$

We recall that
$$|x| = \begin{cases} x, & \text{if } x \text{ is nonnegative} \\ -x, & \text{if } x \text{ is negative} \end{cases}$$

From this definition we see that for $x \geq 0$, $y = |x|$ is equivalent to $y = x$. For $x < 0$, $y = |x|$ is equivalent to $y = -x$. If we graph

$$y = x, \, x \geq 0$$
and
$$y = -x, \, x < 0$$

on the same plane we will have the graph of the absolute value function. This graph is shown in Figure 4.23.

The **bracket function** is specified by the equation

$$y = [x]$$

The symbol $[x]$ is used to denote the largest integer n such that $n \leq x < n+1$. Thus

$$[2] = 2$$
$$[4.05] = 4$$
$$[\pi] = 3$$
$$[\tfrac{15}{8}] = 1$$
$$[-1] = -1$$
$$[-1.1] = -2$$

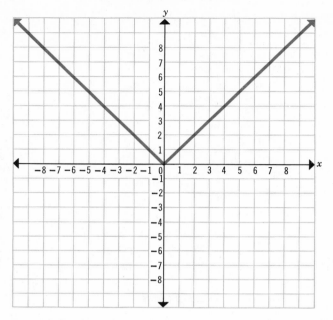

Figure 4.23

To graph the bracket function, specified by $y = [x]$, we consider unit intervals in the domain. If $0 \leq x < 1$, then $[x] = 0$. If $1 \leq x < 2$, then $[x] = 1$. If $2 \leq x < 3$, then $[x] = 2$; and so forth. If $-1 \leq x < 0$, then $[x] = -1$. If $-2 \leq x < -1$, then $[x] = -2$, and so forth. The graph of $y = [x]$ is shown in Figure 4.24.

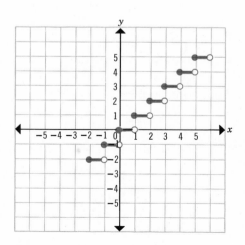

Figure 4.24

EXAMPLE 1. Graph the function specified by $y = |x + 2|$.

SOLUTION. From the definition of absolute value we see that for $x + 2 \geq 0$, $y = |x + 2|$ is equivalent to $y = x + 2$. For $x + 2 < 0$, $y = |x + 2|$ is equivalent to $y = -(x + 2)$. If we graph

$$y = x + 2, x + 2 \geq 0 \quad \text{or} \quad x \geq -2$$

and
$$y = -(x + 2), x + 2 < 0 \quad \text{or} \quad x < -2$$

on the same plane we have the graph of $y = |x + 2|$. This graph is shown in Figure 4.25.

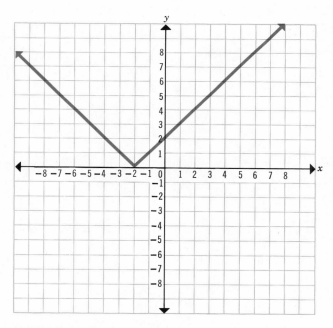

Figure 4.25

EXAMPLE 2. Graph the function specified by $y - 2 = |3x|$.

SOLUTION. From the definition of absolute value we see that for $3x \geq 0$, $y - 2 = |3x|$ is equivalent to the equation $y - 2 = 3x$. For $3x < 0$, $y - 2 = |3x|$ is equivalent to the equation $y - 2 = -3x$. If we graph

$$y - 2 = 3x, \, x \geq 0$$

and $$y - 2 = -3x, \, x < 0$$

on the same plane, we have the graph of $y - 2 = |3x|$. This graph is shown in Figure 4.26.

EXAMPLE 3. Graph the function specified by $y = [\![x + 1]\!]$.

SOLUTION. To graph the equation $y = [\![x + 1]\!]$ we consider unit intervals in the domain of the function.

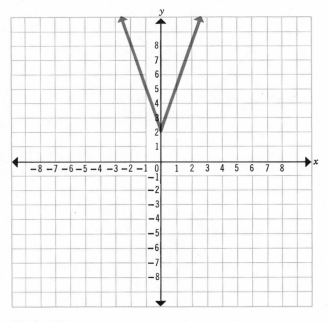

Figure 4.26

If $0 \leq x < 1$, then $[\![x + 1]\!] = 1$.
If $1 \leq x < 2$, then $[\![x + 1]\!] = 2$.
If $2 \leq x < 3$, then $[\![x + 1]\!] = 3$, and so forth.
If $-1 \leq x < 0$, then $[\![x + 1]\!] = 0$
If $-2 \leq x < -1$, then $[\![x + 1]\!] = -1$, and so forth.

The graph of the function is shown in Figure 4.27.

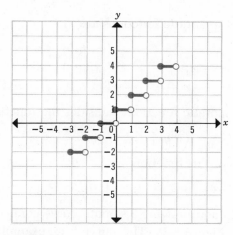

Figure 4.27

EXERCISES 4.6

Graph the functions specified by the given equations (Exercises 1–15).

1. $y = |x| + 2$
2. $y = |x - 1|$
3. $y + 2 = |x + 3|$
4. $2y = |x|$
5. $3y = |x - 1|$
6. $y - 1 = |2x|$
7. $y - 2 = |2x - 3|$
8. $y = [\![x]\!] + 1$

9. $y = [\![x + 1]\!]$
10. $y = 2[\![x]\!]$
11. $y = 2[\![x]\!] - 3$
12. $y = [\![2x]\!]$
13. $y = [\![2x - 4]\!]$
14. $y = [\![2x - 1]\!]$
15. $y = [\![x]\!] + |x|$

4.7 ⏷ Quadratic Functions

Another important function is the function specified by the quadratic equation

$$y = ax^2 + bx + c$$

where $a \neq 0$. Such a function is called a **quadratic function.**

DEFINITION 4.6. A function is a **quadratic function** if and only if it is of
the form

$$\{(x, y) \mid y = ax^2 + bx + c\}$$

and $a \neq 0$.

Some examples of quadratic functions are

$$\{(x, y) \mid y = x^2\} \qquad\qquad \{(x, f(x)) \mid f(x) = -x^2\}$$
$$\{(x, y) \mid y = 4x^2 - x\} \qquad\qquad \{(x, f(x)) \mid f(x) = 1 - 4x - 2x^2\}$$
$$\{(x, y) \mid y = 3x^2 - 4x + 1\} \qquad \{(x, f(x)) \mid f(x) = 2x - 3x^2\}$$

Let us graph a quadratic function. Consider the function

$$\{(x, y) \mid y = x^2 + 6x + 5\}$$

Some ordered pairs of this function are

$$(-5, 0), (-3, -4), (-2, -3), (-1, 0), (0, 5), (1, 12), (2, 21), (3, 32)$$

Plotting these points and drawing a smooth curve connecting them we have
the graph shown in Figure 4.28. This curve is an example of a **parabola.**

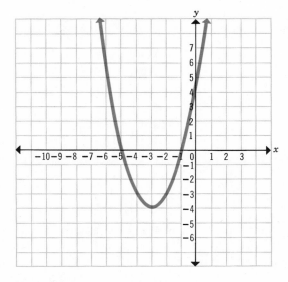

Figure 4.28

In general, the graph of the solution set of the quadratic equation of the form

$$y = ax^2 + bx + c, \ a \neq 0$$

is a parabola. Since for each value of x this equation determines one and only one value of y, this equation gives us a function whose domain is the set of all real numbers and whose range is some subset of the set of real numbers. If we observe the graph in Figure 4.28, we see that the range of the function specified by $y = x^2 + 6x + 5$ is the set

$$\{y \mid y \geq -4\}$$

To graph a quadratic function and to observe some of its properties we express the quadratic function in terms that involve the square of a linear expression in x. To do this, we use the process discussed in Section 3.6 known as **completing the square.** For the quadratic function specified by

$$y = ax^2 + bx + c, \ a \neq 0$$

we have the following equivalent equations.

(1) $$y = a\left(x^2 + \frac{b}{a}x + \frac{c}{a}\right)$$

(2) $$y = a\left(x^2 + \frac{b}{a}x + \frac{b^2}{4a^2} - \frac{b^2}{4a^2} + \frac{c}{a}\right)$$

(3) $$y = a\left(x^2 + \frac{b}{a}x + \frac{b^2}{4a^2}\right) - \left(\frac{b^2}{4a} - c\right)$$

(4) $$y = a\left(x + \frac{b}{2a}\right)^2 - \frac{b^2 - 4ac}{4a}$$

Since the square of any real number is positive or zero, the least value of $\left(x + \dfrac{b}{2a}\right)^2$ will occur when x has value $-\dfrac{b}{2a}$, that is, when $\left(x + \dfrac{b}{2a}\right)$ is equal to zero. If $a > 0$, then the first term in the right member of Equation 4 is positive or zero so that when $\left(x + \dfrac{b}{2a}\right)$ takes on the value 0, y is $-\dfrac{b^2 - 4ac}{4a}$, and this value of y will be the y-coordinate of the lowest point of the graph. This lowest point thus has coordinates:

$$\left(-\frac{b}{2a}, \ -\frac{b^2 - 4ac}{4a}\right)$$

On the other hand, if $a < 0$ then the first term in the right member of Equation 4 is negative or zero, so that $-\dfrac{b^2 - 4ac}{4a}$ will be the y-coordinate of the highest point of the graph. Thus we see that the point with coordinates

$$\left(-\frac{b}{2a}, -\frac{b^2 - 4ac}{4a}\right)$$

is either the highest (when $a < 0$) or the lowest (when $a > 0$) point on the parabola. This point is called the **vertex** of the parabola.

When $ax^2 + bx + c$ has value 0, then the quadratic function

$$\{(x, y) \mid y = ax^2 + bx + c\}$$

has value 0. Since any point on the x-axis has y-coordinate 0, the solutions of the quadratic equation

$$ax^2 + bx + c = 0$$

will be the abscissas of the points where the graph of the function crosses the x-axis. In general, a value of x for which $f(x)$ is 0 is called a **zero** of f.

We summarize the above discussion by the following.

1. The graph of $\{(x, y) \mid y = ax^2 + bx + c\}$ is a parabola.
2. The vertex of this parabola is the point with coordinates

$$\left(-\frac{b}{2a}, -\frac{b^2 - 4ac}{4a}\right)$$

3. If $a > 0$ the parabola opens upward; if $a < 0$ the parabola opens downward.
4. The x-intercepts of the graph are the zeros of the quadratic function, which are the real solutions (if any) of the quadratic equation $ax^2 + bx + c = 0$.

EXAMPLE 1. Graph the function $\{(x, y) \mid y = x^2 - 5x + 4\}$.

SOLUTION. The solution set of the equation $x^2 - 5x + 4 = 0$ is $\{1, 4\}$. Thus the points $(1, 0)$ and $(4, 0)$ are on the graph. Completing the square we obtain

$$\begin{aligned} y &= x^2 - 5x + 4 \\ &= x^2 - 5x + \tfrac{25}{4} - \tfrac{25}{4} + 4 \\ &= (x - \tfrac{5}{2})^2 - \tfrac{9}{4} \end{aligned}$$

The function has minimum value of $-\tfrac{9}{4}$ which occurs when x is $\tfrac{5}{2}$. Therefore the minimum point, which is the vertex of the parabola, has coordinates $(\tfrac{5}{2}, -\tfrac{9}{4})$. We can find several other ordered pairs which belong to this function and complete the graph (Figure 4.29).

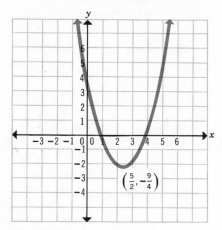

Figure 4.29

EXAMPLE 2. Graph the function $\{(x, y) \mid y = -x^2 - 4x - 8\}$. Find the coordinates of the vertex of the parabola.

SOLUTION. Completing the square we have

$$y = -(x^2 + 4x + 4) - 4$$
$$= -(x + 2)^2 - 4$$

The vertex is the point whose coordinates are $(-2, -4)$. Some other points which are on the graph are $(0, -8)$, $(-3, -5)$, and $(-1, -5)$. Plotting these points and drawing a smooth curve connecting them, we obtain the graph shown in Figure 4.30.

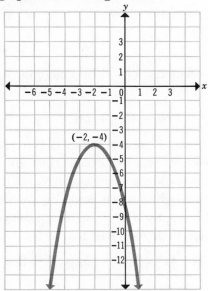

Figure 4.30

Graph and find the coordinates of the vertex (Exercises 1–10).

1. $\{(x, y) \mid y = x^2\}$
2. $\{(x, y) \mid y = x^2 - 4\}$
3. $\{(x, y) \mid y = x^2 + 2x - 8\}$
4. $\{(x, y) \mid y = 2x^2 - 3x + 4\}$
5. $\{(x, y) \mid y = -x^2 + x - 2\}$

6. $\{(x, y) \mid y = x^2 - 3x + 2\}$
7. $\{(x, y) \mid y = -x^2 + 6x + 7\}$
8. $\{(x, y) \mid y = -x^2 - 8x + 9\}$
9. $\{(x, f(x)) \mid f(x) = -\frac{1}{4}x^2 - 4x\}$
10. $\{(x, f(x)) \mid f(x) = -2x^2 - 3x - 4\}$

Solve by completing the square (Exercises 11–13).

11. Find two numbers whose sum is 16 and whose product is a maximum.
12. Find the maximum possible area of a rectangle whose perimeter is 160 feet.
13. Mr. Harris wishes to fence with 32 feet of wire fencing the largest possible pen for his dog. A side of the house will be one side of the pen and will need no fencing. Find the dimensions of the pen.

Find the zeros of the following quadratic functions (Exercises 14–20).

14. $\{(x, y) \mid y = x^2 - 4\}$
15. $\{(x, y) \mid y = x^2 - 8x + 15\}$
16. $\{(x, y) \mid y = 16x^2 - 9\}$
17. $\{(x, y) \mid y = 10x^2 - x - 21\}$

18. $\{(x, y) \mid y = 5x^2 - 13x - 6\}$
19. $\{(x, y) \mid y = 2x^2 + 3x - 2\}$
20. $\{(x, y) \mid y = 5x^2 - 2x - 7\}$

4.8 ◆ The Conics

The graphs of the relations defined by quadratic equations in x and y of the form

(1) $$Ax^2 + Bxy + Cy^2 + Dx + Ey + F = 0$$

in which A, B, C, D, E, and F are real numbers and A, B, and C are not all 0, are called **conic sections** or simply **conics**. They are called conics because the graphs of all equations of this form are curves that result from the intersection of a plane with a right circular cone (Figure 4.31).

We shall discuss the graphs of some special instances of Equation 1. If B and C in Equation 1 are both zero, we have

$$Ax^2 + Dx + Ey + F = 0$$

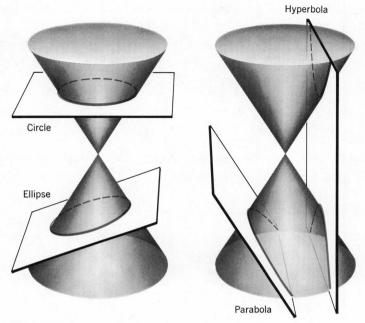

Figure 4.31

which is equivalent to

$$y = -\frac{A}{E}x^2 - \frac{D}{E}x - \frac{F}{E}$$

$$y = ax^2 + bx + c$$

where $a = -\dfrac{A}{E}$, $b = -\dfrac{D}{E}$, and $c = -\dfrac{F}{E}$. This equation was discussed in Section 4.7. The graph of $y = ax^2 + bx + c$ is a **parabola,** one of the conics.

In Equation 1, if B, D, and E are all zero, we have

(2) $$Ax^2 + Cy^2 + F = 0 \qquad A \neq 0, C \neq 0$$

We now consider this form of Equation 1. Let us write Equation 2 in the form

(3) $$Ax^2 + Cy^2 = K$$

If $A = C$ and A, C, and K all have the same sign, we may write Equation 3 as

(4) $$x^2 + y^2 = R \qquad R = \frac{K}{A}$$

The graph of Equation 4 is a **circle** whose center is at the origin and whose radius is $\sqrt{R}$.

If $A \neq C$ and A, C, and K all have the same sign and $K \neq 0$, the graph of Equation 3 is an **ellipse** that has center at the origin and intersects the x-axis at the points $\left(\dfrac{\sqrt{K}}{A}, 0\right)$ and $\left(-\dfrac{\sqrt{K}}{A}, 0\right)$ and the y-axis at the points $\left(0, \dfrac{\sqrt{K}}{C}\right)$ and $\left(0, -\dfrac{\sqrt{K}}{C}\right)$.

If A and C have opposite signs and $K \neq 0$, the graph of Equation 3 is a **hyperbola** whose center is at the origin.

EXAMPLE 1. Graph the relation $\{(x, y) \mid x^2 + y^2 = 4\}$.

SOLUTION. Since $x^2 + y^2 = 4$ is of the form $x^2 + y^2 = R$, we know that the graph is a circle with center at the origin and radius $\sqrt{4} = 2$. The graph of this relation is shown in Figure 4.32.

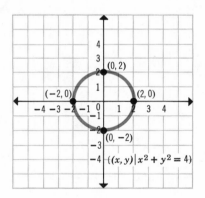

Figure 4.32

EXAMPLE 2. Graph the relation $\{(x, y) \mid 4x^2 + 9y^2 = 36\}$.

SOLUTION. Since this equation is of the form $Ax^2 + Cy^2 = K$ with $A \neq C$ and where A, C, and K all have the same sign, the graph is an ellipse with center at the origin. When y is 0, then x is either 3 or -3, so that the points $(3, 0)$ and $(-3, 0)$ are on the ellipse. When x is 0, then y is 2 or -2, so the points $(0, 2)$ and $(0, -2)$ are on the ellipse. Assigning values of x between -3 and 3 or values of y between -2 and 2 will give us other points on the ellipse and allow us to sketch the curve (Figure 4.33).

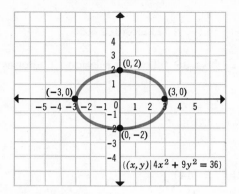

Figure 4.33

EXAMPLE 3. Graph the relation $\{(x, y) \mid x^2 - y^2 = 1\}$.

SOLUTION. Since $x^2 - y^2 = 1$ is of the form $Ax^2 + Cy^2 = K$ with $A \neq C$ and with opposite signs, $K \neq 0$, the graph is a hyperbola. If x is any number between -1 and 1, then y^2 would have to be negative, which is impossible in the field of real numbers. By assigning values of x which are greater than 1 or less than -1 we can find additional points on the hyperbola and sketch the curve (Figure 4.34).

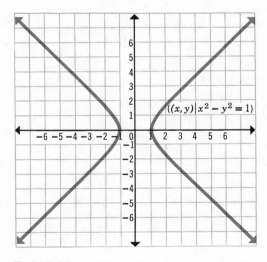

Figure 4.34

EXERCISES 4.8

Sketch the graph of each of the following relations (Exercises 1–10).

1. $\{(x, y) \mid x^2 + y^2 = 25\}$.
2. $\{(x, y) \mid x^2 + y^2 - 36 = 0\}$.
3. $\{(x, y) \mid 5x^2 + y^2 = 25\}$.
4. $\{(x, y) \mid 4x^2 - 9y^2 = 36\}$.
5. $\{(x, y) \mid x^2 + 4y^2 = 8\}$.
6. $\{(x, y) \mid 16x^2 - 9y^2 = 144\}$.
7. $\{(x, y) \mid 10x^2 - 2y^2 = 5\}$.
8. $\{(x, y) \mid 16x^2 + 9y^2 - 144 = 0\}$.
9. $\{(x, y) \mid x^2 - y^2 - 3 = 0\}$.
10. $\{(x, y) \mid y^2 = 4 - 2x^2\}$.

CHAPTER REVIEW

1. If $A = \{1, 2, 5\}$ and $B = \{0, 2, 6\}$, find $A \times B$ and $B \times A$.
2. Sketch the graph of the compound sentence
$$|y| < 1 \qquad \text{and} \qquad x > 2$$
3. Sketch the graph of the following relations.
   (a) $\{(x, y) \mid |x| = |y|\}$ (b) $\{(x, y) \mid y < x\}$.
4. Find the domain and range of each of the relations.
   (a) $\{(x, y) \mid x^2 + y^2 \leq 1\}$ (c) $\{(0, 2), (1, 1), (1, 4), (3, 3)\}$
   (b) $\{(x, y) \mid y = x^2 + 3\}$
5. Which of the following relations are functions?
   (a) $\{(x, y) \mid 2x^2 + 3y^2 = 4\}$ (d) $\{(x, y) \mid x = \sqrt{y + 3}\}$
   (b) $\{(x, y) \mid 2x + 3y = 4\}$ (e) $\{(x, y) \mid x^2 = y + 3\}$
   (c) $\{(x, y) \mid y = \sqrt{x + 2}\}$
6. The function f is specified by

$$f(x) = \frac{x^2 - 1}{x + 1}$$

Find the following:

   (a) $f(0)$ (e) $f(-1)$
   (b) $f(1)$ (f) $f(-2)$
   (c) $f(2)$ (g) $f(-3)$
   (d) $f(3)$

Sketch the graphs of the following functions (Exercises 7–8).

7. $\{(x, y) \mid 3x + 4y = 12\}$ 8. $\{(x, y) \mid y = -3x + 5\}$
9. Find the distance between the following pairs of points.
    (a) $(0, 1)$ and $(3, 5)$ (c) $(0, 1)$ and $(0, 7)$
    (b) $(2, 0)$ and $(3, 5)$ (d) $(3, 0)$ and $(0, 4)$
10. What is the distance between $(2, 1)$ and the midpoint of the line segment joining $(0, 8)$ to $(0, 4)$?
11. What is the slope of each of the following?
    (a) The line passing through $(3, 2)$ and $(5, 1)$.
    (b) The line joining $(2, 1)$ and $(7, 1)$.
    (c) The line passing through $(1, 1)$ and $(1, 5)$.

Find a linear equation whose graph is (Exercises 12–14):

12. The line passing through $(-4, 1)$ and $(5, -2)$.
13. The line passing through $(-2, 4)$ with slope 3.
14. The line with slope -2 and y-intercept 4.

Sketch the graphs (Exercises 15–18).

15. (a) $\{(x, y) \mid y = |x - 1|\}$.   (b) $\{(x, y) \mid y = [\![2x]\!]\}$.
16. $\{(x, y) \mid y = x^2 + 2x\}$.
17. $\{(x, y) \mid y = 2x^2 - 3x\}$.
18. $\{(x, y) \mid y = -3x^2 + 5x\}$.
19. A rectangular field is to be fenced off with 800 feet of wire. What is the largest area which can be fenced off?
20. Sketch the graph of:
    (a) $\{(x, y) \mid x^2 + y^2 = 9\}$.
    (b) $\{(x, y) \mid 9x^2 + 25y^2 = 225\}$.
    (c) $\{(x, y) \mid 2x^2 - y^2 = 3\}$.

CHAPTER 5

Exponential and Logarithmic Functions

5.1 ◆ Exponential Functions

Before we can construct an exponential function whose domain is the set of real numbers, we need to construct a definition for

$$b^x$$

where b is some fixed positive real number, $b \neq 1$, and x is an irrational number. (Recall that an irrational number is a real number which cannot be expressed as the quotient of two integers.) For example, we must be able to have such exponents as occur in

$$b^{\sqrt{2}}, \; b^{\pi}, \; b^{-\sqrt[3]{5}}$$

To do this we use the fact that any irrational number may be approximated to a desired degree of accuracy by rational numbers. For example, we know that

$$1 < \sqrt{2} < 2$$
$$1.4 < \sqrt{2} < 1.5$$
$$1.41 < \sqrt{2} < 1.42$$
$$1.414 < \sqrt{2} < 1.415$$
$$1.4142 < \sqrt{2} < 1.4143, \text{ and so forth}$$

Thus, for any definition we make, $2^{\sqrt{2}}$ must be a number with the following properties:

$$2^1 < 2^{\sqrt{2}} < 2^2$$
$$2^{1.4} < 2^{\sqrt{2}} < 2^{1.5}$$
$$2^{1.41} < 2^{\sqrt{2}} < 2^{1.42}$$
$$2^{1.414} < 2^{\sqrt{2}} < 2^{1.415}$$
$$2^{1.4142} < 2^{\sqrt{2}} < 2^{1.4143}, \text{ and so forth}$$

We can continue this process indefinitely, so that the difference between the left and right members of the string of inequalities is as small as we desire. It can be proved, but we shall accept it without proof, that there is one number, $2^{\sqrt{2}}$, that will satisfy the sequence of inequalities above no matter how long we continue the process.

We can make a similar argument for every irrational number exponent and hence we shall assume that for $b > 0$, $b \neq 1$, if x is any real number—rational or irrational—then b^x names one and only one real number.

It is possible to show that the laws of exponents for rational number exponents are also satisfied by irrational number exponents. We shall accept this without proof, and merely state these laws.

THEOREM 5.1. If $a > 0$, $a \neq 1$, $b > 0$, $b \neq 1$, and m and n are any real numbers, then

(a) $a^m \cdot a^n = a^{m+n}$

(b) $(a^m)^n = a^{mn}$

(c) $\dfrac{a^m}{a^n} = a^{m-n}$

(d) $(ab)^m = a^m b^m$

(e) $\left(\dfrac{a}{b}\right)^n = \dfrac{a^n}{b^n}$

(f) $a^0 = 1$

Given $b > 0$, for all real values of x there is one and only one number of the form b^x. Therefore the equation

$$y = b^x$$

specifies the function

$$\{(x, y) \mid y = b^x\}$$

If $b = 1$, this equation specifies the constant function, since $1^x = 1$ for all x. If $b \neq 1$ (but $b > 0$), then the equation $y = b^x$ specifies a nonconstant function called an **exponential function**.

Let us consider the graph of the exponential function specified by the equation

$$y = 2^x$$

Some ordered pairs that are solutions of this equation are

$$(-3, \tfrac{1}{8}), (-2, \tfrac{1}{4}), (-1, \tfrac{1}{2}), (0, 1), (1, 2), (2, 4), (3, 8)$$

Plotting the points with these coordinates and connecting them with a smooth curve gives us the graph in Figure 5.1.

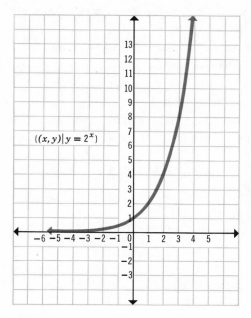

Figure 5.1

In a similar way we can find the graph of the exponential function specified by the equation

$$y = (\tfrac{1}{2})^x$$

This graph is shown in Figure 5.2.

We now state, without proof, the properties of the function specified by an equation of the form

$$y = b^x, \text{ where } b > 0$$

1. The function has only positive values.
2. If $b > 1$, the function has increasing values as x increases. In this case the function is an **increasing function.**
3. If $0 < b < 1$, the function decreases as x increases. In this case the function is a **decreasing function.**
4. Whatever positive number b is, b^0 is 1.
5. This function has no zeros. (That is, there is no value of x for which $b^x = 0$.)

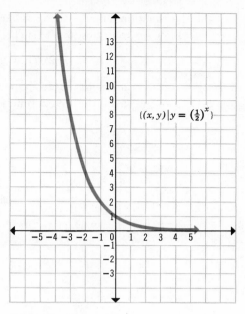

$\{(x, y) \mid y = \left(\tfrac{1}{2}\right)^x\}$

Figure 5.2

EXAMPLE 1. Using the laws of exponents show that $\left(\tfrac{1}{2}\right)^{4x}$ and $\left(\tfrac{1}{16}\right)^x$ are identical.

SOLUTION
$$\left(\tfrac{1}{2}\right)^{4x} = [\left(\tfrac{1}{2}\right)^4]^x = \left(\tfrac{1}{16}\right)^x$$

EXAMPLE 2. Find the solution set of $\left(\tfrac{1}{3}\right)^x = 81$.

SOLUTION
$$81 = 3^4 = \frac{1}{3^{-4}} = \frac{1^{-4}}{3^{-4}} = \left(\frac{1}{3}\right)^{-4}$$

The solution set is $\{-4\}$.

EXAMPLE 3. Determine an integer n such that $n < x < n + 1$ and $5^x = 26.9$

SOLUTION

$5^2 = 25$ and $5^3 = 125$, hence $2 < x < 3$, and $n = 2$.

EXAMPLE 4. Sketch the graph of the exponential function specified by $y = 4^x$.

SOLUTION. Since $4 > 0$, the function has increasing values as x increases. Some ordered pairs which are solutions of $y = 4^x$ are

$$(-2, \tfrac{1}{16}), (-\tfrac{3}{2}, \tfrac{1}{8}), (-1, \tfrac{1}{4}), (-\tfrac{1}{2}, \tfrac{1}{2}), (0, 1), (\tfrac{1}{2}, 2), (1, 4), (2, 16)$$

The graph of the given function has the same shape as the graph of $y = 2^x$ (Figure 5.1). The graph of the given function is shown in Figure 5.3.

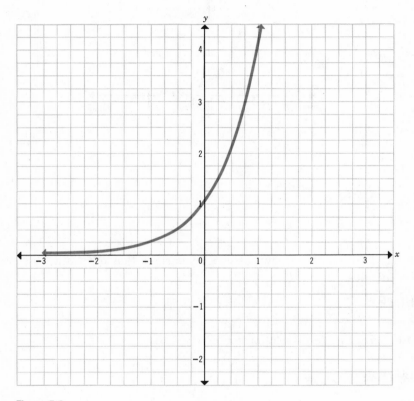

Figure 5.3

EXERCISES 5.1

Tell whether the exponential functions specified by the equations below are decreasing or increasing functions (Exercises 1–10).

1. $y = 3^x$
2. $y = (\tfrac{1}{3})^x$
3. $y = 5^x$
4. $y = (\tfrac{1}{2})^{-x}$
5. $y = 4^{-x}$
6. $y = (3)^{-x}$
7. $y = (\tfrac{1}{4})^x$
8. $y = (\tfrac{2}{3})^{-x}$
9. $y = 4^{2x}$
10. $y = (\tfrac{1}{10})^{3x}$

Sketch the graphs of the exponential functions specified by the equations below (Exercises 11–20).

11. $y = 3^x$ 16. $y = 3^{-x}$
12. $y = 10^x$ 17. $y = 10^{-x}$
13. $y = (\frac{1}{2})^{-x}$ 18. $y = (\frac{1}{3})^{-x}$
14. $y = (\frac{1}{10})^x$ 19. $y = 4^{2x}$
15. $y = (\frac{1}{4})^x$ 20. $y = (\frac{1}{2})^{2x}$

21. Using laws of exponents, show that $(\frac{1}{4})^x$ and 4^{-x} are identical.
22. Using laws of exponents, show that 3^{3x} may be written 27^x.
23. Using laws of exponents, show that 3^{x-2} may be written $\frac{1}{9} \cdot 3^x$.

Find the solution set of each of the following equations. Solve for x by inspection (Exercises 24–29).

24. $10^x = 100$ 26. $3^x = \frac{1}{27}$ 28. $10^x = \frac{1}{1000}$
25. $(\frac{1}{2})^x = 16$ 27. $16^x = 4$ 29. $4^{-x} = \frac{1}{64}$

30. Determine an integer n such that $n < x < n + 1$ and x is a solution of:
     (a) $2^x = 18.3$ (c) $10^x = 1119$
     (b) $4^x = 24.9$ (d) $3^x = 14.2$

5.2 ◆ Inverse Relations and Functions

A relation S in the set R of real numbers is a subset of $R \times R$. Consider the relation

$$S = \{(2, 1), (3, 2), (4, 3), (5, 4)\}$$

whose domain is $\{2, 3, 4, 5\}$ and whose range is $\{1, 2, 3, 4\}$.

If we interchange the first and second components of each ordered pair of S, we have the relation

$$P = \{(1, 2), (2, 3), (3, 4), (4, 5)\}$$

whose domain is $\{1, 2, 3, 4\}$ and whose range is $\{2, 3, 4, 5\}$.

We observe that the domain of S is the range of P, and that the range of S is the domain of P.

We call P the **inverse relation** of S and write

$$P = S^{-1}$$

DEFINITION 5.1. If S is a relation, the **inverse relation** of S denoted by S^{-1} read "S inverse" is defined to be the relation obtained when the components of each ordered pair of S are interchanged.

Using set-builder notation, we write

$$S^{-1} = \{(y, x) \mid (x, y) \in S\}$$

EXAMPLE 1. Find the inverse relation of

$$R = \{(1, 1), (2, 3), (3, 1), (1, 4), (-1, 0)\}$$

SOLUTION. Interchanging the first and second components of each ordered pair of R, we have

$$R^{-1} = \{(1, 1), (3, 2), (1, 3), (4, 1), (0, -1)\}$$

Since every function is a relation, every function has an inverse relation. The inverse relation of a function may or may not be a function. For example, consider the two functions:

$$f = \{(1, 2), (2, 3), (3, 4), (5, 7)\}$$
$$g = \{(1, 4), (2, 4), (3, 7)\}$$

If we interchange the first and second components of each ordered pair in each function above, we get the inverse relations of functions f and g:

$$f^{-1} = \{(2, 1), (3, 2), (4, 3), (7, 5)\}$$
$$g^{-1} = \{(4, 1), (4, 2), (7, 3)\}$$

In the first case, the inverse relation f^{-1} of the function f is also a function, but the inverse relation g^{-1} of the function g is not a function because the ordered pairs $(4, 1)$ and $(4, 2)$ have the same first component but different second components.

If the inverse relation of a function is also a function, it is called the **inverse function** of the given function. We see, then, that every function has an inverse relation, but every function does not necessarily have an inverse function.

When a relation is specified by an equation, the inverse relation is found by interchanging the variables in the equation. If, for example,

$$f = \{(x, y) \mid y = 3x + 2\}$$

then

$$f^{-1} = \{(y, x) \mid x = 3y + 2\}$$

If we solve the equation $x = 3y + 2$ explicitly for y we get $y = \dfrac{x - 2}{3}$. The inverse relation can then be written

$$f^{-1} = \left\{ (x, y) \mid y = \frac{x - 2}{3} \right\}$$

The inverse relation written in this form is convenient for graphing.

EXAMPLE 2. Find the inverse relation of

$$f = \left\{ (x, y) \mid y = \frac{5x - 3}{4} \right\}$$

Does f have an inverse function?

SOLUTION. Interchanging the variables in the equation $y = \dfrac{5x - 3}{4}$, we

obtain $x = \dfrac{5y - 3}{4}$. The inverse relation of f is

$$f^{-1} = \left\{ (y, x) \mid x = \frac{5y - 3}{4} \right\}$$

Solving $x = \dfrac{5y - 3}{4}$ explicitly for y, we obtain $y = \dfrac{4x + 3}{5}$. Thus
the inverse relation of f may be written

$$f^{-1} = \left\{ (x, y) \mid y = \frac{4x + 3}{5} \right\}$$

Since every value of x gives one and only one value of y we see that
f^{-1} is a function and f has an inverse function.

EXAMPLE 3. Find the inverse relation of $f = \{(x, y) \mid y = x^2 + 1\}$. Is f^{-1}
a function?

SOLUTION. Interchanging the variables in $y = x^2 + 1$, we obtain $x = y^2 + 1$.
Thus the inverse relation of function f is

$$f^{-1} = \{(y, x) \mid x = y^2 + 1\}$$

Solving $x = y^2 + 1$ explicitly for y, we obtain

$$y = \pm \sqrt{x - 1}$$

and we may write

$$f^{-1} = \{(x, y) \mid y = \pm \sqrt{x - 1}\}$$

Since every value of x in $y = \pm \sqrt{x - 1}$ gives two values of y, f^{-1} is
not a function and function f does not have an inverse function.

EXERCISES 5.2

Find the inverse relation of each function given below. Determine whether each of the functions has an inverse function (Exercises 1–20).

1. $\{(1, 2), (2, 4), (3, 5), (4, 8)\}$.
2. $\{(1, 3), (2, 1), (3, 5), (4, 3)\}$.
3. $\{(0, -2), (1, -1), (2, 2), (3, 3)\}$.
4. $\{(-1, 0), (0, 1), (1, 0), (2, 1)\}$.
5. $\{(-2, -2), (-1, -1), (0, 0), (1, 0)\}$.
6. $\{(-3, 5), (-1, 1), (2, 7), (3, -2)\}$.
7. $\{(1, 1), (2, 2), (3, 3), (15, 15)\}$.
8. $\{(-5, \frac{1}{2}), (-3, -3), (0, 1), (1, 8), (5, 1)\}$.
9. $\{(-1, 0), (-\frac{1}{2}, 1), (-\frac{1}{3}, 2), (-\frac{1}{4}, 4), (-\frac{1}{10}, 11)\}$.
10. $\{(x, y) \mid y = 3x\}$.
11. $\{(x, y) \mid y = x^2\}$.
12. $\{(x, y) \mid y = \sqrt{x}\}$.
13. $\{(x, y) \mid y = x^3\}$.
14. $\{(x, y) \mid y = x^4\}$.
15. $\{(x, y) \mid y = x^2 - 2x + 2\}$.
16. $\{(x, y) \mid y = x^3 + 1\}$.
17. $\left\{(x, y) \mid y = \dfrac{1}{x}\right\}$.
18. $\left\{(x, y) \mid y = \dfrac{1}{x - 2}\right\}$.
19. $\left\{(x, y) \mid y + 1 = \dfrac{1}{x + 2}\right\}$.
20. $\{(x, y) \mid x^2 + y^2 = 0\}$.

5.3 ◆ Logarithmic Functions

Figures 5.1 and 5.2 (Section 5.1) show the possible forms of the graph of the exponential function

$$\{(x, y) \mid y = b^x\}$$

where b is a positive real number not equal to 1. In either case, for each value of x there is paired one and only one value of y. It is also true that in either case, a horizontal line through a given point on the positive y-axis will intersect the graph at exactly one point (Figure 5.4). That is, $b^p = b^q$ if and only if $p = q$. Hence the inverse relation of the exponential function is also a function. That is, the exponential function has an inverse function

$$\{(y, x) \mid x = b^y\}$$

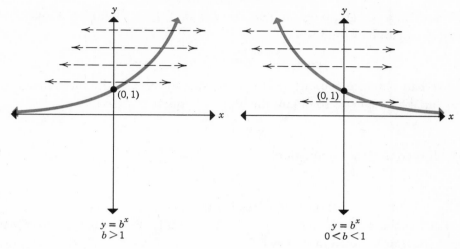

$$y = b^x$$
$$b > 1$$

$$y = b^x$$
$$0 < b < 1$$

Figure 5.4

We would like to have a way to solve the equation $x = b^y$ explicitly for y. The exponent that satisfies the equation $x = b^y$ is called the **logarithm of x to the base b,** which we write

$$y = \log_b x$$

We read this: y equals the logarithm of x to the base b. We state this formally by Definition 5.2.

DEFINITION 5.2 The equation $y = \log_b x$ is equivalent to the equation

$$x = b^y$$

We also assume that b is positive and not equal to 1.

We see that the inverse function of an exponential function is

$$\{(x, y) \mid y = \log_b x\}$$

and is called a **logarithmic function.**

We obtain the graph of the logarithmetic function with base b by graphing the equation $y = \log_b x$. Since equation

$$y = \log_b x$$

is equivalent to equation

$$x = b^y$$

this simplifies our graphing problem. In Section 5.1 we found the graph of a typical equation of the form

$$y = b^x$$

We can obtain equation $x = b^y$ from equation $y = b^x$ by interchanging the variables x and y. To obtain the graph of equation

$$y = \log_b x$$

which is equivalent to equation

$$x = b^y$$

from equation $y = b^x$, we merely relabel the axes.

Let us look at the graph of $y = 2^x$ (Figure 5.1). Relabeling the axes we have the graph shown in Figure 5.5.

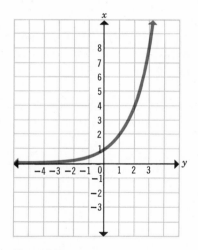

Figure 5.5

To get the axes in their customary position we rotate the figure 90° about the origin in a counterclockwise direction and then turn the whole graph over (that is rotate it about the y axis). Having done this we have the graph of $x = 2^y$, shown in Figure 5.6. Since $x = 2^y$ and $y = \log_2 x$ are two ways of writing the same equation, Figure 5.6 is the graph of $y = \log_2 x$.

Although Figure 5.6 shows the graph of the logarithmic function for $b = 2$, it is representative of the logarithmic function for any real number $b > 1$.

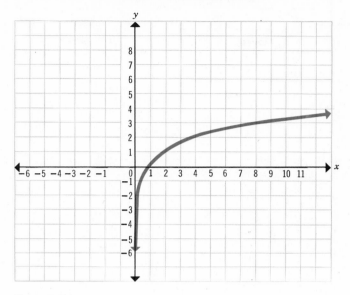

Figure 5.6

We now list without proof some properties of logarithmic functions.

1. $\log_b 1 = 0$, hence the only zero of the function is when $x = 1$.
2. $\log_b x$ is *less than* zero when $0 < x < 1$ and $b > 1$ and also when $x > 1$ and $0 < b < 1$.
3. $\log_b x$ is *greater than* zero when $x > 1$ and $b > 1$ and also when $x < 1$ and $0 < b < 1$.

The graph of the logarithmic function for $b = \frac{1}{2}$ is shown in Figure 5.7 (See page 148).

It should be observed that, for $b > 0$ and $b \neq 1$, equations

$$y = \log_b x \text{ and } x = b^y$$

are equivalent, both specifying the same function. Thus an exponential equation may be written in logarithmic form and a logarithmic equation may be written in exponential form. Some examples are:

Exponential form	Logarithmic form
$3^2 = 9$	$\log_3 9 = 2$
$10^2 = 100$	$\log_{10} 100 = 2$
$(\frac{1}{2})^{-3} = 8$	$\log_{\frac{1}{2}} 8 = -3$

We now prove some theorems called the **laws of logarithms**.

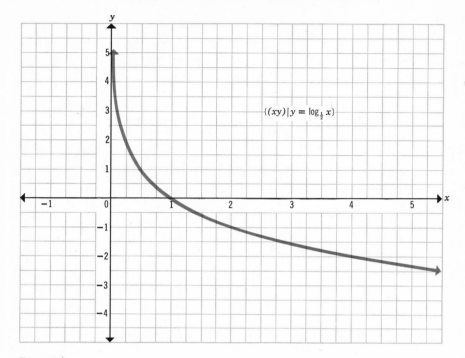

Figure 5.7

THEOREM 5.1. If $b > 0$, $b \neq 1$, and x_1 and x_2 are positive real numbers, then $\log_b (x_1 \cdot x_2) = \log_b x_1 + \log_b x_2$.

PROOF. Let $M = \log_b x_1$ and $N = \log_b x_2$. Then $b^M = x_1$ and $b^N = x_2$, by the definition of a logarithm. By the laws of exponents,

$$x_1 \cdot x_2 = b^M \cdot b^N = b^{M+N}$$

By the definition of a logarithm,

$$\log_b (x_1 \cdot x_2) = M + N$$

Hence, since $M = \log_b x_1$ and $N = \log_b x_2$, we have

$$\log_b (x_1 \cdot x_2) = \log_b x_1 + \log_b x_2$$

THEOREM 5.2. If $b > 0$, $b \neq 1$, and x_1 and x_2 are positive real numbers, then

$$\log_b \frac{x_1}{x_2} = \log_b x_1 - \log_b x_2$$

PROOF. Since x_1 and x_2 are both positive, neither is 0 and we have

$$x_2 \cdot \frac{x_1}{x_2} = x_1$$

Thus, by Theorem 5.1,

$$\log_b x_1 = \log_b \left(x_2 \cdot \frac{x_1}{x_2} \right) = \log_b x_2 + \log_b \frac{x_1}{x_2}$$

and

$$\log_b \frac{x_1}{x_2} = \log_b x_1 - \log_b x_2$$

THEOREM 5.3. If $b > 0$, $b \neq 1$, x is a positive number, and p is any real number, then

$$\log_b x^p = p \log_b x$$

PROOF. Let $M = \log_b x$. Then, by definition of a logarithm,

$$b^M = x$$

By the laws of exponents,

$$x^p = (b^M)^p = b^{Mp} = b^{pM}$$

Thus, by the definition of a logarithm,

$$\log_b x^p = pM = p \cdot \log_b x$$

Theorem 5.3 assures us that:

$$\log_b 9 = \log_b 3^2 = 2 \log_b 3$$
$$\log_b \sqrt{5} = \log_b 5^{1/2} = \tfrac{1}{2} \log_b 5$$

EXAMPLE 1. Express $\tfrac{1}{4} \log_b x + \tfrac{1}{2} \log_b y$ as a single logarithm with coefficient 1, using the laws of logarithms.

SOLUTION

$$\begin{aligned} \tfrac{1}{4} \log_b x + \tfrac{1}{2} \log_b y &= \log_b x^{1/4} + \log_b y^{1/2} && \text{Theorem 5.3} \\ &= \log_b (x^{1/4} \cdot y^{1/2}) && \text{Theorem 5.1} \end{aligned}$$

EXAMPLE 2. Express $\log_b \sqrt[3]{\dfrac{x+1}{x^2}}$ as the sum or difference of simpler logarithmic quantities using the laws of logarithms.

SOLUTION

$$\log_b \sqrt[3]{\frac{x+1}{x^2}} = \frac{1}{3} \left[\log_b \frac{x+1}{x^2} \right] \qquad\qquad \text{Theorem 5.3}$$

$$= \frac{1}{3} [\log_b (x + 1) - \log_b x^2] \qquad \text{Theorem 5.2}$$

$$= \frac{1}{3} [\log_b (x + 1) - 2 \log_b x] \qquad \text{Theorem 5.3}$$

EXERCISES 5.3

Express in logarithmetic notation (Exercises 1–9).

1. $2^3 = 8$ 4. $(\frac{1}{2})^2 = \frac{1}{4}$ 7. $5^1 = 5$
2. $3^4 = 81$ 5. $(\frac{1}{4})^2 = \frac{1}{16}$ 8. $2^{-6} = \frac{1}{64}$
3. $5^2 = 25$ 6. $27^{2/3} = 9$ 9. $(10)^{-2} = \frac{1}{100}$

Express in exponential notation (Exercises 10–17).

10. $\log_{10} 100 = 2$ 14. $\log_{1/2} 16 = -4$
11. $\log_2 32 = 5$ 15. $\log_{1/3} 27 = -3$
12. $\log_{10} \frac{1}{100} = -2$ 16. $\log_4 4 = 1$
13. $\log_5 1 = 0$ 17. $\log_8 16 = \frac{4}{3}$

Find by inspection the value of x, b, or y (Exercises 18–25).

18. $y = \log_2 4$ 22. $\log_{10} 0.01 = y$
19. $y = \log_8 1$ 23. $\log_{10} x = -3$
20. $\log_b 16 = 4$ 24. $\log_{1/2} x = -4$
21. $\log_b 100 = 2$ 25. $\log_5 x = 3$

Express each of the following as a single logarithm (Exercises 26–31).

26. $\log_b x + \log_b y$
27. $\log_b x - \log_b z$
28. $\log_b 2x - \log_b 3y$
29. $\frac{1}{2} \log_b x - \frac{2}{3} \log_b y$
30. $\log_b x - \log_b y + \log_b z$
31. $\frac{3}{4} \log_b x - \frac{3}{4} \log_b y$

Express as the sum or difference of simpler logarithmic quantities (Exercises 32–41).

32. $\log_b (xyz)$ 35. $\log_b x^{1/2}$

33. $\log_b \frac{xy}{z}$ 36. $\log_b x^3 y^2$

34. $\log_b \frac{x^2}{y}$ 37. $\log_b \sqrt{\frac{x}{y}}$

38. $\log_b \sqrt[4]{\dfrac{x^3}{y^2}}$ 40. $\log_b \sqrt{x(x-a)(x-b)}$

39. $\log_b 2\sqrt{x^3 y^3}$ 41. $\log_b \sqrt[5]{\dfrac{(x-1)^2}{x^2}}$

Graph the functions specified by the following equations (Exercises 42–45).

42. $y = \log_{10} x$ 44. $y = \log_2 x$

43. $y = \log_3 x$ 45. $y = \log_{1/2} x$

5.4 ✦ Common Logarithms

There are two logarithmic functions that are commonly used in mathematics. One of these is specified by the equation

$$y = \log_{10} x$$

and the other by

$$y = \log_e x$$

where e is a specific irrational number for which 2.718281828459 is an approximation accurate to twelve decimal places. We shall discuss first the function specified by

$$y = \log_{10} x$$

Values of this function are called **logarithms to the base 10** or **common logarithms.** For computation that involves scientific notation this system of logarithms is very convenient. The advantages will become apparent in the discussion that follows.

It is easy to find values of $\log_{10} x$ when x is an integer power of 10. For example:

since $1000 = 10^3$, $\log_{10} 1000 = \log_{10} 10^3 = 3$

since $100 = 10^2$, $\log_{10} 100 = \log_{10} 10^2 = 2$

since $10 = 10^1$; $\log_{10} 10 = \log_{10} 10^1 = 1$

since $1 = 10^0$; $\log_{10} 1 = \log_{10} 10^0 = 0$

since $0.1 = \dfrac{1}{10^1} = 10^{-1}$; $\log_{10} 0.1 = \log_{10} 10^{-1} = -1$

since $0.01 = \dfrac{1}{10^2} = 10^{-2}$; $\log_{10} 0.01 = \log_{10} 10^{-2} = -2$

and so forth.

Table II of the appendix is a table of common logarithms, called a "log table," and is used to find $\log_{10} x$ when $1 \leq x < 10$. A portion of such a table is shown in Figure 5.8. It should be noted that the logarithms given are

Table II—Common Logarithms.

N	0	1	2	3	4	5	6	7	8	9
55	7404	7412	7419	7427	7435	7443	7451	7459	7466	7474
56	7482	7490	7497	7505	7513	7520	7528	7536	7543	7551
57	7559	7566	7574	7582	7589	7597	7604	7612	7619	7627
58	7634	7642	7649	7657	7664	7672	7679	7686	7694	7701
59	7709	7716	7723	7731	7738	7745	7752	7760	7767	7774
60	7782	7789	7796	7803	7810	7818	7825	7832	7839	7846
61	7853	7860	7868	7875	7882	7889	7896	7903	7910	7917
62	7924	7931	7938	7945	7952	7959	7966	7973	7980	7987
63	7993	8000	8007	8014	8021	8028	8035	8041	8048	8055
64	8062	8069	8075	8082	8089	8096	8102	8109	8116	8122
65	8129	8136	8142	8149	8156	8162	8169	8176	8182	8189
66	8195	8202	8209	8215	8222	8228	8235	8241	8248	8254
67	8261	8267	8274	8280	8287	8293	8299	8306	8312	8319
68	8325	8331	8338	8344	8351	8357	8363	8370	8376	8382
69	8388	8395	8401	8407	8414	8420	8426	8432	8439	8445
70	8451	8457	8463	8470	8476	8482	8488	8494	8500	8506
71	8513	8519	8525	8531	8537	8543	8549	8555	8561	8567
72	8573	8579	8585	8591	8597	8603	8609	8615	8621	8627
73	8633	8639	8645	8651	8657	8663	8669	8675	8681	8686
74	8692	8698	8704	8710	8716	8722	8727	8733	8739	8745
75	8751	8756	8762	8768	8774	8779	8785	8791	8797	8802
76	8808	8814	8820	8825	8831	8837	8842	8848	8854	8859
77	8865	8871	8876	8882	8887	8893	8899	8904	8910	8915
78	8921	8927	8932	8938	8943	8949	8954	8960	8965	8971
79	8976	8982	8987	8993	8998	9004	9009	9015	9020	9025
80	9031	9036	9042	9047	9053	9058	9063	9069	9074	9079
81	9085	9090	9096	9101	9106	9112	9117	9122	9128	9133
82	9138	9143	9149	9154	9159	9165	9170	9175	9180	9186
83	9191	9196	9201	9206	9212	9217	9222	9227	9232	9238
84	9243	9248	9253	9258	9263	9269	9274	9279	9284	9289
85	9294	9299	9304	9309	9315	9320	9325	9330	9335	9340
86	9345	9350	9355	9360	9365	9370	9375	9380	9385	9390
87	9395	9400	9405	9410	9415	9420	9425	9430	9435	9440
88	9445	9450	9455	9460	9465	9469	9474	9479	9484	9489
89	9494	9499	9504	9509	9513	9518	9523	9528	9533	9538
90	9542	9547	9552	9557	9562	9566	9571	9576	9581	9586
91	9590	9595	9600	9605	9609	9614	9619	9624	9628	9633
92	9638	9643	9647	9652	9657	9661	9666	9671	9675	9680
93	9685	9689	9694	9699	9703	9708	9713	9717	9722	9727
94	9731	9736	9741	9745	9750	9754	9759	9763	9768	9773
95	9777	9782	9786	9791	9795	9800	9805	9809	9814	9818
96	9823	9827	9832	9836	9841	9845	9850	9854	9859	9863
97	9868	9872	9877	9881	9886	9890	9894	9899	9903	9908
98	9912	9917	9921	9926	9930	9934	9939	9943	9948	9952
99	9956	9961	9965	9969	9974	9978	9983	9987	9991	9996
N	0	1	2	3	4	5	6	7	8	9

Figure 5.8

merely rational number approximations which are accurate for the number of decimal places given in the table.

In this table all decimal points have been omitted. In the column headed "N," the decimal point should occur after the first digit in each numeral. In the other columns, a decimal point should occur before the first digit in each numeral. Each number named in the column headed "N" represents the first two significant digits of N, while each number in the top row of the table contains the third digit. For example, to find $\log_{10} 8.64$, we look down the column headed "N" for 86 and then move to the right to the numeral which appears in the column headed "4." Finding 9365 in the table, we have

$$\log_{10} 8.64 = 0.9365$$

This means that

$$10^{0.9365} = 8.64$$

When we write $\log_{10} 8.64 = 0.9365$, we are using the equality sign in an inexact sense. Since the rational number given in the table is only an approximation of the actual logarithm, we should write $\log_{10} 8.64 \doteq 0.9365$ (recall that the symbol $\doteq$ means approximately equal to). We shall follow customary usage, however, and use $=$ instead of $\doteq$.

To find $\log_{10} x$ for values outside the range of the table, that is, for $x > 10$ or $0 < x < 1$, we represent x in **scientific notation.** That is, we represent the number whose logarithm we wish to find as a product of a number between 1 and 10 and an appropriate power of 10. We then use Theorem 5.1 to find the logarithm. Some examples are given below:

$$
\begin{aligned}
\log_{10} 87.2 &= \log_{10} (8.72 \times 10^1) \\
&= \log_{10} 8.72 + \log_{10} 10^1 \\
&= 0.9405 + 1 \\
&= 1.9405 \\
\log_{10} 872 &= \log_{10} (8.72 \times 10^2) \\
&= \log_{10} 8.72 + \log_{10} 10^2 \\
&= 0.9405 + 2 \\
&= 2.9405 \\
\log_{10} 8720 &= \log_{10} (8.72 \times 10^3) \\
&= \log_{10} 8.72 + \log_{10} 10^3 \\
&= 0.9405 + 3 \\
&= 3.9405.
\end{aligned}
$$

In each of the above examples the decimal fraction part of the logarithm is the same and the integer part of the logarithm is the exponent of 10 which occurs when the number is written in scientific notation.

Finding common logarithms is quite mechanical when we consider $\log_{10} x$ to consist of two parts: an integer part, called the **characteristic,** and a decimal fraction part, called the **mantissa.** The log table can then be seen to be a table of mantissas.

We now consider $\log_{10} x$ for $0 < x < 1$. As an example,

$$\begin{aligned}
\log_{10} 0.872 &= \log_{10} (8.72 \times 10^{-1}) \\
&= \log 8.72 + \log_{10} 10^{-1} \\
&= 0.9405 + (-1)
\end{aligned}$$

It is customary to write characteristics which are negative integers in the form $n - 10$, or $n - 20$, or $n - 30$, and so forth, where n is a positive integer. Thus we write

$$\begin{aligned}
\log_{10} 0.872 &= 0.9405 + (-1) \\
&= 0.9405 + (9 - 10) \\
&= 9.9405 - 10
\end{aligned}$$

We also have

$$\begin{aligned}
\log_{10} 0.0000000000872 &= \log_{10} (8.72 \times 10^{-11}) \\
&= 0.9405 + (-11) \\
&= 0.9405 + (9 - 20) \\
&= 9.9405 - 20
\end{aligned}$$

If we are given $\log_{10} x$ it is possible to use the log table to find x. In this case x is called the **antilogarithm** (**antilog$_{10}$**) **of $\log_{10} x$.** For example, antilog$_{10}$ 1.5955 can be found by locating the mantissa, 0.5955, in the body of the log table and observing that $\log_{10} 3.94 = 0.5955$. Thus

$$\begin{aligned}
1.5955 &= 1 + 0.5955 \\
&= \log_{10} 10^1 + \log_{10} 3.94 \\
&= \log_{10} (3.94 \times 10^1) \\
&= \log_{10} 39.4.
\end{aligned}$$

Therefore

$$\text{antilog}_{10} 1.5955 = 39.4$$

Since the table gives us common logarithms of numbers with three significant figures, when we want to find the logarithm of a number with four or more significant figures (for example, 3826) we use a process called **linear interpolation.**

A function is a set of ordered pairs of numbers. A table of logarithms is just such a set of ordered pairs of the logarithmic function. For each number,

x, there is associated the number $\log_{10} x$ to form the ordered pair $(x, \log_{10} x)$. The table of common logarithms (Table II, page 487) gives three digits for the number x and four digits for the number $\log_{10} x$. By using a process called linear interpolation, we can use this table to find approximations to logarithms of numbers which have four digit numerals.

Suppose we want to find $\log_{10} 2.412$. A magnified portion of the graph of $y = \log_{10} x$ is shown in Figure 5.9.

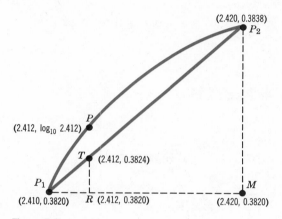

Figure 5.9

We use the line segment connecting P_1, whose coordinates are $(2.410, 0.3820)$, and P_2, whose coordinates are $(2.420, 0.3838)$, as an approximation to the curve which passes through these points. The number $\log_{10} 2.412$ is the y-coordinate of the point P, which has x-coordinate 2.412. Since there is no way of finding this y-coordinate from the table, we shall use in its place the y-coordinate of the point T—that point of the line segment with x-coordinate 2.412.

Observing Figure 5.9, we can see that the triangles P_1TR and P_1P_2M are similar because they are both right triangles with a common angle with vertex at P_1. Since the sides of similar triangles are proportional, we have

$$\frac{P_1R}{P_1M} = \frac{TR}{P_2M}$$

Since we know three of these numbers, we have

$$\frac{2.412 - 2.410}{2.420 - 2.410} = \frac{TR}{0.3838 - 0.3820}$$

or

$$\frac{0.002}{0.010} = \frac{TR}{0.0018}$$

$$\frac{1}{5} = \frac{TR}{0.0018}$$

$TR = 0.00036 \doteq 0.0004$ (rounding to four decimal places)

Since the y-coordinate of R is 0.3820, we see that the y-coordinate of T must be 0.3820 + 0.0004, or 0.3824.

The example below shows a shortened form of the procedure for linear interpolation.

EXAMPLE 1. Find $\log_{10} 48.24$.

SOLUTION

$$0.10 \begin{cases} 0.04 \begin{cases} \log_{10} 48.20 = 1.6830 \\ \log_{10} 48.24 = ? \end{cases} y \\ \log_{10} 48.30 = 1.6839 \end{cases} 0.0009$$

$$\frac{4}{10} = \frac{y}{0.0009}$$

$$y = \frac{4}{10}(0.0009) = 0.00036 \doteq 0.0004$$

$$\log_{10} 48.24 = 1.6830 + 0.0004 = 1.6834$$

Example 2 shows the procedure used for finding antilog$_{10}$ of a number whose mantissa is not in the table.

EXAMPLE 2. Find antilog$_{10}$ 2.7638.

SOLUTION

$$0.0008 \begin{cases} 0.0004 \begin{cases} \text{antilog}_{10}\ 2.7634 = 580.0 \\ \text{antilog}_{10}\ 2.7638 = ? \end{cases} y \\ \text{antilog}_{10}\ 2.7642 = 581.0 \end{cases} 1.0$$

$$\frac{4}{8} = \frac{y}{1}$$

$$y = \frac{1}{2} = 0.5$$

$$\text{antilog}_{10}\ 2.7638 = 580.0 + 0.5 = 580.5$$

EXERCISES 5.4

Find each logarithm. Interpolate as required (Exercises 1–20).

1. $\log_{10} 3.86$
2. $\log_{10} 5.09$
3. $\log_{10} 17.8$
4. $\log_{10} 786$
5. $\log_{10} 0.0932$
6. $\log_{10} 7.683$
7. $\log_{10} 0.09342$
8. $\log_{10} 86,710$
9. $\log_{10} 0.0008431$
10. $\log_{10} 0.1111$
11. $\log_{10} 786,300$
12. $\log_{10} 48.23$
13. $\log_{10} (5.78 \times 10^{-3})$
14. $\log_{10} (1.76 \times 10^{4})$
15. $\log_{10} (1.234 \times 10^{5})$
16. $\log_{10} (7.816 \times 10^{-3})$
17. $\log_{10} (5.141 \times 10^{3})$
18. $\log_{10} (3.142 \times 10^{-2})$
19. $\log_{10} (2.718 \times 10^{-5})$
20. $\log_{10} (1.414 \times 10^{-28})$

Find each antilogarithm. Interpolate as required (Exercises 21–40).

21. $\text{antilog}_{10} 0.6021$
22. $\text{antilog}_{10} 0.7042$
23. $\text{antilog}_{10} 3.9777$
24. $\text{antilog}_{10} 2.9666$
25. $\text{antilog}_{10} (7.8414 - 10)$
26. $\text{antilog}_{10} (9.5276 - 10)$
27. $\text{antilog}_{10} (8.7388 - 10)$
28. $\text{antilog}_{10} 4.6609$
29. $\text{antilog}_{10} 0.8237$
30. $\text{antilog}_{10} 1.6606$
31. $\text{antilog}_{10} 1.2092$
32. $\text{antilog}_{10} (8.2804 - 10)$
33. $\text{antilog}_{10} (7.4248 - 20)$
34. $\text{antilog}_{10} 2.6064$
35. $\text{antilog}_{10} 1.9246$
36. $\text{antilog}_{10} (7.5591 - 10)$
37. $\text{antilog}_{10} 3.6725$
38. $\text{antilog}_{10} (7.7814 - 20)$
39. $\text{antilog}_{10} (1.3027)$
40. $\text{antilog}_{10} (8.9957)$

5.5 ◆ Computations Using Logarithms

Numerical computations which involve multiplication, division, raising to a power, or extracting roots can be simplified through the use of logarithms. For convenience, we restate the laws of logarithms (Theorems 7.1, 7.2, and 7.3) here, using base ten.

(i) $\log_{10} (x_1 \cdot x_2) = \log_{10} x_1 + \log_{10} x_2$.

(ii) $\log_{10} \dfrac{x_1}{x_2} = \log_{10} x_1 - \log_{10} x_2$.

(iii) $\log_{10} x^p = p \log_{10} x$.

We shall also need the following assumptions.

ASSUMPTION 1. If $M = N$, then $\log_{10} M = \log_{10} N$.
ASSUMPTION 2. If $\log_{10} M = \log_{10} N$, then $M = N$.
ASSUMPTION 3. If $M = N$, then $10^M = 10^N$.

EXAMPLE 1. Using logarithms, compute $\dfrac{(8.32)^{1/2}(17.6)^3}{(3.21)^{1/4}}$.

SOLUTION. Setting

$$N = \frac{(8.32)^{1/2}(17.6)^3}{(3.21)^{1/4}}$$

we obtain

$$\log_{10} N = \log_{10} \frac{(8.32)^{1/2}(17.6)^3}{(3.21)^{1/4}}$$
$$= \tfrac{1}{2} \log_{10} 8.32 + 3 \log_{10} 17.6 - \tfrac{1}{4} \log_{10} 3.21$$
$$= \tfrac{1}{2}(0.9201) + 3(1.2455) - \tfrac{1}{4}(0.5065)$$
$$= 0.4601 + 3.7365 - 0.1266$$
$$= 4.0700$$
$$\text{antilog}_{10}\ 4.0700 = 11{,}750$$
$$N \doteq 11{,}750$$

EXAMPLE 2. Using logarithms, compute $\sqrt[5]{0.0364}$.

SOLUTION. Setting

$$N = \sqrt[5]{0.0364}$$

we obtain

$$\log_{10} N = \log_{10} \sqrt[5]{0.0364}$$
$$= \tfrac{1}{5} \log_{10} 0.0364$$
$$= \tfrac{1}{5}(8.5611 - 10)$$
$$= \tfrac{1}{5}(48.5611 - 50)$$
$$= 9.7122 - 10$$
$$\text{antilog}_{10}\ 9.7122 - 10 = 0.5155$$
$$N \doteq 0.5155$$

EXAMPLE 3. Using logarithms, compute $\dfrac{3.21}{\sqrt[3]{-4.12}}$.

SOLUTION. Since the cube root of a negative number is negative and the quotient of a positive number divided by a negative number is negative, the result will be negative. In doing the computation we ignore the sign. Setting

$$N = \frac{3.21}{\sqrt[3]{4.12}}$$

we have

$$\log_{10} N = \log_{10} 3.21 - \tfrac{1}{3} \log 4.12$$
$$= 0.5065 - \tfrac{1}{3}(0.6149)$$
$$= 0.5065 - 0.2050$$
$$= 0.3015$$
$$N = \text{antilog}_{10}\, 0.3015 \doteq 2.002.$$

Since the result is negative, the quotient is -2.002.

EXERCISES 5.5

Compute using logarithms (Exercises 1–20).

1. $(7.86)(149)$

2. $(80.3)(0.672)$

3. $\dfrac{3.27}{0.864}$

4. $\dfrac{(2.76)(18.2)}{11.4}$

5. $\sqrt{1.82}$

6. $(0.931)^2$

7. $\left(\dfrac{1.23}{4.11}\right)^3$

8. $\left(\dfrac{2.17}{4.03}\right)^{1/3}$

9. $\left[\dfrac{(1.04)(7.19)}{(2.18)}\right]^3$

10. $\left[\dfrac{2.31}{(8.05)(5.11)}\right]^4$

11. $\left[\dfrac{(1.25)(5.72)}{(3.15)(6.61)}\right]^{1/2}$

12. $\dfrac{(2.71)^2(5.05)^3}{(1.13)^3(8.71)^2}$

13. $\dfrac{(0.913)^2(-1.24)^5}{(1.03)^4(0.0123)^3}$

14. $\dfrac{(0.00315)(35.7)^2}{(218)^3(0.0141)^2}$

15. $\dfrac{(10.7)^3(0.00216)^4}{(0.000111)^2(3.33)^5}$

16. $\dfrac{(31.08)(4.213)}{(8.207)(10.05)}$

17. $\dfrac{(0.03679)(-1.003)}{(21.77)(0.0004034)}$

18. $\dfrac{(26.43)(1.892)^3}{(0.842)\sqrt{14.34}}$

19. $\sqrt[6]{\dfrac{(22.39)(6.619)^5}{9007}}$

20. $\dfrac{(0.03582)(0.8768)}{0.004144}$

5.6 ◆ Exponential and Logarithmic Equations

An equation in which a variable appears in an exponent is called an **exponential equation**. Solution sets of some exponential equations may be found by using the common logarithm function.

EXAMPLE 1. Find the solution set of $3^x = 30$.

SOLUTION. Since $3^x > 0$ for all x, this equation is equivalent to

$$\log_{10} 3^x = \log_{10} 30$$
$$x \log_{10} 3 = \log_{10} 30$$
$$x = \frac{\log_{10} 30}{\log_{10} 3} = \frac{1.4771}{0.4771} = 3.095996 \doteq 3.096$$

The solution set is $\{3.096\}$.

EXAMPLE 2. Find the solution set of $3^{x+1} = 4^{x-7}$.

SOLUTION. This equation is equivalent to

$$\log_{10} 3^{x+1} = \log_{10} 4^{x-7}$$
$$(x + 1) \log_{10} 3 = (x - 7) \log_{10} 4$$
$$x(\log_{10} 4 - \log_{10} 3) = \log_{10} 3 + 7 \log_{10} 4$$
$$x = \frac{\log_{10} 3 + 7 \log_{10} 4}{\log_{10} 4 - \log_{10} 3}$$
$$= \frac{0.4771 + 4.2147}{0.6021 - 0.4771} = 37.5344 \doteq 37.53$$

The solution set is $\{37.53\}$.

A logarithmic equation is one in which the logarithm of a variable occurs. Solution sets of some logarithmic equations can be found by using the laws of logarithms.

EXAMPLE 3. Find the solution set of

$$\log_{10} x = 2 \log_{10} 4 + \log_{10} 32$$

SOLUTION. This equation is equivalent to

$$\log_{10} x = \log_{10} (4^2)(32)$$
$$x = 4^2 \cdot 32 = 512$$

The solution set is $\{512\}$.

EXAMPLE 4. Find the solution set of $\log_{10} (3x + 2) = \log_{10} (x - 1) + 1$.

SOLUTION. This equation is equivalent to

$$\log_{10} (3x + 2) - \log_{10} (x - 1) = 1$$

$$\log_{10} \frac{3x + 2}{x - 1} = 1 = \log_{10} 10$$

$$\frac{3x + 2}{x - 1} = 10$$

$$3x + 2 = 10x - 10$$

$$7x = 12$$

$$x = \tfrac{12}{7}$$

The solution set is $\{\tfrac{12}{7}\}$.

EXERCISES 5.6

Find the solution sets (Exercises 1–20).

1. $2^x = 32$	11. $\log_{10} 2x = 3$
2. $3^x = 27$	12. $\log_{10} (3x + 5) = 2$
3. $2^x = 27$	13. $\log_{10} 2x - \log_{10} 2 = 5$
4. $3^x = 36$	14. $3 \log_{10} x = \log_{10} 8$
5. $3^{x+1} = 2^{2x-14}$	15. $2 \log_{10} x = 1$
6. $4^{x-1} = 3^{3x}$	16. $\log_{10} (2x - 1) = \log_{10} (x + 3) - 2$
7. $2^{4x+3} = 4^{9x-1}$	17. $\log_{10} (3x + 5) + 3 = \log_{10} (2x + 1)$
8. $3^{2x-5} = 2^{3x+5}$	18. $\log_{10} (x + 2) - \log_{10} (x + 1) = 3$
9. $5^x = 3$	19. $\log_{10} (2x - 1) + \log_{10} (x - 3) = 2$
10. $9^x = 3^{2x}$	20. $\log_{10} (x - 1) + \log_{10} (x + 1) = \log_{10} (2x + 1)$

5.7 ◆ Logarithms to Other Bases

The number e ($e = 2.718281828459 \ldots$) is of great importance in mathematics. This number (which was named after the mathematician Leonard Euler) is the base of a system of logarithms called **natural logarithms** or **Napierian logarithms** (after the mathematician John Napier). Tables of natural logarithms exist, but it is desirable to be able to determine such logarithms by the use of a table of common logarithms. To do this we develop a formula for converting logarithms from one base to another.

We consider the equation

$$(1) \qquad\qquad y = \log_b x$$

where $b > 0$. This equation, by the definition of logarithm to the base b, is equivalent to

$$(2) \qquad\qquad x = b^y$$

If we take the logarithm of each member of Equation 2 to the base a, where $a > 0$, we have

(3) $$\log_a x = y \log_a b$$

which is equivalent to

(4) $$y = \frac{\log_a x}{\log_a b}$$

Substituting the value of y from Equation 1 in Equation 4, we have the **change of base formula**

(5) $$\log_b x = \frac{\log_a x}{\log_a b}$$

This is the formula used for converting from a logarithm to the base a to a logarithm to the base b. In particular, then, to convert from a logarithm to the base 10 (common logarithm) to a logarithm to the base e (natural logarithm), the formula becomes

(6) $$\log_e x = \frac{\log_{10} x}{\log_{10} e}$$

($\log_e x$ is frequently written $\ln x$.) Interpolation in the common logarithm table tells us that

$$\log_{10} e = \log_{10} 2.718$$
$$= 0.4343$$

so we have
$$\log_e x = \frac{\log_{10} x}{0.4343}$$

EXAMPLE 1. Find (a) $\log_e 72$; (b) $\log_3 0.324$; (c) $\log_3 2.17$.

SOLUTION

(a) $\log_e 72 = \dfrac{\log_{10} 72}{0.4343} = \dfrac{1.8573}{0.4343} = 4.277$

(b) $\log_e 0.324 = \dfrac{\log_{10} 0.324}{0.4343} = \dfrac{9.5105 - 10}{0.4343} = \dfrac{-0.4895}{0.4343} = -1.1271$

(c) $\log_3 2.17 = \dfrac{\log_{10} 2.17}{\log_{10} 3} = \dfrac{0.3365}{0.4771} = 0.7053$

EXAMPLE 2. Given $\log_2 16 = 4$, find $\log_{16} 2$.

SOLUTION. Since

$$\log_b x = \frac{\log_a x}{\log_a b}$$

we have, letting $b = 16$, $a = 2$, and $x = 2$,

$$\log_{16} 2 = \frac{\log_2 2}{\log_2 16}$$

$$= \tfrac{1}{4}$$

EXERCISES 5.7

Find the following logarithms (Exercises 1–10).

1. $\log_e 728$	6. $\log_e 926$
2. $\log_e 76$	7. $\log_e 99$
3. $\log_e 423$	8. $\log_e 107$
4. $\log_e 87$	9. $\log_e 88$
5. $\log_e 114$	10. $\log_e 27$

Find the following logarithms (Exercises 11–16).

11. $\log_2 321$	14. $\log_{12} 17$
12. $\log_4 163$	15. $\log_9 89$
13. $\log_7 32$	16. $\log_{16} 763$

Compute the following (Exercises 17–20).

17. If $\log_3 9 = 2$, find $\log_9 3$.
18. If $\log_4 64 = 3$. find $\log_{64} 4$.
19. If $\log_{10} 10{,}000 = 4$, find $\log_{10000} 10$.
20. If $\log_5 125 = 3$, find $\log_{125} 5$.

CHAPTER REVIEW

Find the inverse function for each of the following (Exercises 1–2).

1. $\{(0, 0), (1, 5), (2, 3), (4, 4)\}$.
2. $\{(x, y) \mid y = 3x + 2\}$.

Sketch the graphs of each of the following functions (Exercises 3–4).

3. $\{(x, y) \mid y = 3^x\}$.
4. $\{(x, y) \mid y = \log_3 x\}$.
5. Express in logarithmic form:
   (a) $16 = 2^4$ (b) $x = 2^y$
6. Express in exponential form:
   (a) $\log_{1/2} 8 = -3$ (b) $y = \log_5 x$

Find the value of the following (Exercises 7–14).

7. (a) $\log_{10} 2.07$ (b) $\log_{10} 8.91$ (c) $\log_{10} 5.33$
8. (a) $\log_{10} 21.5$ (b) $\log_{10} 354,000$ (c) $\log_{10} 0.00417$
9. (a) $\log_{10} 3.182$ (b) $\log_{10} 7.033$ (c) $\log_{10} 5.003$
10. (a) $\log_{10} 0.08231$ (b) $\log_{10} 18,300,000$ (c) $\log_{10} 333$
11. (a) $\text{antilog}_{10} 0.9410$ (b) $\text{antilog}_{10} 0.4409$ (c) $\text{antilog}_{10} 0.7308$
12. (a) $\text{antilog}_{10} 3.9085$ (b) $\text{antilog}_{10} (8.1271 - 10)$ (c) $\text{antilog}_{10} 1.7466$
13. (a) $\text{antilog}_{10} 0.6183$ (b) $\text{antilog}_{10} 0.8334$ (c) $\text{antilog}_{10} 0.1415$
14. (a) $\text{antilog}_{10} (7.8244 - 10)$ (b) $\text{antilog}_{10} (9.5780 - 10)$
    (c) $\text{antilog}_{10} 0.9347$

Use logarithms to evaluate the following (Exercises 15–16).

15. $\dfrac{(3.18)^4(0.0141)^2(4.18)^2}{(385)^2(0.00817)^3(1.24)^4}$ 16. $\dfrac{\sqrt[3]{18.31}\,\sqrt[4]{4,105}}{(8.103)^2\,\sqrt[5]{2.003}}$

Find the solution sets (Exercises 17–18).

17. $2^{2x+1} = 3^{x-2}$ 18. $\log(2x + 1) = 3 - \log_{10}(x - 2)$

Evaluate (Exercises 19–20).

19. $\log_3 5$ 20. $\log_e 25$

CHAPTER 6

Trigonometric Functions

6.1 ◆ The Unit Circle

In this chapter we shall consider moves along arcs of a circle. We shall call the course traveled in such a move a **path**. For the circle in Figure 6.1 we have taken the radius to be 4 units. The circumference of this circle is thus 8π units ($C = 2\pi r$).

This circle has been cut into eight arcs of equal length by the points $A, B, \ldots, H$, spaced equally around the circle. Each of these arcs has length π. A path along this circle from point A to point E, passing through points B, C, and D, is 4π units in length. We agree to consider a counterclockwise motion as positive and a clockwise motion as negative. We use an ordered pair to denote a path. The first component of the ordered pair gives the point at

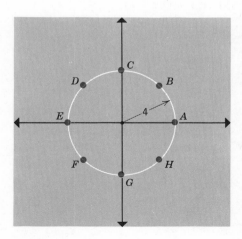

Figure 6.1

which the path begins. The second component of the ordered pair tells the length and direction of the path by using a positive number to denote counterclockwise motion and a negative number to denote clockwise motion. Thus the path from A to E passing through B, C, and D is denoted by the ordered pair $(A, 4\pi)$. In this system, the ordered pair $(A, -\pi)$ denotes the path traveled by a move from A to H in a clockwise direction.

A circle whose radius is 1 unit is called a **unit circle.** For the unit circle in Figure 6.2, the points A, B, C, and D are spaced equally along the circle and divide the circle into four arcs, each of length $\dfrac{\pi}{2}$. If we use ordered pairs to denote paths along this circle, we see that many paths have the same initial point and the same terminal point, but are given by different ordered pairs. For example, the paths listed below have initial point A and terminal point B.

$$\left(A, \frac{\pi}{2}\right) \qquad \left(A, \frac{\pi}{2} - 2\pi\right)$$

$$\left(A, \frac{\pi}{2} + 2\pi\right) \qquad \left(A, \frac{\pi}{2} - 4\pi\right)$$

$$\left(A, \frac{\pi}{2} + 4\pi\right) \qquad \left(A, \frac{\pi}{2} - 6\pi\right)$$

$$\left(A, \frac{\pi}{2} + 6\pi\right) \qquad \left(A, \frac{\pi}{2} - 8\pi\right)$$

$$\vdots \qquad\qquad \vdots$$

$$\left(A, \frac{\pi}{2} + 2n\pi\right) \qquad (A, -2n\pi)$$

where n is a nonnegative integer.

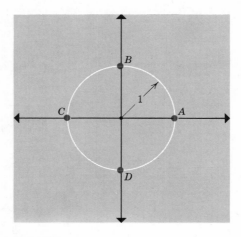

Figure 6.2

We see that for $n = 0, 1, 2, 3, \ldots$, the path

$$\left(A, \frac{\pi}{2} \pm 2n\pi\right)$$

has initial point A and terminal point B.

In a similar manner it can be seen that for $n = 0, 1, 2, \ldots$, any path of the form $\left(A, \frac{3\pi}{2} \pm 2n\pi\right)$ has initial point A and terminal point D, and that any path of the form $(A, \pi \pm 2n\pi)$ has initial point A and terminal point C.

In general, if P is any point on the unit circle and θ is any real number, and if path (A, θ) has initial point A and terminal point P, then for $n = 0, 1, 2, 3, \ldots, (A, \theta \pm 2n\pi)$ will also have initial point A and terminal point P (Figure 6.3).

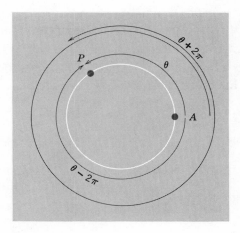

Figure 6.3

6.2 ✦ The Unit Circle with a Coordinate System

We now place a unit circle on the Cartesian plane in such a manner that the center of the circle falls at the origin. Further, we shall consider only those paths with initial point A whose coordinates are $(1, 0)$ (Figure 6.4).

As before, counterclockwise paths will be denoted by positive numbers and clockwise paths by negative numbers. Thus, for any real number θ, the terminal point of the path denoted by (A, θ) is some point P on the unit circle. If the point P has coordinates (x, y), we must have $x^2 + y^2 = 1$ (since the distance from P to the origin must be 1, the radius of the circle). Some points on the circle have coordinates that are easy to determine. These points are

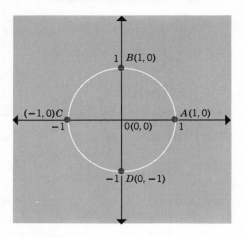

Figure 6.4

shown in Figure 6.4. Since points A, B, C, and D are points on the axes, we see that the angles AOB, BOC, COD, and DOA all have degree measures of $90°$, that is, they are right angles. Since the radius of the unit circle is 1, the circumference has length 2π units. We then see that:

The terminal point of $(A, 0)$ is A.

The terminal point of $\left(A, \dfrac{\pi}{2}\right)$ is B.

The terminal point of (A, π) is C.

The terminal point of $\left(A, \dfrac{3\pi}{2}\right)$ is D.

EXAMPLE 1. Consider the point P which bisects the path $\left(A, \dfrac{\pi}{2}\right)$, that is, is the path with initial point A and terminal point P and which moves counterclockwise. Find the coordinates of P.

SOLUTION. We denote the coordinates of P by (x, y). Because of the symmetry of a circle about its center, the coordinates of point R in Figure 6.5 must be $(x, -y)$. Since P bisects the path which moves counterclockwise from A to B, the length of the path (arc) from A to P must be $\dfrac{\pi}{4}$. The length of the path from A to R must also be $\dfrac{\pi}{4}$. Thus the arc from P to R is $\dfrac{\pi}{2}$ units in length. But the length of the path

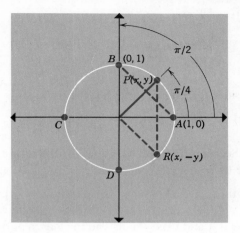

Figure 6.5

from A to B is also $\frac{\pi}{2}$ units in length. Since, in a given circle, arcs of equal length subtend chords of equal length, the chord PR and the chord AB have the same length. Using the distance formula we have

$$\sqrt{(x-x)^2 + [y-(-y)]^2} = \sqrt{(1-0)^2 + (0-1)^2}$$
$$(x-x)^2 + [y-(-y)]^2 = (1-0)^2 + (0-1)^2$$
$$(2y)^2 = 1^2 + (-1)^2$$
$$4y^2 = 2$$
$$y^2 = \tfrac{1}{2}$$

Hence y is $\dfrac{1}{\sqrt{2}}$ or $-\dfrac{1}{\sqrt{2}}$. Since P is in the first quadrant, we know that its y-coordinate is positive, hence y is $\dfrac{1}{\sqrt{2}}$. We can now find the x-coordinate by replacing y by $\dfrac{1}{\sqrt{2}}$ in the equation $x^2 + y^2 = 1$, which gives

$$x^2 + \left(\frac{1}{\sqrt{2}}\right)^2 = 1$$
$$x^2 + \tfrac{1}{2} = 1$$
$$x^2 = \tfrac{1}{2}$$

so that x is $\dfrac{1}{\sqrt{2}}$ or $-\dfrac{1}{\sqrt{2}}$. Again, since P is in the first quadrant, we know that its x-coordinate must be positive so that we conclude that the coordinates of P are

$$\left(\frac{1}{\sqrt{2}}, \frac{1}{\sqrt{2}}\right) \quad \text{or} \quad \left(\frac{\sqrt{2}}{2}, \frac{\sqrt{2}}{2}\right)$$

EXAMPLE 2. Consider the point Q which is the terminal point of the path $\left(A, \frac{\pi}{6}\right)$. Find the coordinates of Q.

SOLUTION. Let the coordinates of Q be (x, y). Because a circle is symmetric about its center, the coordinates of point R must be $(x, -y)$. That is, R is the terminal point of the path $\left(A, -\frac{\pi}{6}\right)$. The length of the arc between A and Q is $\frac{\pi}{6}$ units and the length of the arc between A and R is also $\frac{\pi}{6}$ units. (Figure 6.6) Thus the length of the arc QR is $\frac{\pi}{6} + \frac{\pi}{6} = \frac{\pi}{3}$ units. It is also true that the length of the arc between Q and B is $\frac{\pi}{2} - \frac{\pi}{6} = \frac{\pi}{3}$ units. Since equal arcs subtend chords of equal length, we must have chord RQ equal in length to chord BQ. Using the distance formula, we have

$$\sqrt{(x-0)^2 + (y-1)^2} = \sqrt{(x-x)^2 + [y-(-y)]^2}$$
$$(x-0)^2 + (y-1)^2 = (x-x)^2 + [y-(-y)]^2$$
$$x^2 + y^2 - 2y + 1 = 4y^2$$

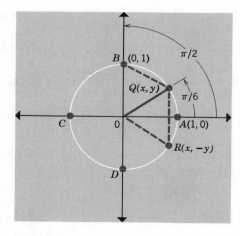

Figure 6.6

since $x^2 + y^2 = 1$ (Q is on the unit circle), we can replace $x^2 + y^2$ in the left member by 1, giving

$$1 - 2y + 1 = 4y^2$$
$$4y^2 + 2y - 2 = 0$$
$$2y^2 + y - 1 = 0$$
$$(2y - 1)(y + 1) = 0$$

Thus y is -1 or $\frac{1}{2}$. Since Q is in the first quadrant, its y-coordinate must be positive so that y is $\frac{1}{2}$. Since $x^2 + y^2 = 1$, we find x by replacing y by $\frac{1}{2}$ in this equation:

$$x^2 + \tfrac{1}{4} = 1$$
$$x^2 = \tfrac{3}{4}$$

Thus x is $\dfrac{\sqrt{3}}{2}$ or $-\dfrac{\sqrt{3}}{2}$. Since Q is in the first quadrant, its x-coordinate must be positive, so that the coordinates of Q are

$$\left(\frac{\sqrt{3}}{2}, \frac{1}{2}\right)$$

EXERCISES 6.1

1. The unit circle below is cut into eight arcs of equal length by the points shown. Determine the length of each of the eight arcs.

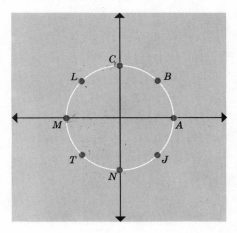

Use the figure in Exercise 1 and determine the terminal points of the arcs given below (Exercises 2–12).

2. $\left(A, \dfrac{\pi}{4}\right)$

3. $\left(A, \dfrac{\pi}{2}\right)$

4. $\left(A, \dfrac{3\pi}{4}\right)$

5. $(A, -\pi)$

6. $\left(A, -\dfrac{7\pi}{4}\right)$

7. $\left(A, \dfrac{5\pi}{4}\right)$

8. $\left(A, \dfrac{9\pi}{4}\right)$

9. $\left(A, -\dfrac{15\pi}{4}\right)$

10. $\left(A, \dfrac{\pi}{2} + 2n\pi\right)$, $n = 0, \pm 1, \pm 2, \ldots$

11. $\left(A, -\dfrac{3\pi}{4} + 2n\pi\right)$, $n = 0, \pm 1, \pm 2, \ldots$

12. $\left(A, -\dfrac{5\pi}{4} + 2n\pi\right)$, $n = 0, \pm 1, \pm 2, \ldots$

The unit circle below is divided into twelve arcs of equal length by the points shown. What are the terminal points of the following arcs (Exercises 13–22)?

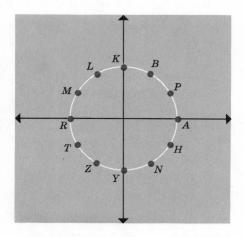

13. $\left(A, \dfrac{\pi}{6}\right)$

14. $\left(A, \dfrac{\pi}{3}\right)$

15. $\left(A, \dfrac{2\pi}{3}\right)$

16. $\left(A, -\dfrac{4\pi}{3}\right)$

17. $\left(A, -\dfrac{5\pi}{6}\right)$

18. $\left(A, \dfrac{7\pi}{6}\right)$

19. $\left(A, \dfrac{7\pi}{3}\right)$

20. $\left(A, -\dfrac{13\pi}{6}\right)$

21. $\left(A, \dfrac{3\pi}{2} + 2n\pi\right)$, $n = 0, \pm 1, \pm 2, \ldots$

22. $\left(A, -\dfrac{\pi}{3} + 2n\pi\right)$, $n = 0, \pm 1, \pm 2, \ldots$

Using the unit circle divided into twelve equal arcs, determine the values of θ for which (A, θ) has each of the following terminal points (Exercises 23–40).

23. B, with $0 < \theta < 2\pi$.
24. B, with $-2\pi < \theta < 0$.
25. B, with $2\pi < \theta < 4\pi$.
26. M, with $0 < \theta < 2\pi$.
27. M, with $-2\pi < \theta < 0$.
28. M, with $2\pi < \theta < 4\pi$.
29. Z, with $0 < \theta < 2\pi$.
30. Z, with $4\pi < \theta < 6\pi$.
31. Z, with $-6\pi < \theta < -4\pi$.
32. Y, with $0 < \theta < 2\pi$.
33. Y, with $-2\pi < \theta < 0$.
34. Y, with $10\pi < \theta < 12\pi$.
35. N, with $0 < \theta < 2\pi$.
36. N, with $-2\pi < \theta < 0$.
37. N with $-14\pi < \theta < -12\pi$.
38. Determine the coordinates of the point T.
39. Determine the coordinates of the point H.
40. Determine the coordinates of the point B.

6.3 ✦ The Circle Function

The **unit circle** in the Cartesian plane is the graph of the equation

$$x^2 + y^2 = 1$$

For certain paths—given by ordered pairs of the form (A, θ), where A is the point with coordinates $(1, 0)$ and θ is a real number giving the length and direction of the path—we have determined the coordinates of the terminal points. We now define the **circle function**, C, whose range will be the set of ordered pairs that are coordinates of points on the unit circle.

DEFINITION 6.1. The **circle function**, C, is given by $C = \{(\theta, (x, y)) \mid (x, y)$ are the coordinates of the terminal point of the path (A, θ) and A has coordinates $(1, 0)\}$.

The circle function, C, associates with each real number, θ, an ordered pair, (x, y), whose components are the coordinates of the terminal point of the path given by (A, θ), where A is the point $(1, 0)$ and the path is along the unit circle. Values of $C(\theta)$ are easily found for certain special values of θ. Since the path from A to A has 0 length, the path $(A, 0)$ terminates at A, so that

$$C(0) = (1, 0)$$

Since the circumference of the unit circle has length 2π, we see that for $n = 0, 1, 2, 3, \ldots,$

$$C(0 \pm 2n\pi) = C(0) = (1, 0)$$

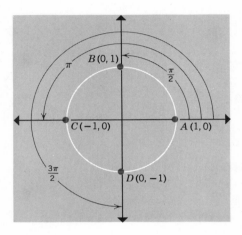

Figure 6.7

Figure 6.7 shows the coordinates of the points B: $(0, 1)$; C: $(-1, 0)$; and D: $(0, -1)$. Since the points A, B, C, and D divide the circle into four arcs of equal length, we see that

$$C\left(\frac{2\pi}{4}\right) = C\left(\frac{\pi}{2}\right) = (0, 1)$$

$$C\left(\frac{2\pi}{2}\right) = C(\pi) = (-1, 0)$$

$$C\left(\frac{3}{4} \cdot 2\pi\right) = C\left(\frac{3\pi}{2}\right) = (0, -1)$$

In Section 6.2 we found some other values of the circle functions:

$$C\left(\frac{\pi}{4}\right) = \left(\frac{1}{\sqrt{2}}, \frac{1}{\sqrt{2}}\right) = \left(\frac{\sqrt{2}}{2}, \frac{\sqrt{2}}{2}\right)$$

$$C\left(\frac{\pi}{6}\right) = \left(\frac{\sqrt{3}}{2}, \frac{1}{2}\right)$$

EXAMPLE 1. Find the value of $C\left(\dfrac{\pi}{3}\right)$.

SOLUTION. Figure 6.8 shows the unit circle with $P(x, y)$ denoting the terminal point of the path $\left(A, \dfrac{\pi}{3}\right)$. We know that $C\left(\dfrac{\pi}{6}\right) = \left(\dfrac{\sqrt{3}}{2}, \dfrac{1}{2}\right)$ (see Section 6.2). Let us draw a line through the origin with slope 1. We see that the circle is symmetric with respect to this particular one of its diameters. The points $P(x, y)$ and $P_2\left(\dfrac{\sqrt{3}}{2}, \dfrac{1}{2}\right)$ are equidistant from this line so that the x-coordinate of one is the y-coordinate of the other, and vice versa. Thus

$$C\left(\frac{\pi}{3}\right) = \left(\frac{1}{2}, \frac{\sqrt{3}}{2}\right)$$

In a similar manner, we can find $C\left(\dfrac{2\pi}{3}\right)$, $C\left(\dfrac{4\pi}{3}\right)$, and $C\left(\dfrac{5\pi}{3}\right)$.

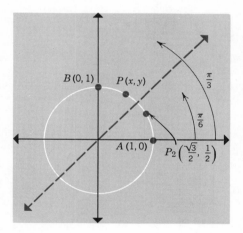

Figure 6.8

EXERCISES 6.2

Find values for the following (Exercises 1–12).

1. $C\left(\dfrac{3\pi}{4}\right)$ 3. $C\left(\dfrac{7\pi}{4}\right)$

2. $C\left(\dfrac{5\pi}{4}\right)$ 4. $C\left(\dfrac{2\pi}{3}\right)$

5. $C\left(\dfrac{4\pi}{3}\right)$ 9. $C\left(\dfrac{11\pi}{6}\right)$

6. $C\left(\dfrac{5\pi}{3}\right)$ 10. $C\left(-\dfrac{3\pi}{4}\right)$

7. $C\left(\dfrac{5\pi}{6}\right)$ 11. $C\left(-\dfrac{5\pi}{3}\right)$

8. $C\left(\dfrac{7\pi}{6}\right)$ 12. $C\left(-\dfrac{7\pi}{4}\right)$

13. Complete the tables below.

θ	0	$\dfrac{\pi}{6}$	$\dfrac{\pi}{4}$	$\dfrac{\pi}{3}$	$\dfrac{\pi}{2}$	$\dfrac{2\pi}{3}$	$\dfrac{3\pi}{4}$	$\dfrac{5\pi}{6}$	π
$C(\theta)$	$(1,0)$				$(0,-1)$				$(-1,0)$

θ	$\dfrac{7\pi}{6}$	$\dfrac{5\pi}{4}$	$\dfrac{4\pi}{3}$	$\dfrac{3\pi}{2}$	$\dfrac{5\pi}{3}$	$\dfrac{7\pi}{4}$	$\dfrac{11\pi}{6}$	2π
$C(\theta)$				$(0,-1)$				$(1,0)$

Use the table in Exercise 13 to find the values of the following (Exercises 14–23).

14. $C(12\pi)$ 19. $C\left(-\dfrac{5\pi}{6}\right)$

15. $C(11\pi)$ 20. $C\left(\dfrac{9\pi}{4}\right)$

16. $C(17\pi)$ 21. $C\left(\dfrac{29\pi}{3}\right)$

17. $C(28\pi)$ 22. $C\left(\dfrac{13\pi}{6}\right)$

18. $C\left(\dfrac{7\pi}{6}\right)$ 23. $C\left(\dfrac{15\pi}{4}\right)$

24. Which of the following are true statements?

(a) $C\left(\dfrac{\pi}{4}\right) = C\left(-\dfrac{\pi}{4}\right)$ (d) $C\left(\dfrac{\pi}{2}\right) = C\left(-\dfrac{3\pi}{2}\right)$

(b) $C(2\pi) = C(-2\pi)$ (e) $C\left(\dfrac{\pi}{4}\right) = C\left(-\dfrac{3\pi}{4}\right)$

(c) $C\left(\dfrac{\pi}{6}\right) = C\left(\dfrac{5\pi}{6}\right)$ (f) $C(\pi) = C(3\pi)$

(g) $C\left(\dfrac{2\pi}{3}\right) = C\left(-\dfrac{3\pi}{4}\right)$ (i) $C(2\pi) = C(0)$

(h) $C\left(\dfrac{\pi}{4} + 2\pi\right) = C\left(-\dfrac{\pi}{4}\right)$ (j) $C\left(\dfrac{5\pi}{4}\right) = C\left(-\dfrac{\pi}{4}\right)$

6.4 ◆ Periodic Functions

We have observed that the circle function

$C = \{(\theta, (x, y)) \mid (x, y)$ are the coordinates of the terminal point of the path
(A, θ) and A has coordinates $(1, 0)\}$

has the property

$$C(\theta) = C(\theta + 2\pi) = C(\theta + 4\pi) = \cdots = C(\theta \pm 2n\pi)$$

where n is a nonnegative integer. A function with such a property is called a **periodic function** and the numbers 2π, 4π, ..., $2n\pi$, are called periods of the function.

DEFINITION 6.2. A function f is said to be **periodic** with period p, $p > 0$, if for all real numbers x

$$f(x + p) = f(x)$$

The period, p, is called the **fundamental period** of f provided that p is the least positive number that is also a period of f.

We shall refer to the fundamental period of a periodic function as *the* period of that function. If a function is periodic with fundamental period p, then $-p, 2p, -2p, 3p, -3p$, and so forth are also periods of the function.

6.5 ◆ The Trigonometric Functions Cosine and Sine

In Section 6.3 we introduced the circle function:

$C = \{(\theta, (x, y)) \mid (x, y)$ are the coordinates of the terminal point of the path (A, θ) and A has coordinates $(1, 0)\}$

The domain of C is the set of all real numbers and the range of C is the set

$$\{(x, y) \mid x^2 + y^2 = 1\}$$

We now consider separately the first and second components of the ordered pairs in the range of the circle function. We construct two new functions from the circle function by doing this.

For any given real number θ, we call the first component of the ordered pair $C(\theta)$ the **cosine** of θ, which we abbreviate $\cos \theta$. We call the second component of $C(\theta)$ the **sine** of θ which we abbreviate $\sin \theta$.

DEFINITION 6.3. For any real number θ the **cosine** of θ, abbreviated $\cos \theta$, is the abscissa of $C(\theta)$ and the **sine** of θ, abbreviated $\sin \theta$, is the ordinate of $C(\theta)$:

$$\cos \theta = x \ [\text{the abscissa of } C(\theta)]$$
$$\sin \theta = y \ [\text{the ordinate of } C(\theta)]$$

We note that Definition 6.3 gives rise to sets of ordered pairs $(\theta, \cos \theta)$ and $(\theta, \sin \theta)$ and thus defines two functions. The domain of each of these functions is the set of all real numbers θ. However, since every point $C(\theta)$ lies on the unit circle, neither of its coordinates (x, y) can exceed 1 in absolute value. Therefore

$$|\cos \theta| \leq 1 \qquad \text{and} \qquad |\sin \theta| \leq 1$$

In other words, the ranges of these functions are restricted by the requirements $-1 \leq \cos \theta \leq 1$ and $-1 \leq \sin \theta \leq 1$ for all values of $\theta \in R$.

DEFINITION 6.4. The **cosine function** is the set of all ordered pairs $(\theta, \cos \theta)$ where $\cos \theta$ is the x-coordinate of the terminal point of the path (A, θ). The **sine function** is the set of all ordered pairs $(\theta, \sin \theta)$ where $\sin \theta$ is the y-coordinate of the terminal point of the path (A, θ).

Then

cosine function $= \{(\theta, \cos \theta) \mid \cos \theta \text{ is the } x\text{-coordinate of the terminal point}$
of the path $(A, \theta)\}$

sine function $= \{(\theta, \sin \theta) \mid \sin \theta \text{ is the } y\text{-coordinate of the terminal point of}$
the path $(A, \theta)\}$

Since the x-coordinates of points in the first and fourth quadrants are positive, we see that $\cos \theta$ is positive for paths with terminal points in these quadrants. Similarly, $\cos \theta$ is negative for paths with terminal points in the second and third quadrants. Figure 6.9 shows the signs of $\cos \theta$ and $\sin \theta$ in the four quadrants.

Since the circle function is periodic with period 2π, the cosine function

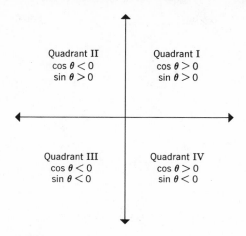

Figure 6.9

and the sine function are also periodic with period 2π. Hence, for $n = 0$, $1, 2, \ldots$:

$$\cos (\theta \pm 2\pi n) = \cos \theta$$
$$\sin (\theta \pm 2\pi n) = \sin \theta$$

By definition

$$C(\theta) = (x, y) = (\cos \theta, \sin \theta)$$

Hence we can find certain values of $\cos \theta$ and $\sin \theta$ by using the points whose coordinates we have found. These values are given in Table 6.1.

We can use the fact that the cosine and sine functions both have period 2π to find the values of these functions for some numbers not given in Table 6.1.

EXAMPLE 1. Find $\cos \dfrac{15\pi}{6}$ and $\sin \dfrac{15\pi}{6}$.

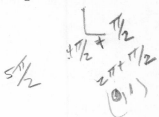

SOLUTION. We first observe that

$$\frac{15\pi}{6} = 2\pi + \frac{\pi}{2}$$

Thus

$$\cos \frac{15\pi}{6} = \cos \left(\frac{\pi}{2} + 2\pi\right) = \cos \frac{\pi}{2} = 0$$

$$\sin \frac{15\pi}{6} = \sin \left(\frac{\pi}{2} + 2\pi\right) = \sin \frac{\pi}{2} = 1$$

Table 6.1

θ	$\cos\theta$	$\sin\theta$
0	1	0
$\dfrac{\pi}{6}$	$\dfrac{1}{2}\sqrt{3}$	$\dfrac{1}{2}$
$\dfrac{\pi}{4}$	$\dfrac{1}{2}\sqrt{2}$	$\dfrac{1}{2}\sqrt{2}$
$\dfrac{\pi}{3}$	$\dfrac{1}{2}$	$\dfrac{1}{2}\sqrt{3}$
$\dfrac{\pi}{2}$	0	1
$\dfrac{2\pi}{3}$	$-\dfrac{1}{2}$	$\dfrac{1}{2}\sqrt{3}$
$\dfrac{3\pi}{4}$	$-\dfrac{1}{2}\sqrt{2}$	$\dfrac{1}{2}\sqrt{2}$
$\dfrac{5\pi}{6}$	$-\dfrac{1}{2}\sqrt{3}$	$\dfrac{1}{2}$
π	-1	0
$\dfrac{7\pi}{6}$	$-\dfrac{1}{2}\sqrt{3}$	$-\dfrac{1}{2}$
$\dfrac{5\pi}{4}$	$-\dfrac{1}{2}\sqrt{2}$	$-\dfrac{1}{2}\sqrt{2}$
$\dfrac{4\pi}{3}$	$-\dfrac{1}{2}$	$-\dfrac{1}{2}\sqrt{3}$
$\dfrac{3\pi}{2}$	0	-1
$\dfrac{5\pi}{3}$	$\dfrac{1}{2}$	$-\dfrac{1}{2}\sqrt{3}$
$\dfrac{7\pi}{4}$	$\dfrac{1}{2}\sqrt{2}$	$-\dfrac{1}{2}\sqrt{2}$
$\dfrac{11\pi}{6}$	$\dfrac{1}{2}\sqrt{3}$	$-\dfrac{1}{2}$
2π	1	0

EXAMPLE 2. Find $\cos\dfrac{17\pi}{3}$ and $\sin\dfrac{17\pi}{3}$.

SOLUTION

$$\frac{17\pi}{3} = 4\pi + \frac{5\pi}{3}$$

$$\cos\frac{17\pi}{3} = \cos\left(\frac{5\pi}{3} + 4\pi\right) = \cos\frac{5\pi}{3} = \frac{1}{2}$$

$$\sin\frac{17\pi}{3} = \sin\left(\frac{5\pi}{3} + 4\pi\right) = \sin\frac{5\pi}{3} = -\frac{\sqrt{3}}{2} = -\frac{1}{2}\sqrt{3}$$

EXAMPLE 3. Find $\cos\left(-\dfrac{17\pi}{4}\right)$ and $\sin\left(-\dfrac{17\pi}{4}\right)$.

SOLUTION

$$-\frac{17\pi}{4} = \frac{7\pi}{4} - 6\pi$$

$$\cos\left(-\frac{17\pi}{4}\right) = \cos\left(\frac{7\pi}{4} - 6\pi\right) = \cos\frac{7\pi}{4} = \frac{1}{\sqrt{2}} = \frac{\sqrt{2}}{2} = \frac{1}{2}\sqrt{2}$$

$$\sin\left(-\frac{17\pi}{4}\right) = \sin\left(\frac{7\pi}{4} - 6\pi\right) = \sin\frac{7\pi}{4} = -\frac{1}{\sqrt{2}} = -\frac{\sqrt{2}}{2} = -\frac{1}{2}\sqrt{2}$$

EXERCISES 6.3

Find the value of each of the following (Exercises 1–40).

1. $\sin\dfrac{\pi}{3}$

2. $\cos\dfrac{\pi}{4}$

3. $\cos\pi$

4. $\sin\dfrac{\pi}{6}$

5. $\sin\dfrac{3\pi}{2}$

6. $\cos\dfrac{2\pi}{3}$

7. $\sin\dfrac{5\pi}{3}$

8. $\cos\dfrac{7\pi}{4}$

9. $\sin\dfrac{11\pi}{6}$

10. $\sin\dfrac{7\pi}{4}$

11. $\cos\dfrac{13\pi}{6}$

12. $\sin\dfrac{7\pi}{3}$

13. $\sin\left(-\dfrac{7\pi}{3}\right)$

14. $\cos\dfrac{5\pi}{2}$

15. $\cos\dfrac{11\pi}{4}$

16. $\cos5\pi$

17. $\cos6\pi$

18. $\cos\left(-7\pi\right)$

19. $\sin\dfrac{13\pi}{4}$

20. $\sin\dfrac{7\pi}{2}$

21. $\sin\left(-\dfrac{15\pi}{4}\right)$

22. $\sin\left(-\dfrac{16\pi}{3}\right)$

23. $\cos\dfrac{17\pi}{4}$

24. $\cos\dfrac{17\pi}{3}$

25. $\sin\dfrac{21\pi}{4}$

26. $\sin\dfrac{22\pi}{3}$

27. $\sin \dfrac{32\pi}{3}$ 34. $\sin \dfrac{107\pi}{3}$

28. $\sin \left(-\dfrac{31\pi}{3} \right)$ 35. $\cos \dfrac{107\pi}{4}$

29. $\cos \left(-\dfrac{11\pi}{3} \right)$ 36. $\cos \dfrac{107\pi}{2}$

30. $\cos \left(-\dfrac{11\pi}{2} \right)$ 37. $\cos 107\pi$

31. $\cos \dfrac{41\pi}{4}$ 38. $\sin \left(-\dfrac{215\pi}{4} \right)$

32. $\cos \left(-\dfrac{31\pi}{4} \right)$ 39. $\sin \left(-\dfrac{1001\pi}{3} \right)$

33. $\sin \left(-\dfrac{22\pi}{3} \right)$ 40. $\cos \left(-\dfrac{2851\pi}{2} \right)$

6.6 ◆ Some Theorems for the Cosine and Sine Functions

Whenever $(\theta, (x, y))$ is one of the ordered pairs of the circle function, the ordered pair (x, y), of the range, must satisfy the equation

$$x^2 + y^2 = 1$$

since these are the coordinates of a point on the unit circle. Since, in this case, $\cos \theta = x$ and $\sin \theta = y$, we can use the substitution property of equality to obtain

$$(\cos \theta)^2 + (\sin \theta)^2 = 1$$

It is customary to write $(\cos \theta)^2$ as $\cos^2 \theta$, read: "cosine squared theta," and $(\sin \theta)^2$ as $\sin^2 \theta$, read: "sine squared theta." Thus the above equation becomes

(1) $$\cos^2 \theta + \sin^2 \theta = 1$$

This equation is called the **basic identity** of trigonometry.

Since the x-coordinates of points in the first and fourth quadrants are positive, and those of points in the second and third quadrants are negative, equation 1 is equivalent to

$$\cos \theta = \begin{cases} \sqrt{1 - \sin^2 \theta} & \text{in Quadrants I and IV} \\ -\sqrt{1 - \sin^2 \theta} & \text{in Quadrants II and III} \end{cases}$$

In a similar manner we find

$$\sin \theta = \begin{cases} \sqrt{1 - \cos^2 \theta} & \text{in Quadrants I and II} \\ -\sqrt{1 - \cos^2 \theta} & \text{in Quadrants III and IV} \end{cases}$$

We now prove some theorems about the cosine and sine functions. We begin by considering four paths on the unit circle (Figure 6.10).

We observe in Figure 6.10 that:

B is the terminal point of the path (A, θ_3).

C is the terminal point of the paths (A, θ_1) and (D, θ_3).

D is the terminal point of the path (A, θ_2).

$\theta_1 = \theta_2 + \theta_3$.

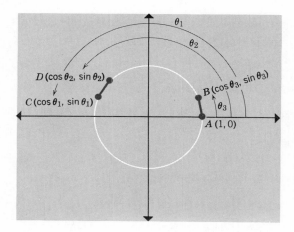

Figure 6.10

Since the length of the arc between A and B is equal in length to the arc between D and C, and since arcs of equal length subtend chords of equal length, the line segments AB and CD are equal in length. Using the distance formula, we have

$$\sqrt{(1 - \cos \theta_3)^2 + (0 - \sin \theta_3)^2} = \sqrt{(\cos \theta_1 - \cos \theta_2)^2 + (\sin \theta_1 - \sin \theta_2)^2}$$

Squaring both sides of the above equation, we obtain

$$(1 - \cos \theta_3)^2 + (0 - \sin \theta_3)^2 = (\cos \theta_1 - \cos \theta_2)^2 + (\sin \theta_1 - \sin \theta_2)^2$$

Simplifying, we have

$$1 - 2 \cos \theta_3 + \cos^2 \theta_3 + \sin^2 \theta_3 = \cos^2 \theta_1 - 2 \cos \theta_1 \cos \theta_2 \\ + \cos^2 \theta_2 + \sin^2 \theta_1 - 2 \sin \theta_1 \sin \theta_2 + \sin^2 \theta_2$$

Regrouping terms, we obtain

$$1 + (\cos^2 \theta_3 + \sin^2 \theta_3) - 2 \cos \theta_3 = (\cos^2 \theta_1 + \sin^2 \theta_1) \\ + (\cos^2 \theta_2 + \sin^2 \theta_2) - 2 \cos \theta_1 \cos \theta_2 - 2 \sin \theta_1 \sin \theta_2$$

But the basic identity says that $\cos^2 \theta + \sin^2 \theta = 1$, so that we have

$$2 - 2 \cos \theta_3 = 2 - 2 \cos \theta_1 \cos \theta_2 - 2 \sin \theta_1 \sin \theta_2$$

that is,

$$\cos \theta_3 = \cos \theta_1 \cos \theta_2 + \sin \theta_1 \sin \theta_2$$

Since $\theta_3 = \theta_1 - \theta_2$, we can write

(2) $$\cos (\theta_1 - \theta_2) = \cos \theta_1 \cos \theta_2 + \sin \theta_1 \sin \theta_2$$

A similar diagram can be drawn on the unit circle to illustrate the case in which θ_1 or θ_2 are both negative real numbers. The same equation (2) will result. This proves:

THEOREM 6.1. For all real numbers θ_1 and θ_2,

$$\cos (\theta_1 - \theta_2) = \cos \theta_1 \cos \theta_2 + \sin \theta_1 \sin \theta_2$$

We use this theorem to prove several other theorems.

THEOREM 6.2. For all real numbers, θ,

$$\cos (-\theta) = \cos \theta$$

PROOF. In Theorem 6.1, let $\theta_1 = 0$ and $\theta_2 = \theta$. Then

$$\cos (0 - \theta) = \cos 0 \cos \theta + \sin 0 \sin \theta$$
$$\cos (-\theta) = (1) \cos \theta + (0) \sin \theta$$
$$= \cos \theta$$

THEOREM 6.3. For all real numbers, θ,

$$\cos \left(\frac{\pi}{2} - \theta \right) = \sin \theta$$

PROOF. In Theorem 6.1, let $\theta_1 = \frac{\pi}{2}$ and $\theta_2 = \theta$. Then

$$\cos \left(\frac{\pi}{2} - \theta \right) = \cos \frac{\pi}{2} \cos \theta + \sin \frac{\pi}{2} \sin \theta$$
$$= (0) \cos \theta + (1) \sin \theta$$
$$= \sin \theta$$

THEOREM 6.4. For all real numbers, θ,

$$\sin \left(\frac{\pi}{2} - \theta \right) = \cos \theta$$

PROOF. If we replace θ by $\frac{\pi}{2} - \theta$ in Theorem 6.3, we have

$$\cos\left[\frac{\pi}{2} - \left(\frac{\pi}{2} - \theta\right)\right] = \sin\left(\frac{\pi}{2} - \theta\right)$$

$$\cos\theta = \sin\left(\frac{\pi}{2} - \theta\right)$$

THEOREM 6.5. For all real numbers, θ,

$$\sin(-\theta) = -\sin\theta$$

PROOF. By Theorem 6.3, setting $\theta = -\theta$, we have

$$\cos\left(\frac{\pi}{2} - [-\theta]\right) = \sin(-\theta)$$

That is,

$$\sin(-\theta) = \cos\left(\frac{\pi}{2} + \theta\right)$$

By Theorem 6.2,

$$\cos\left(\frac{\pi}{2} + \theta\right) = \cos\left(-\frac{\pi}{2} - \theta\right)$$

By Theorem 6.1,

$$\cos\left(-\frac{\pi}{2} - \theta\right) = \cos\left(-\frac{\pi}{2}\right)\cos\theta + \sin\left(-\frac{\pi}{2}\right)\sin\theta$$

$$= (0)\cos\theta + (-1)\sin\theta$$

$$= -\sin\theta$$

Thus

$$\sin(-\theta) = -\sin\theta$$

THEOREM 6.6. For all real numbers, θ_1 and θ_2,

$$\cos(\theta_1 + \theta_2) = \cos\theta_1\cos\theta_2 - \sin\theta_1\sin\theta_2$$

PROOF

$$
\begin{aligned}
\cos(\theta_1 + \theta_2) &= \cos(\theta_1 - [-\theta_2]) & \\
&= \cos\theta_1\cos(-\theta_2) + \sin\theta_1\sin(-\theta_2) & \text{Theorem 6.1} \\
&= \cos\theta_1\cos\theta_2 + \sin\theta_1\sin(-\theta_2) & \text{Theorem 6.2} \\
&= \cos\theta_1\cos\theta_2 + \sin\theta_1[-\sin\theta_2] & \text{Theorem 6.5} \\
&= \cos\theta_1\cos\theta_2 - \sin\theta_1\sin\theta_2 &
\end{aligned}
$$

THEOREM 6.7. For all real numbers θ_1 and θ_2,

$$\sin(\theta_1 + \theta_2) = \sin\theta_1\cos\theta_2 + \cos\theta_1\sin\theta_2$$

PROOF

$$\sin(\theta_1 + \theta_2) = \cos\left(\frac{\pi}{2} - [\theta_1 + \theta_2]\right) \qquad\qquad \text{Theorem 6.3}$$

$$= \cos \left(\left[\frac{\pi}{2} - \theta_1 \right] - \theta_2 \right) \qquad \text{Why?}$$

$$= \cos \left(\frac{\pi}{2} - \theta_1 \right) \cos \theta_2 + \sin \left(\frac{\pi}{2} - \theta_1 \right) \sin \theta_2 \qquad \text{Theorem 6.1}$$

$$= \sin \theta_1 \cos \theta_2 + \sin \left(\frac{\pi}{2} - \theta_1 \right) \sin \theta_2 \qquad \text{Theorem 6.3}$$

$$= \sin \theta_1 \cos \theta_2 + \cos \theta_1 \sin \theta_2 \qquad \text{Theorem 6.4.}$$

THEOREM 6.8. For all real numbers, θ_1 and θ_2,

$$\sin (\theta_1 - \theta_2) = \sin \theta_1 \cos \theta_2 - \cos \theta_1 \sin \theta_2$$

PROOF

$$\begin{aligned}
\sin (\theta_1 - \theta_2) &= \sin (\theta_1 + [-\theta]) \\
&= \sin \theta_1 \cos (-\theta_2) + \cos \theta_1 \sin (-\theta_2) \qquad \text{Theorem 6.7} \\
&= \sin \theta_1 \cos \theta_2 + \cos \theta_1 \sin (-\theta_2) \qquad \text{Theorem 6.2} \\
&= \sin \theta_1 \cos \theta_2 + \cos \theta_1[-\sin \theta_2] \qquad \text{Theorem 6.5} \\
&= \sin \theta_1 \cos \theta_2 - \cos \theta_1 \sin \theta_2
\end{aligned}$$

At this point we list the theorems we have proved. These equations are called **trigonometric formulas.**

1. $\cos^2 \theta + \sin^2 \theta = 1.$
2. $\cos (\theta_1 + \theta_2) = \cos \theta_1 \cos \theta_2 - \sin \theta_1 \sin \theta_2.$
3. $\cos (\theta_1 - \theta_2) = \cos \theta_1 \cos \theta_2 + \sin \theta_1 \sin \theta_2.$
4. $\sin (\theta_1 + \theta_2) = \sin \theta_1 \cos \theta_2 + \cos \theta_1 \sin \theta_2.$
5. $\sin (\theta_1 - \theta_2) = \sin \theta_1 \cos \theta_2 - \cos \theta_1 \sin \theta_2.$
6. $\cos (-\theta) = \cos \theta.$
7. $\sin (-\theta) = -\sin \theta.$
8. $\cos \left(\frac{\pi}{2} - \theta \right) = \sin \theta.$
9. $\sin \left(\frac{\pi}{2} - \theta \right) = \cos \theta.$

Formulas 2, 3, 4, and 5 are called the **sum and difference formulas** for the cosine and sine functions. Formulas 8 and 9 are called **reduction formulas.**

EXAMPLE 1. Given $\cos \theta = \frac{2}{3}$ and $0 < \theta < \frac{\pi}{2}$, find $\sin \theta$.

SOLUTION. Since $0 < \theta < \frac{\pi}{2}$, we are in the first quadrant, so that

$$\sin \theta = \sqrt{1 - \cos^2 \theta}$$

$$= \sqrt{1 - (\tfrac{2}{3})^2}$$
$$= \sqrt{1 - \tfrac{4}{9}}$$
$$= \sqrt{\tfrac{5}{9}}$$
$$= \frac{\sqrt{5}}{3}$$

EXAMPLE 2. Given $\sin \theta = \dfrac{1}{3}$ and $\dfrac{\pi}{2} < \theta < \dfrac{3\pi}{2}$, find $\cos \theta$.

SOLUTION. Since $\dfrac{\pi}{2} < \theta < \dfrac{3\pi}{2}$, $\cos \theta$ is negative, hence

$$\cos \theta = -\sqrt{1 - \sin^2 \theta}$$
$$= -\sqrt{1 - (\tfrac{1}{3})^2}$$
$$= -\sqrt{1 - \tfrac{1}{9}}$$
$$= -\sqrt{\tfrac{8}{9}}$$
$$= -\frac{2\sqrt{2}}{3}$$

EXAMPLE 3. Find $\cos\left(\dfrac{5\pi}{12}\right)$.

SOLUTION. Since

$$\frac{5\pi}{12} = \frac{3\pi}{12} + \frac{2\pi}{12} = \frac{\pi}{4} + \frac{\pi}{6}$$

Theorem 6.6 gives us

$$\cos\left(\frac{5\pi}{12}\right) = \cos\left(\frac{\pi}{4} + \frac{\pi}{6}\right)$$
$$= \cos\frac{\pi}{4}\cos\frac{\pi}{6} - \sin\frac{\pi}{4}\sin\frac{\pi}{6}$$
$$= \frac{\sqrt{2}}{2} \cdot \frac{\sqrt{3}}{2} - \frac{\sqrt{2}}{2} \cdot \frac{1}{2}$$
$$= \frac{\sqrt{6} - \sqrt{2}}{4}$$

EXAMPLE 4. Find $\sin\left(-\dfrac{5\pi}{6}\right)$.

SOLUTION. By Theorem 6.5

$$\sin\left(-\frac{5\pi}{6}\right) = -\sin\frac{5\pi}{6} = -\frac{1}{2}$$

EXERCISES 6.4

Find the value of the following (Exercises 1–20).

1. $\sin \dfrac{7\pi}{12}$; $\left(\dfrac{7\pi}{12} = \dfrac{\pi}{4} + \dfrac{\pi}{3}\right)$

11. $\sin \dfrac{13\pi}{12}$

2. $\cos \dfrac{7\pi}{12}$

12. $\cos \dfrac{5\pi}{12}$

3. $\sin \dfrac{11\pi}{12}$ $\left(\dfrac{11\pi}{12} = \dfrac{\pi}{4} + \dfrac{2\pi}{3}\right)$

13. $\sin \left(-\dfrac{3\pi}{4}\right)$

4. $\cos \dfrac{11\pi}{12}$

14. $\cos \left(-\dfrac{5\pi}{6}\right)$

5. $\sin \dfrac{5\pi}{12}$

15. $\sin \left(-\dfrac{7\pi}{6}\right)$

6. $\sin \dfrac{\pi}{12}$

16. $\sin \left(-\dfrac{7\pi}{4}\right)$

7. $\cos \dfrac{\pi}{12}$

17. $\cos \left(-\pi\right)$

8. $\cos \dfrac{7\pi}{2}$

18. $\cos \left(-\dfrac{5\pi}{12}\right)$

9. $\cos \left(-\dfrac{\pi}{12}\right)$

19. $\sin \left(-\dfrac{11\pi}{12}\right)$

10. $\sin \left(-\dfrac{\pi}{12}\right)$

20. $\cos \left(-\dfrac{\pi}{12}\right)$

21. If $\cos \theta = \dfrac{5}{13}$ and $\dfrac{3\pi}{2} < \theta < 2\pi$, find $\sin \theta$.

22. If $\cos \theta = -\dfrac{12}{13}$ and $\dfrac{\pi}{2} < \theta < \pi$, find $\sin \theta$.

23. If $\sin \theta = -\dfrac{3}{5}$ and $\dfrac{3\pi}{2} < \theta < 2\pi$, find $\cos \theta$.

24. If $\sin \theta = -\dfrac{4}{5}$ and $\pi < \theta < \dfrac{3\pi}{2}$, find $\cos \theta$.

25. If $\cos \theta = -\dfrac{4}{5}$ and $\dfrac{\pi}{2} < \theta < \pi$, find $\sin \theta$.

26. If $\cos \theta = \dfrac{3}{5}$ and $-\dfrac{\pi}{2} < \theta < 0$, find $\sin \theta$.

27. If $\sin \theta = \dfrac{12}{13}$ and $-\dfrac{3\pi}{2} < \theta < -\pi$, find $\cos \theta$.

28. If $\cos \theta = \dfrac{5}{13}$ and $4\pi < \theta < 5\pi$, find $\sin \theta$.

29. If $\sin \theta = -\dfrac{3}{5}$ and $-\pi < \theta < -\dfrac{\pi}{2}$, find $\cos \theta$.

30. If $\cos \theta = \dfrac{4}{5}$ and $\dfrac{3\pi}{2} < \theta < 2\pi$, find $\sin \theta$.

6.7 ◆ More Trigonometric Formulas

The following formulas are true for all real numbers, θ:

1. $\sin^2 \theta + \cos^2 \theta = 1$
2. $\cos (\theta_1 + \theta_2) = \cos \theta_1 \cos \theta_2 - \sin \theta_1 \sin \theta_2$
3. $\cos (\theta_1 - \theta_2) = \cos \theta_1 \cos \theta_2 + \sin \theta_1 \sin \theta_2$
4. $\sin (\theta_1 + \theta_2) = \sin \theta_1 \cos \theta_2 + \cos \theta_1 \sin \theta_2$
5. $\sin (\theta_1 - \theta_2) = \sin \theta_1 \cos \theta_2 - \cos \theta_1 \sin \theta_2$
6. $\cos (-\theta) = \cos \theta$
7. $\sin (-\theta) = -\sin \theta$
8. $\cos \left(\dfrac{\pi}{2} - \theta\right) = \sin \theta$
9. $\sin \left(\dfrac{\pi}{2} - \theta\right) = \cos \theta$
10. $\cos (\pi - \theta) = -\cos \theta$
11. $\sin (\pi - \theta) = \sin \theta$
12. $\cos \left(\dfrac{3\pi}{2} - \theta\right) = -\sin \theta$
13. $\sin \left(\dfrac{3\pi}{2} - \theta\right) = -\cos \theta$
14. $\cos \left(\dfrac{3\pi}{2} + \theta\right) = \sin \theta$
15. $\sin \left(\dfrac{3\pi}{2} + \theta\right) = -\cos \theta$
16. $\cos (2\pi - \theta) = \cos \theta$
17. $\sin (2\pi - \theta) = -\sin \theta$
18. $\cos 2\theta = \cos^2 \theta - \sin^2 \theta$
19. $\cos 2\theta = 1 - 2 \sin^2 \theta$
20. $\cos 2\theta = 2 \cos^2 \theta - 1$
21. $\sin 2\theta = 2 \sin \theta \cos \theta$
22. $\cos \dfrac{\theta}{2} = \pm \sqrt{\dfrac{1 + \cos\theta}{2}}$
23. $\sin \dfrac{\theta}{2} = \pm \sqrt{\dfrac{1 - \cos \theta}{2}}$

We proved some of these formulas in Section 6.6. We shall accept the others without proof. We now use the formulas given above to solve some problems.

EXAMPLE 1. Find the value of $\sin\left(-\dfrac{\pi}{12}\right)$.

SOLUTION. Since $-\dfrac{\pi}{12} = \dfrac{1}{2}\left(-\dfrac{\pi}{6}\right)$, we have

$$\sin\left(-\frac{\pi}{12}\right) = \sin\frac{1}{2}\left(-\frac{\pi}{6}\right)$$

$$= -\sqrt{\frac{1 - \cos\left(-\frac{\pi}{6}\right)}{2}} \qquad \text{Formula 23; } -\frac{\pi}{6} \text{ is in Q-IV}$$

$$= -\sqrt{\frac{1 - \cos\frac{\pi}{6}}{2}} \qquad \text{Formula 6}$$

$$= -\sqrt{\frac{1 - \frac{\sqrt{3}}{2}}{2}} \qquad \cos\frac{\pi}{6} = \frac{\sqrt{3}}{2}$$

$$= -\frac{\sqrt{2 - \sqrt{3}}}{2}$$

EXAMPLE 2. Simplify $1 - \sin^2 8\theta$ by reducing it to a single term.

SOLUTION. Since $\cos^2\theta = 1 - \sin^2\theta$ from the basic identity, (Formula 1) we have

$$1 - \sin^2 8\theta = \cos^2 8\theta$$

EXAMPLE 3. Find the value of $\cos\dfrac{7\pi}{4}$.

SOLUTION. Since $\dfrac{7\pi}{4} = \dfrac{3\pi}{2} + \dfrac{\pi}{4}$, using Formula 14, we have

$$\cos\frac{7\pi}{4} = \cos\left(\frac{3\pi}{2} + \frac{\pi}{4}\right)$$

$$= \sin\frac{\pi}{4}$$

$$= \frac{\sqrt{2}}{2}$$

EXAMPLE 4. Find the value of $\sin \dfrac{5\pi}{6}$.

SOLUTION. Since $\dfrac{5\pi}{6} = \pi - \dfrac{\pi}{6}$, using Formula 11, we have

$$\sin \frac{5\pi}{6} = \sin \left(\pi - \frac{\pi}{6} \right)$$

$$= \sin \frac{\pi}{6}$$

$$= \frac{1}{2}$$

EXAMPLE 5. Simplify $1 - 2 \sin^2 \dfrac{3\theta}{7}$ by reducing it to a single term.

SOLUTION. Using Formula 19, we have

$$1 - 2 \sin^2 \frac{3\theta}{7} = \cos 2 \left(\frac{3\theta}{7} \right)$$

$$= \cos \frac{6\theta}{7}$$

EXERCISES 6.5

Find the value of each of the following (Exercises 1–6).

1. $\sin \dfrac{\pi}{8} \left(\dfrac{\pi}{8} = \dfrac{1}{2} \cdot \dfrac{\pi}{4} \right)$

2. $\cos \dfrac{\pi}{8}$

3. $\sin \dfrac{\pi}{12}$

4. $\cos \dfrac{\pi}{12}$

5. $\cos \left(-\dfrac{\pi}{8} \right)$

6. $\sin \left(-\dfrac{\pi}{8} \right)$

Simplify by reducing to a single term (Exercises 7–15).

7. $2 \sin \dfrac{\pi}{6} \cos \dfrac{\pi}{6}$

8. $\cos^2 \dfrac{\pi}{4} - \sin^2 \dfrac{\pi}{4}$

9. $\sin 3\theta \cos 2\theta + \cos 3\theta \sin 2\theta$

10. $\cos \dfrac{\theta}{5} \cos \dfrac{\theta}{4} + \sin \dfrac{\theta}{5} \sin \dfrac{\theta}{4}$

11. $2 \sin \dfrac{\pi}{4} \cos \dfrac{\pi}{4}$

12. $\cos^2 3\theta - \sin^2 3\theta$

13. $1 - 2 \sin^2 \dfrac{3\pi}{7}$

14. $-\sqrt{\dfrac{1 + \cos \dfrac{\pi}{4}}{2}}$

15. $2 \sin \dfrac{4\pi}{9} \cos \dfrac{4\pi}{9}$

Given $\cos \theta = a$ and $\sin \theta = $ b. Find each of the following in terms of a and b (Exercises 16-23).

16. $\cos (\pi - \theta)$ 20. $\cos \left(\dfrac{3\pi}{2} + \theta \right)$

17. $\cos 2\theta$ 21. $\cos \left(\dfrac{3\pi}{2} - \theta \right)$

18. $\sin (\pi - \theta)$ 22. $\sin (2\pi - \theta)$

19. $\sin \left(\dfrac{3\pi}{2} + \theta \right)$ 23. $\sin 2\theta$

24. Prove: For all real numbers θ, $\cos (\pi - \theta) = -\cos \theta$.
25. Prove: For all real numbers θ, $\sin (\pi - \theta) = \sin \theta$.
26. Prove: For all real numbers θ, $\sin (2\pi - \theta) = -\sin \theta$.
27. Prove: For all real numbers θ, $\sin 2\theta = 2 \sin \theta \cos \theta$.
28. Prove: For all real numbers θ, $\cos 2\theta = \cos^2 \theta - \sin^2 \theta$.
29. Prove: For all real numbers θ, $\cos 2\theta = 1 - 2 \sin^2 \theta$.
30. Prove: For all real numbers θ, $\cos 2\theta = 2 \cos^2 \theta - 1$.

6.8 ◆ Other Trigonometric Functions

We now introduce four new trigonometric functions that are constructed using the sine and cosine functions. For any value of θ other than one that calls for division by zero, we have the following.

1. The **tangent** of θ, denoted by tan θ, is given by

$$\tan \theta = \frac{\sin \theta}{\cos \theta} = \frac{y}{x}$$

where (x, y) are the coordinates of the terminal point of the path (A, θ).

2. The **cotangent** of θ, denoted by cot θ, is given by

$$\cot \theta = \frac{\cos \theta}{\sin \theta} = \frac{x}{y}$$

where (x, y) are the coordinates of the terminal point of the path (A, θ).

3. The **secant** of θ, denoted by sec θ, is given by

$$\sec \theta = \frac{1}{\cos \theta} = \frac{1}{x}$$

where (x, y) are the coordinates of the terminal point of the path (A, θ).

4. The **cosecant** θ, denoted by csc θ, is given by

$$\csc \theta = \frac{1}{\sin \theta} = \frac{1}{y}$$

where (x, y) are the coordinates of the terminal point of the path (A, θ).
For formal definitions we have:

DEFINITION 6.5. The **tangent** function is the set of all ordered pairs $(\theta, \frac{y}{x})$
where $x = \cos \theta$ and $y = \sin \theta$, $x \neq 0$.

Then

$$\text{tangent function} = \left\{ \left(\theta, \frac{y}{x}\right) \middle| x = \cos \theta \text{ and } y = \sin \theta, x \neq 0 \right\}$$

Since $\tan \theta = \frac{\sin \theta}{\cos \theta}$, the domain of the tangent function must exclude

those values of θ for which $\cos \theta = 0$. These are numbers of the form $\frac{\pi}{2} \pm n\pi$,

where $n = 0, 1, 2, 3, \ldots$. Thus the domain of the tangent function is

$$\left\{ \theta \mid \theta \neq \frac{\pi}{2} \pm n\pi, n = 0, 1, 2, 3, \ldots \right\}$$

The range of the tangent function is the set of all real numbers.

Since the sine and cosine functions have period 2π, the tangent function must also have period 2π. However

$$\begin{aligned}
\tan (\theta + \pi) &= \frac{\sin (\theta + \pi)}{\cos (\theta + \pi)}, \\
&= \frac{\sin \theta \cos \pi + \cos \theta \sin \pi}{\cos \theta \cos \pi - \sin \theta \sin \pi} \\
&= \frac{(\sin \theta)(-1) + (\cos \theta)(0)}{(\cos \theta)(-1) + (\sin \theta)(0)} \\
&= \frac{\sin \theta}{\cos \theta} \\
&= \tan \theta
\end{aligned}$$

Thus the tangent function also has period π. We show that π is, in fact, the fundamental period of the tangent function by supposing that there is a number, p, with $0 < p < \pi$, such that for all θ, $\tan (\theta + p) = \tan \theta$. Then, in particular, for $\theta = 0$, we have

$$\begin{aligned}
\tan p &= \tan (0 + p) \\
&= \tan 0 \\
&= \tfrac{0}{1} \\
&= 0
\end{aligned}$$

Since $\tan \theta = \dfrac{\sin \theta}{\cos \theta}$, we must also have $\sin p = 0$. But there is no value of p, with $0 < p < \pi$, for which $\sin p = 0$. Thus, π is the fundamental period of the tangent function.

DEFINITION 6.6. The **cotangent** function is the set of all ordered pairs $\left(\theta, \dfrac{x}{y} \right)$ where $x = \cos \theta$ and $y = \sin \theta$, $y \neq 0$.

Then

$$\text{cotangent function} = \left\{ \left(\theta, \frac{x}{y} \right) \middle| x = \cos \theta \text{ and } y = \sin \theta, y \neq 0 \right\}$$

Since $\cot \theta = \dfrac{\cos \theta}{\sin \theta}$, the domain of the cotangent function must exclude those values of θ for which $\sin \theta = 0$, that is, numbers of the form $\pm n\pi$, where $n = 0, 1, \ldots$. Thus the domain of the cotangent function is

$$\{ \theta \mid \theta \neq \pm n\pi, n = 0, 1, 2, \ldots \}$$

The range of the cotangent function is the set of all real numbers. Since a value of the cotangent function is the reciprocal of a value of the tangent function (except when either is 0), the cotangent function also has fundamental period π.

DEFINITION 6.7. For $x \neq 0$ and $\theta \neq \frac{\pi}{2} \pm n\pi$ where n is a nonnegative integer, the **secant function** is

$$\text{secant} = \left\{ \left(\theta, \frac{1}{x} \right) \middle| x = \cos \theta \right\}$$

By reasoning as above, the domain of the secant function must exclude all values of θ for which $\cos \theta = 0$, that is numbers of the form $\frac{\pi}{2} \pm n\pi$, where $n = 0, 1, 2, \ldots$. Since $-1 \leq \cos \theta \leq 1$, the range of the secant function is the set of numbers which are reciprocals of numbers of this form, that is,

$$\{z \mid |z| \geq 1\}$$

Since the cosine function has fundamental period 2π, the secant function also has fundamental period 2π.

DEFINITION 6.8. For $y \neq 0$ and $\theta \neq \pm n\pi$, where n is a nonnegative integer, the **cosecant function** is

$$\text{cosecant} = \left\{ \left(\theta, \frac{1}{y} \right) \middle| y = \sin \theta \right\}$$

The domain of the cosecant function must exclude all values of θ for which $\sin \theta = 0$, that is, numbers of the form $\pm n\pi$ where $n = 0, 1, 2, 3, \ldots$. The range of the cosecant function is also $\{z \mid |z| \geq 1\}$.

We can now prove some theorems about the four trigonometric functions that we defined above.

THEOREM 6.9. For all real numbers θ_1 and θ_2, where none of θ_1, θ_2, and $\theta_1 + \theta_2$ has the form $\frac{\pi}{2} \pm n\pi$, n a nonnegative integer,

$$\tan (\theta_1 + \theta_2) = \frac{\tan \theta_1 + \tan \theta_2}{1 - \tan \theta_1 \tan \theta_2}$$

PROOF

$$\tan (\theta_1 + \theta_2) = \frac{\sin (\theta_1 + \theta_2)}{\cos (\theta_1 + \theta_2)} \qquad \text{definition 6.5}$$

$$= \frac{\sin \theta_1 \cos \theta_2 + \cos \theta_1 \sin \theta_2}{\cos \theta_1 \cos \theta_2 - \sin \theta_1 \sin \theta_2} \qquad \text{Theorems 6.6 and 6.7}$$

Dividing both numerator and denominator of the right member of the equation above by $\cos \theta_1 \cos \theta_2$ gives

$$\tan (\theta_1 + \theta_2) = \frac{\dfrac{\sin \theta_1}{\cos \theta_1} + \dfrac{\sin \theta_2}{\cos \theta_2}}{1 - \dfrac{\sin \theta_1}{\cos \theta_2} \dfrac{\sin \theta_2}{\cos \theta_2}}$$

$$= \frac{\tan \theta_1 + \tan \theta_2}{1 - \tan \theta_1 \tan \theta_2}$$

THEOREM 6.10. **For all real numbers, θ_1 and θ_2, where none of θ_1, θ_2, and $\theta_1 - \theta_2$ has the form $\dfrac{\pi}{2} \pm n\pi$, n a nonnegative integer,**

$$\tan (\theta_1 - \theta_2) = \frac{\tan \theta_1 - \tan \theta_2}{1 + \tan \theta_1 \tan \theta_2}$$

The proof of this theorem is left to the reader.

THEOREM 6.11. **For all real numbers θ, where neither θ nor 2θ has the form $\dfrac{\pi}{2} \pm n\pi$, n a nonnegative integer,**

$$\tan 2\theta = \frac{2 \tan \theta}{1 - \tan^2 \theta}$$

PROOF. In Theorem 6.9 we let $\theta_1 = \theta_2 = \theta$. Thus

$$\tan 2\theta = \frac{\tan \theta + \tan \theta}{1 - \tan \theta \tan \theta}$$

$$= \frac{2 \tan \theta}{1 - \tan^2 \theta}$$

THEOREM 6.12. **If θ is a real number, not of the form $n\pi$ (n an integer), then**

$$\tan \frac{\theta}{2} = \frac{1 - \cos \theta}{\sin \theta}$$

PROOF. From the definition of the tangent function, (definition 6.5)

$$\tan \frac{\theta}{2} = \frac{\sin \dfrac{\theta}{2}}{\cos \dfrac{\theta}{2}}$$

We multiply numerator and denominator of the right member by $2 \sin \dfrac{\theta}{2}$ to get

$$\tan \frac{\theta}{2} = \frac{2 \sin^2 \dfrac{\theta}{2}}{2 \sin \dfrac{\theta}{2} \cos \dfrac{\theta}{2}}$$

$$= \frac{1 - \cos \theta}{2 \sin \dfrac{\theta}{2} \cos \dfrac{\theta}{2}} \qquad \sin^2 \frac{\theta}{2} = \frac{1 - \cos \theta}{2} \qquad \text{(formula 23)}$$

$$= \frac{1 - \cos \theta}{\sin \theta} \qquad 2 \sin \frac{\theta}{2} \cos \frac{\theta}{2} = \sin 2 \left(\frac{\theta}{2}\right) \qquad \text{(formula 21)}$$

THEOREM 6.12a. If θ is a real number, not of the form $n\pi$ (n an integer), then

$$\tan \frac{\theta}{2} = \frac{\sin \theta}{1 + \cos \theta}$$

The proof of theorem is left to the reader.

EXAMPLE 1. Given that $0 < \theta < \dfrac{\pi}{2}$ and $\tan \theta = \dfrac{5}{12}$, find the values of $\sin \theta$, $\cos \theta$, $\cot \theta$, $\sec \theta$, and $\csc \theta$.

SOLUTION. From the definition of the tangent function,

$$\frac{\sin \theta}{\cos \theta} = \frac{5}{12}$$

Since θ is in Quadrant I,

$$\cos \theta = \sqrt{1 - \sin^2 \theta}$$

Thus

$$\frac{\sin \theta}{\sqrt{1 - \sin^2 \theta}} = \frac{5}{12}$$
$$12 \sin \theta = 5\sqrt{1 - \sin^2 \theta}$$

Squaring both members produces

$$144 \sin^2 \theta = 25[1 - \sin^2 \theta]$$
$$144 \sin^2 \theta = 25 - 25 \sin^2 \theta$$
$$169 \sin^2 \theta = 25$$
$$\sin^2 \theta = \tfrac{25}{169}$$

Since θ is in Quadrant I, we know that $\sin \theta > 0$, so that

$$\sin \theta = \tfrac{5}{13}$$

Since

$$\cos \theta = \sqrt{1 - \sin^2 \theta}$$

for θ in Quadrant I, we also have

$$\begin{aligned}
\cos \theta &= \sqrt{1 - (\tfrac{5}{13})^2} \\
&= \sqrt{1 - \tfrac{25}{169}} \\
&= \sqrt{\tfrac{144}{169}} \\
&= \tfrac{12}{13}
\end{aligned}$$

Thus

$$\cot \theta = \frac{\cos \theta}{\sin \theta} = \frac{12}{5}$$

$$\sec \theta = \frac{1}{\cos \theta} = \frac{13}{12}$$

and

$$\csc \theta = \frac{1}{\sin \theta} = \frac{13}{5}$$

EXAMPLE 2. Given $\tan \theta = -\dfrac{3}{4}$ and $\dfrac{3\pi}{2} < \theta < 2\pi$, find the values of the five other trigonometric functions.

SOLUTION. From the definition of tangent function,

$$\frac{\sin \theta}{\cos \theta} = -\frac{3}{4}$$

Since θ is in Quadrant IV, $\cos \theta > 0$, so that

$$\cos \theta = \sqrt{1 - \sin^2 \theta}$$

Thus

$$\frac{\sin \theta}{\sqrt{1 - \sin^2 \theta}} = -\frac{3}{4}$$

$$4 \sin \theta = -3\sqrt{1 - \sin^2 \theta}$$

$$16 \sin^2 \theta = 9[1 - \sin^2 \theta]$$

$$16 \sin^2 \theta = 9 - 9 \sin^2 \theta$$

$$25 \sin^2 \theta = 9$$

$$\sin^2 \theta = \tfrac{9}{25}$$

Since θ is in Quadrant IV, $\sin \theta < 0$, so that

$$\sin \theta = -\tfrac{3}{5}$$

Thus

$$\cos \theta = \sqrt{1 - \sin^2 \theta}$$
$$= \sqrt{1 - (-\tfrac{3}{5})^2}$$
$$= \sqrt{1 - \tfrac{9}{25}}$$
$$= \sqrt{\tfrac{16}{25}}$$
$$= \tfrac{4}{5}$$

Finally,

$$\cot \theta = \frac{\cos \theta}{\sin \theta} = -\frac{4}{3}$$

$$\sec \theta = \frac{1}{\cos \theta} = \frac{5}{4}$$

$$\csc \theta = \frac{1}{\sin \theta} = -\frac{5}{3}$$

EXERCISES 6.6

1. Complete the following table:

θ	0	$\dfrac{\pi}{6}$	$\dfrac{\pi}{4}$	$\dfrac{\pi}{3}$	$\dfrac{\pi}{2}$	$\dfrac{2\pi}{3}$	$\dfrac{3\pi}{4}$	$\dfrac{5\pi}{6}$	π
$\tan \theta$	0				not defined				

θ	$\dfrac{7\pi}{6}$	$\dfrac{5\pi}{4}$	$\dfrac{4\pi}{3}$	$\dfrac{3\pi}{2}$	$\dfrac{5\pi}{3}$	$\dfrac{7\pi}{4}$	$\dfrac{11\pi}{6}$	2π
$\tan \theta$				not defined				

Construct a table similar to the table in Exercise 1 for each of the following.

2. $\cot \theta, 0 \leq \theta \leq 2\pi$.
3. $\sec \theta, 0 \leq \theta \leq 2\pi$.
4. $\csc \theta, 0 \leq \theta \leq 2\pi$.

In Exercises 5–20 the value of one trigonometric function is given. Find the values of the other five.

5. $\tan \theta = \dfrac{5}{12}, 0 \leq \theta < \dfrac{\pi}{2}$.

6. $\csc \theta = -\dfrac{6}{5}, \dfrac{\pi}{2} < \theta < \dfrac{3\pi}{2}$.

7. $\sec \theta = -\dfrac{5}{3}, \pi < \theta < \dfrac{3\pi}{2}.$

8. $\tan \theta = -\dfrac{12}{5}, \dfrac{3\pi}{2} < \theta < 2\pi.$

9. $\cot \theta = \dfrac{12}{5}, \pi < \theta < \dfrac{3\pi}{2}.$

10. $\sin \theta = -\dfrac{4}{5}, \pi < \theta < \dfrac{3\pi}{2}.$

11. $\cos \theta = \dfrac{5}{13}, -\dfrac{\pi}{2} < \theta < 0.$

12. $\tan \theta = \dfrac{3}{4}, 2\pi < \theta < 3\pi.$

13. $\csc \theta = -\dfrac{5}{4}, \dfrac{\pi}{2} < \theta < \dfrac{3\pi}{2}.$

14. $\sec \theta = \dfrac{13}{12}, -\dfrac{\pi}{2} < \theta < 0.$

15. $\cot \theta = \dfrac{1}{2}, \pi < \theta < 2\pi.$

16. $\tan \theta = -\dfrac{5}{6}, \dfrac{3\pi}{2} < \theta < 2\pi.$

17. $\cot \theta = \dfrac{4}{3}, \pi < \theta < \dfrac{3\pi}{2}.$

18. $\csc \theta = -5, \pi < \theta < \dfrac{3\pi}{2}.$

19. $\sec \theta = -6, \dfrac{\pi}{2} < \theta < \pi.$

20. $\cot \theta = \dfrac{\sqrt{5}}{3}, \pi < \theta < \dfrac{3\pi}{2}.$

21. Prove Theorem 6.10.

22. Prove Theorem 6.12a.

6.9 ✦ Trigonometric Identities

We have established some basic relations between the six trigonometric functions. Some of these were given in the definitions of the functions and others have been stated as theorems. We call such relations **trigonometric identities.** These equations are true for all values of θ in the domains of the functions involved. For convenience, we restate these identities.

1. $\sin^2 \theta + \cos^2 \theta = 1.$
2. $\sin (-\theta) = -\sin \theta.$
3. $\cos (-\theta) = \cos \theta.$
4. $\tan (-\theta) = -\tan \theta.$
5. $\sin 2\theta = 2 \sin \theta \cos \theta.$
6. $\cos 2\theta = \cos^2 \theta - \sin^2 \theta = 1 - 2 \sin^2 \theta = 2 \cos^2 \theta - 1.$
7. $\tan 2\theta = \dfrac{2 \tan \theta}{1 - \tan^2 \theta}.$
8. $\sin \dfrac{\theta}{2} = \pm \sqrt{\dfrac{1 - \cos \theta}{2}}.$
9. $\cos \dfrac{\theta}{2} = \pm \sqrt{\dfrac{1 + \cos \theta}{2}}.$
10. $\tan \dfrac{\theta}{2} = \dfrac{1 - \cos \theta}{\sin \theta} = \dfrac{\sin \theta}{1 + \cos \theta}.$

11. $\sin(\theta_1 + \theta_2) = \sin\theta_1 \cos\theta_2 + \cos\theta_1 \sin\theta_2.$

12. $\sin(\theta_1 - \theta_2) = \sin\theta_1 \cos\theta_2 - \cos\theta_1 \sin\theta_2.$

13. $\cos(\theta_1 + \theta_2) = \cos\theta_1 \cos\theta_2 - \sin\theta_1 \sin\theta_2.$

14. $\cos(\theta_1 - \theta_2) = \cos\theta_1 \cos\theta_2 + \sin\theta_1 \sin\theta_2.$

15. $\tan(\theta_1 + \theta_2) = \dfrac{\tan\theta_1 + \tan\theta_2}{1 - \tan\theta_1 \tan\theta_2}.$

16. $\tan(\theta_1 - \theta_2) = \dfrac{\tan\theta_1 - \tan\theta_2}{1 + \tan\theta_1 \tan\theta_2}.$

17. $\tan\theta = \dfrac{\sin\theta}{\cos\theta}.$

18. $\cot\theta = \dfrac{\cos\theta}{\sin\theta}.$

19. $\sec\theta = \dfrac{1}{\cos\theta}.$

20. $\csc\theta = \dfrac{1}{\sin\theta}.$

We use the above identities to prove further identities. For anyone who takes more mathematics courses or who takes courses involving mathematics, it is important to be able to prove identities.

It is most desirable to prove an identity by transforming one member of the equation to the other member which should be left unaltered. The member with which we work is usually the more complicated one. There are no set rules for making these transformations. The following suggestions, however, should prove helpful.

1. If one member involves only one function, express the other member in terms of this function.

2. If one member is factorable, factor it.

3. If one member is a sum and the other is a product, work with the sum.

4. When in doubt, express the more complicated member in terms of sines and cosines and simplify.

5. When working with one member, look at the other member to see which transformations will more easily reduce to the other member, the member on which you are working.

6. Avoid, if possible, introducing radicals.

If an identity is extremely complicated, we sometimes reduce each member of the equation to a common third expression. In this book, we shall always transform one member of the identity to the other member which is left unaltered.

The following examples illustrate some techniques for solving identities.

EXAMPLE 1. Prove that

$$\tan^2 \theta + 1 = \sec^2 \theta$$

SOLUTION. Working with the left member, we have

$$
\begin{aligned}
\tan^2 \theta + 1 &= \frac{\sin^2 \theta}{\cos^2 \theta} + 1 \\
&= \frac{\sin^2 \theta + \cos^2 \theta}{\cos^2 \theta} \\
&= \frac{1}{\cos^2 \theta} \\
&= \sec^2 \theta
\end{aligned}
$$

EXAMPLE 2. Prove that

$$\cot^2 \theta + 1 = \csc^2 \theta$$

SOLUTION. Working with the left member we have

$$
\begin{aligned}
\cot^2 \theta + 1 &= \frac{\cos^2 \theta}{\sin^2 \theta} + 1 \\
&= \frac{\cos^2 \theta + \sin^2 \theta}{\sin^2 \theta} \\
&= \frac{1}{\sin^2 \theta} \\
&= \csc^2 \theta
\end{aligned}
$$

EXAMPLE 3. Prove that

$$3 \sin^2 \theta + 4 \cos^2 \theta = 3 + \cos^2 \theta$$

SOLUTION. Working with the left member, we have

$$
\begin{aligned}
3 \sin^2 \theta + 4 \cos^2 \theta &= 3 \sin^2 \theta + (3 \cos^2 \theta + \cos^2 \theta) \\
&= (3 \sin^2 \theta + 3 \cos^2 \theta) + \cos^2 \theta \\
&= 3(\sin^2 \theta + \cos^2 \theta) + \cos^2 \theta \\
&= 3 \cdot 1 + \cos^2 \theta \\
&= 3 + \cos^2 \theta
\end{aligned}
$$

EXAMPLE 4. Prove that

$$\frac{1 + \cos \theta}{1 - \cos \theta} - \frac{1 - \cos \theta}{1 + \cos \theta} = 4 \cot \theta \csc \theta$$

SOLUTION

$$\frac{1 + \cos \theta}{1 - \cos \theta} - \frac{1 - \cos \theta}{1 + \cos \theta} = \frac{(1 + \cos \theta)(1 + \cos \theta)}{(1 - \cos \theta)(1 + \cos \theta)} - \frac{(1 - \cos \theta)(1 - \cos \theta)}{(1 + \cos \theta)(1 - \cos \theta)}$$

$$= \frac{(1 + \cos \theta)^2 - (1 - \cos \theta)^2}{(1 - \cos \theta)(1 + \cos \theta)}$$

$$= \frac{1 + 2 \cos \theta + \cos^2 \theta - 1 + 2 \cos \theta - \cos^2 \theta}{1 - \cos^2 \theta}$$

$$= \frac{4 \cos \theta}{1 - \cos^2 \theta}$$

$$= \frac{4 \cos \theta}{\sin^2 \theta}$$

$$= 4 \cdot \frac{\cos \theta}{\sin \theta} \cdot \frac{1}{\sin \theta}$$

$$= 4 \cot \theta \csc \theta$$

EXERCISES 6.7

Prove that the following are identities for all allowable values of the variables (Exercises 1–40).

1. $\cot \theta = \dfrac{1}{\tan \theta}.$

2. $\csc^2 \theta(1 - \cos^2 \theta) = 1.$

3. $\csc \theta \tan \theta = \sec \theta.$

4. $\cos^2 \theta(\sec^2 \theta - 1) = \sin^2 \theta.$

5. $2 + \tan^2 \theta + \cot^2 \theta = \sec^2 \theta + \csc^2 \theta.$

6. $\cot^2 \theta - \cos^2 \theta = \cos^2 \theta \cot^2 \theta.$

7. $\dfrac{\cos^2 \theta}{1 - \sin \theta} = 1 + \sin \theta.$

8. $3 \cos^4 \theta + 6 \sin^2 \theta = 3 + 3 \sin^4 \theta.$

9. $\sec^2 \theta \csc^2 \theta = \sec^2 \theta + \csc^2 \theta.$

10. $\sec^2 \theta + \tan^2 \theta = \sec^4 \theta - \tan^4 \theta.$

11. $3 \sin^2 \theta + 4 \cos^2 \theta = 3 + \cos^2 \theta.$

12. $\dfrac{\tan^3 \theta - \cot^3 \theta}{\tan \theta - \cot \theta} = \sec^2 \theta + \cot^2 \theta.$

13. $\dfrac{\tan \theta + 2 \cot \theta}{\tan \theta + 3 \cot \theta} = \dfrac{\tan^2 \theta + 2}{\tan^2 \theta + 3}.$

14. $\sec^2 \theta \csc^2 \theta = 2 + \dfrac{\sin^4 \theta + \cos^4 \theta}{\sin^2 \theta \cos^2 \theta}.$

15. $(1 + \sin \theta)(1 + \csc \theta) = 2 + \sin \theta + \csc \theta.$

16. $(\csc\theta - \cot\theta)(1 + \cos\theta) = \sin\theta.$

17. $\dfrac{\sin^3\theta - \cos^3\theta}{\sin^2\theta - \cos^2\theta} = \dfrac{1 + \sin\theta\cos\theta}{\sin\theta + \cos\theta}.$

18. $\sec 2\theta = \dfrac{1}{2\cos^2\theta - 1}.$

19. $2\csc 2\theta = \sec\theta\csc\theta.$

20. $\cos^2 2\theta - \cos^2\theta = \sin^2\theta - \sin^2 2\theta.$

21. $\sin^4\dfrac{\theta}{2} - \cos^4\dfrac{\theta}{2} = -\cos\theta.$

22. $\dfrac{\sin\theta + \sin 2\theta}{1 + \cos\theta + \cos 2\theta} = \tan\theta.$

23. $\cos 4\theta = 1 - 8\sin^2\theta\cos^2\theta.$

24. $\cos^2\theta_1 - \cos^2\theta_2 = \sin(\theta_2 + \theta_1)\sin(\theta_2 - \theta_1).$

25. $\dfrac{\sin\theta_1\cos\theta_2}{\cos\theta_1\sin\theta_2} = \dfrac{\tan\theta_1}{\tan\theta_2}.$

26. $\sin\theta\sec\theta = \tan\theta.$

27. $\sqrt{\dfrac{1 + \cos\theta}{1 - \cos\theta}} = \csc\theta + \cot\theta.$

28. $\dfrac{\cot\theta + \csc\theta}{\sin\theta + \tan\theta} = \csc\theta\cot\theta.$

29. $\dfrac{\tan\theta_1 + \tan\theta_2}{\cot\theta_1 + \cot\theta_2} + \dfrac{1 - \tan\theta_1\tan\theta_2}{1 - \cot\theta_1\cot\theta_2} = 0.$

30. $(\sin\theta_1\cos\theta_2 + \cos\theta_1\sin\theta_2)^2 + (\cos\theta_1\cos\theta_2 - \sin\theta_1\sin\theta_2)^2 = 1.$

31. $\dfrac{\cos\theta}{1 - \tan\theta} + \dfrac{\sin\theta}{1 - \cot\theta} = \sin\theta + \cos\theta.$

32. $\cos^2\theta\cot\theta - \sin^2\theta\tan\theta = \cot\theta - \tan\theta.$

33. $1 - \dfrac{2\tan^2\theta}{\sec^2\theta - 2} = \dfrac{1}{1 - 2\sin^2\theta}.$

34. $\dfrac{\tan^3\theta - \cot^3\theta}{\tan\theta - \cot\theta} = \tan^2\theta + \cot^2\theta + 1.$

35. $\dfrac{1}{\tan\theta - 2\sin\theta} = \dfrac{1 + \sec\theta}{\sec\theta\tan\theta - 2\sin\theta - \tan\theta}.$

36. $(1 - \cos\theta)(1 + \sec\theta) = \dfrac{\sin^2\theta}{\cos\theta}.$

37. $\dfrac{1 - \sin^2\theta\sec^2\theta}{1 - \cos^2\theta\csc^2\theta} = -\tan^2\theta.$

38. $\dfrac{\sec\theta - 1}{\sec\theta + 1} = \sec^2\dfrac{\theta}{2} - 1.$

39. $\dfrac{\sec\theta\cot\theta - \csc\theta\tan\theta}{\cos\theta - \sin\theta} = \csc\theta\sec\theta.$

40. $\dfrac{\cos\theta\cot\theta - \sin\theta\tan\theta}{\csc\theta - \sec\theta} = 1 + \sin\theta\cos\theta.$

6.10 ◆ Trigonometric Equations

Trigonometric equations are equations in which the various trigonometric functions occur. Some examples are:

$$2\sin\theta + 1 = 0$$
$$\tan^2\theta - 1 = 0$$
$$\cos^2\theta + 2\cos\theta + 1 = 0$$

The solution set of a trigonometric equation is the set of all values of the variables which satisfy the equation. Since trigonometric functions are periodic, the solution set of a trigonometric equation will usually be an infinite set.

The process of solving a trigonometric equation involves algebraic as well as trigonometric methods. There is no general rule for solving trigonometric equations, but the following suggestions should prove helpful.

(1) *If only one function of a single variable is involved, solve for the values of the function. Then determine the solutions.*

EXAMPLE 1. Find the solution set of $2\cos\theta = 1$.

SOLUTION. This equation is equivalent to

$$\cos\theta = \tfrac{1}{2}$$

There are two points on the unit circle with y-coordinate $\tfrac{1}{2}$. Thus the values of θ between 0 and 2π which satisfy this equation are $\dfrac{\pi}{3}$ and $\dfrac{5\pi}{3}$. Since the cosine function has period 2π, we may add any integer multiple of 2π to these solutions to get the solution set

$$\left\{\frac{\pi}{3} \pm 2n\pi \mid n = 0, 1, 2, \ldots\right\} \cup \left\{\frac{5\pi}{3} \pm 2n\pi \mid n = 0, 1, 2, \ldots\right\}$$

EXAMPLE 2. Find the solution set of $4\cos\theta + 1 = 2\cos\theta$.

SOLUTION. This equation is equivalent to

$$2\cos\theta = -1$$
$$\cos\theta = -\tfrac{1}{2}$$

The two values of θ between 0 and 2π, which satisfy this last equation are $\dfrac{2\pi}{3}$ and $\dfrac{4\pi}{3}$. Hence the solution set is

$$\left\{\frac{2\pi}{3} \pm 2\pi n \mid n = 0, 1, 2, 3\right\} \cup \left\{\frac{4\pi}{3} \pm 2\pi n \mid n = 1, 2, 3, \ldots\right\}$$

EXAMPLE 3. Find the solution set of

$$\sin 2\theta = \tfrac{1}{2}, 0 \le \theta < 2\pi$$

SOLUTION. In order for $\sin 2\theta$ to be $\tfrac{1}{2}$, 2θ must be an element of the set

$$\left\{\frac{\pi}{6} \pm 2n\pi \mid n = 0, 1, 2, \ldots\right\} \cup \left\{\frac{5\pi}{6} \pm 2n\pi \mid n = 0, 1, 2, \ldots\right\}$$

Since these are the possible values of 2θ, we must have θ one of the numbers in the set

$$\left\{\frac{\pi}{12} \pm n\pi \mid n = 0, 1, 2, \ldots\right\} \cup \left\{\frac{5\pi}{12} \pm n\pi \mid n = 0, 1, 2, \ldots\right\}$$

Since the solution set can only contain values of θ between 0 and 2π, it must be

$$\left\{\frac{\pi}{12}, \frac{\pi}{12} + \pi, \frac{5\pi}{12}, \frac{5\pi}{12} + \pi\right\} = \left\{\frac{\pi}{12}, \frac{5\pi}{12}, \frac{13\pi}{12}, \frac{17\pi}{12}\right\}$$

EXAMPLE 4. Find the solution set of

$$3 \sin \theta = 5$$

SOLUTION. Since this equation is equivalent to

$$\sin \theta = \tfrac{5}{3}$$

and since $\sin \theta$ is the y-coordinate of some point on the unit circle, we must have $-1 \le \sin \theta \le 1$. Since $\tfrac{5}{3} > 1$, there are no values of θ which satisfy the given equation. Hence the solution set of the given equation is $\varnothing$.

(2) *If one member of the equation is zero, and the other is factorable, set each factor equal to zero and solve the resulting equations.*

EXAMPLE 5. Find the solution set of

$$2 \sin^2 \theta - \sin \theta - 1 = 0; 0 \le \theta < 2\pi$$

SOLUTION. The given equation is equivalent to

$$(2 \sin \theta + 1)(\sin \theta - 1) = 0$$
$$2 \sin \theta + 1 = 0 \quad \text{or} \quad \sin \theta - 1 = 0$$
$$\sin \theta = -\tfrac{1}{2} \quad \text{or} \quad \sin \theta = 1$$

There are two values of θ between 0 and 2π for which $\sin \theta = -\frac{1}{2}$, namely, $\dfrac{7\pi}{6}$ and $\dfrac{11\pi}{6}$. There is only one value of θ between 0 and 2π for which $\sin \theta = 1$, namely, $\dfrac{\pi}{2}$. The solution set is

$$\left\{ \frac{\pi}{2}, \frac{7\pi}{6}, \frac{11\pi}{6} \right\}$$

(3) *If several functions of a single variable are involved, use the fundamental identities to express both members in terms of a single function. Then proceed as in (1) or (2).*

EXAMPLE 6. Find the solution set of

$$\csc^2 \theta - \cot \theta = 1, 0 \leq \theta < 2\pi$$

SOLUTION. The given equation is equivalent to

$$1 + \cot^2 \theta - \cot \theta = 1$$

since $\csc^2 \theta = 1 + \cot^2 \theta$. Now $1 + \cot^2 \theta - \cot \theta = 1$ is equivalent to

$$\cot^2 \theta - \cot \theta = 0$$
$$\cot \theta (\cot \theta - 1) = 0$$

and

$$\cot \theta = 0 \quad \text{or} \quad \cot \theta = 1$$

There are two values of θ between 0 and 2π for which $\cot \theta = 0$, namely, $\dfrac{\pi}{2}$ and $\dfrac{3\pi}{2}$. There are two values of θ between 0 and 2π for which $\cot \theta = 1$, namely, $\dfrac{\pi}{4}$ and $\dfrac{5\pi}{4}$. The solution set is

$$\left\{ \frac{\pi}{4}, \frac{\pi}{2}, \frac{5\pi}{4}, \frac{3\pi}{2} \right\}$$

(4) *If several variables are involved, use the fundamental identities to express both members in terms of a single variable. Then proceed as in (3).*

EXAMPLE 7. Find the solution set of

$$\sin 2\theta = \sqrt{2}\cos\theta, 0 \le \theta < 2\pi$$

SOLUTION. This equation involves two variables, θ and 2θ. We replace $\sin 2\theta$ by $2\sin\theta\cos\theta$ to obtain the equivalent equations,

$$2\sin\theta\cos\theta = \sqrt{2}\cos\theta$$
$$2\sin\theta\cos\theta - \sqrt{2}\cos\theta = 0$$
$$\cos\theta\,(2\sin\theta - \sqrt{2}) = 0$$
$$\cos\theta = 0 \text{ or } \sin\theta = \frac{\sqrt{2}}{2}$$

There are two values of θ between 0 and 2π for which $\cos\theta$ is 0, namely, $\dfrac{\pi}{2}$ and $\dfrac{3\pi}{2}$. There are two values of θ between 0 and 2π for which $\sin\theta = \dfrac{\sqrt{2}}{2}$, namely, $\dfrac{\pi}{4}$ and $\dfrac{3\pi}{4}$. The solution set is

$$\left\{\frac{\pi}{4}, \frac{\pi}{2}, \frac{3\pi}{4}, \frac{3\pi}{2}\right\}$$

EXERCISES 6.8

Find the solution sets of the following equations, $0 \le \theta < 2\pi$.

1. $\cos\theta = \frac{1}{2}\sqrt{3}$
2. $4\sin^2\theta = 1$
3. $2\cos^2\theta = 1$
4. $4\sin^2\theta = 3$
5. $(\cos\theta - 1)(2\sin\theta + 1) = 0$
6. $4\cos^2\theta = 3$
7. $\sec^2\theta = 1$
8. $\cot^2\theta = 1$
9. $(\tan\theta)(2\sin\theta - 1) = 0$
10. $(2\cos\theta - \sqrt{3})(\sqrt{2}\sin\theta - 1) = 0$
11. $\sin\theta(\csc 3\theta - 1) = 0$
12. $(\sec 2\theta + 1)(2\sin 2\theta - 1) = 0$
13. $\tan^2\theta + \tan\theta = 0$
14. $\sin^2\theta - 3\cos^2\theta = 0$
15. $\cos 2\theta = 1$
16. $\cos 2\theta + \sin\theta = 0$
17. $\sin\theta + \sin\dfrac{\theta}{2} = 0$
18. $2\sin^2\theta + 5\sin\theta + 2 = 0$

19. $2\sin^2\theta + 9\sin\theta + 7 = 0$
20. $\cos 2\theta = 3\sin\theta + 2$
21. $\sin^2 2\theta + 3\cos^2\theta = 3$
22. $\sec^2\theta + 3\csc^2\theta = 8$
23. $\sin 4\theta = \sin 2\theta$
24. $\cot^2\theta + \csc\theta = -4$
25. $\csc^2\theta - \cot\theta = 1$
26. $\sin^2\theta + 3\cos\theta = 1$
27. $\tan^2\theta = \sec\theta - 1$
28. $\cos^2\theta + 3\sin\theta - 3 = 0$
29. $\tan^2\theta = 3$
30. $\sec^2\theta - 2\tan\theta = 0$
31. $\dfrac{\cos\theta}{\cos\theta - \sin\theta} + \dfrac{\tan\theta}{\tan\theta - 1} = 0$
32. $3\tan\theta = 2\cos\theta$
33. $4\sin^2\theta - 4\sin\theta + 1 = 0$
34. $\sin^2\theta - \cos^2\theta - \cos\theta = 1$
35. $\dfrac{1}{3 + \cos\theta} = \dfrac{1}{4 - \cos\theta}$

CHAPTER REVIEW

1. Find the terminal points of the following paths [A is $(1, 0)$].

   (a) $\left(A, \dfrac{3\pi}{4}\right)$ (b) $(A, -\pi)$ (c) $\left(A, \dfrac{7\pi}{6}\right)$ (d) $\left(A, \dfrac{35\pi}{4}\right)$

2. Find, $C(\theta)$, for the following values of θ:

   (a) $\dfrac{\pi}{4}$ (b) $\dfrac{5\pi}{6}$ (c) $\dfrac{7\pi}{4}$ (d) -3π

Evaluate (Exercises 3–6).

3. $\sin\left(\dfrac{7\pi}{3}\right)$ 4. $\cos\left(-\dfrac{3\pi}{4}\right)$ 5. $\sin\left(\dfrac{5\pi}{6}\right)$ 6. $\cos 13\pi$

7. Find the periods of the following functions.
   (a) sine function
   (b) tangent function
   (c) secant function

Evaluate (Exercises 8–11).

8. $\cos\left(\dfrac{\pi}{3} - \dfrac{\pi}{4}\right)$ 9. $\sin\left(\dfrac{3\pi}{4} - \dfrac{1\pi}{3}\right)$ 10. $\sin(2x + 2y)$

11. $\cos(2x - 2y)$

Find the values of the remaining trigonometric functions, given the following (Exercises 12–14).

12. $\cot\theta = 1$, $\dfrac{\pi}{2} < \theta < \dfrac{3\pi}{2}$.

13. $\sec\theta = 2$, $-\dfrac{\pi}{2} < \theta < 0$.

14. $\csc\theta = \dfrac{2\sqrt{3}}{3}$, $0 < \theta < \dfrac{\pi}{2}$.

Show that the following equations are identities (Exercises 15–18).

15. $\csc^2\theta - 1 = \cot^2\theta$.
16. $\cos 3\theta = 4\cos^3\theta - 3\cos\theta$.

17. $\dfrac{1}{\tan\theta + \cot\theta} = \sin\theta\cos\theta$.

18. $\sin^4\theta - \cos^4\theta + 2\dfrac{\cot^2\theta}{\csc^2\theta} = 1$.

Find the solution sets ($0 \leq \theta < 2\pi$) (Exercises 19 and 20).

19. $\cot\theta = -1$ 20. $3\sin^2\theta + \sin\theta - 4 = 0$

CHAPTER 7

Trigonometric Ratios

7.1 ◆ Angles

In Chapter 6 we defined each of the six trigonometric functions as a function whose domain is the set R of real numbers (with the exceptions of the cases where division by zero is required). Specifically, with each real number θ, which represented the directed length of an arc on the unit circle, we associated the coordinates (x, y) of the end point P of the arc. We now return to the more traditional viewpoint and consider trigonometric ratios of angles.

In geometry an **angle** is defined to be the union of two rays with a common end point. In trigonometry we generalize this definition by introducing the measure of an angle, which is the amount of rotation required to move from the position of the initial ray of an angle to the terminal ray. The designation of the measure of an angle thus requires an **initial side**, such as $\overrightarrow{OB}$ in Figure 7.1, a **terminal** side, such as $\overrightarrow{OA}$ in the figure, and an amount of rotation. The common point of the two rays forming an angle, the point O in Figure 7.1, is called the **vertex** of that angle. There are two possible directions of rotation that result in moving from the position of ray $\overrightarrow{OB}$ to position of ray $\overrightarrow{OA}$. We say that the measure of an angle is positive if the rotation is counterclockwise and negative if the rotation is clockwise. We shall usually use a curved arrow, as in Figure 7.1, to represent the direction of rotation.

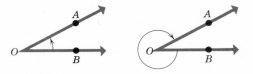

Figure 7.1

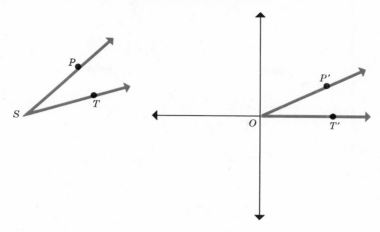

Figure 7.2

When the rays forming an angle are in the Cartesian plane, each angle in the plane is congruent (has the same measure) to an angle with vertex at the origin and initial side along the positive *x*-axis. An angle so located is said to be in **standard position.** Thus, in Figure 7.2, angle *P'OT'* is in standard position and is congruent to angle *PST*. If the terminal side of an angle in standard position is in a given quadrant, the angle is said to be in that quadrant. Thus, in Figure 7.3, angle *AOB* is in Quadrant I, angle *ROT* is in Quadrant II, angle *MON* is in Quadrant III, and angle *KOL* is in Quadrant IV. Note that angle *KOL* has a negative measure, while the other three angles have positive measures.

7.2 ◆ The Measure of an Angle

The most reasonable unit to be used in measuring an angle is the number of complete rotations or portions of complete rotations that are needed to move from the initial side to the terminal side of the angle. If we construct a

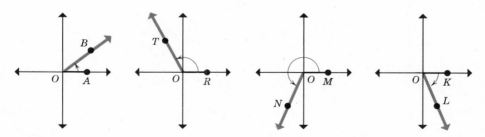

Figure 7.3

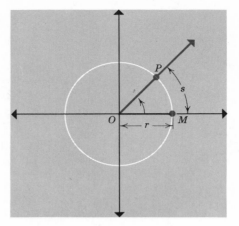

Figure 7.4

circle of radius r and center at the vertex of the angle (Figure 7.4), then the number of complete rotations of an angle is determined by the ratio of the length of the intercepted arc to the length, $2\pi r$, of the circumference of that circle. For angle MOP, in Figure 7.4, with initial side $\overrightarrow{OM}$ and terminal side $\overrightarrow{OP}$, if the arc MP has length s, we can define the measure of angle MOP in rotations as

$$\frac{s}{2\pi r}$$

We now consider two concentric circles with centers at O (Figure 7.5) the inner circle with radius r and the outer circle with radius r'. For A and A', the

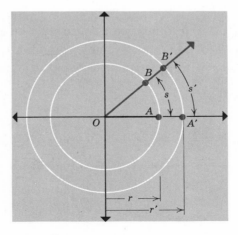

Figure 7.5

intersection points of these circles with the initial side of the angle AOB, and B and B', the intersection points of these circles with the terminal side of the angle, we have an arc AB of length s on the inner circle and an arc $A'B'$ of length s' on the outer circle. We recall from geometry that the lengths of arc cut off by a given central angle on concentric circles are proportional to the circumferences of the circles. Thus

$$\frac{s}{2\pi r} = \frac{s'}{2\pi r'}$$

and we see that the measure in rotations of any angle is independent of the radius of the measuring circle.

One of the most common units used in measuring angles is the degree. We define a **degree** (using the symbol °) as $\frac{1}{360}$ of one complete rotation; a **minute** (using the symbol ′) as $\frac{1}{60}$ of a degree; and a **second** (using the symbol ″) as $\frac{1}{60}$ of a minute. Thus the measure of an angle in degrees is the measure in rotations, multiplied by 360°. For an angle which subtends an arc of length s on a circle of radius r, the degree measure of the angle becomes

$$\pm \frac{s}{2\pi r}(360°)$$

where the measure is positive if the rotation is counterclockwise and negative if the rotation is clockwise.

An angle of one complete rotation in a counterclockwise direction has a degree measure of 360°. An angle of one-half a complete rotation in a counterclockwise direction has a degree measure of 180°. An angle of one-fourth a complete rotation in a counterclockwise direction has a degree measure of 90°.

Similarly, an angle of one complete rotation in a clockwise direction has a degree measure of −360°. An angle of one-half a complete rotation in a clockwise direction has a degree measure of −180°. An angle of one-fourth a complete rotation in a clockwise direction has a degree measure of −90°.

An angle of degree measure of ±90° is called a **right angle**—its two sides are perpendicular to each other. An angle of degree measure of ±180° is called a **straight angle**—its sides are, in fact, parts of one straight line.

We also allow angles to have measures of more than one complete rotation. Thus an angle of two complete rotations in a counterclockwise direction has a degree measure of 720°.

The degree is one of the two common units used in measuring angles. The other unit of angular measure is called the **radian.** To define a radian begin with a circle of radius r and center at the origin (Figure 7.6). We mark on

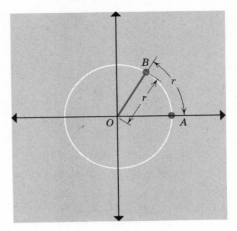

Figure 7.6

the circumference the arc AB, starting at point A and moving in the positive direction toward B, so that the length of this arc is equal to r, the radius of the circle. The angle AOB, with vertex at the center of the circle, initial side $\overrightarrow{OA}$ and terminal side $\overrightarrow{OB}$ will be measured as one **radian** (abbreviated rad). In this way we can measure any angle in radian units.

The precise relation between the degree measure and the radian measure of an angle will now be determined. For a circle of radius r, the length, C, of the circumference is $C = 2\pi r$. Thus an arc of length r needs to be laid off in the positive direction 2π times in order to cover the complete circumference. Hence, one complete rotation requires an angle of radian measure 2π. Since one complete rotation also requires an angle with degree measure $360°$, we have, as the relation between the two measures:

$$(1) \qquad\qquad 2\pi \text{ radians} = 360°$$

From Equation 1 we obtain

$$(2) \qquad\qquad 1 \text{ radian} = \frac{360°}{2\pi} = \frac{180°}{\pi}$$

$$(2a) \qquad\qquad 1° = \frac{2\pi}{360} = \frac{\pi}{180} \text{ radians}$$

Equations 2 and 2a are called **conversion formulas**.

When using radian measure for angles, it is customary to omit the indication "radian" so that an angle of 5 is understood to be an angle of 5 radians.

We see that 1 radian is much larger than $1°$. In fact, 1 radian equals approximately $57° \ 17' \ 44''$.

EXAMPLE 1. Find the degree measure of an angle whose radian measure is

(a) $\dfrac{\pi}{2}$;   (b) $\dfrac{\pi}{3}$;   (c) $-\dfrac{\pi}{4}$

SOLUTION

(a) $\dfrac{\pi}{2} = \left(\dfrac{\pi}{2} \cdot \dfrac{180}{\pi}\right)^{\circ} = 90^{\circ}$

(b) $\dfrac{\pi}{3} = \left(\dfrac{\pi}{3} \cdot \dfrac{180}{\pi}\right)^{\circ} = 60^{\circ}$

(c) $-\dfrac{\pi}{4} = \left(-\dfrac{\pi}{4} \cdot \dfrac{180}{\pi}\right)^{\circ} = -45^{\circ}$

EXAMPLE 2. Find the radian measure of an angle whose degree measure is

(a) 30°;   (b) 180°;   (c) -315°.

SOLUTION

(a) $30^{\circ} = 30(1^{\circ}) = 30\left(\dfrac{\pi}{180}\right) = \dfrac{\pi}{6}$

(b) $180^{\circ} = \left(180 \cdot \dfrac{\pi}{180}\right) = \pi$

(c) $-315^{\circ} = \left(-315 \cdot \dfrac{\pi}{180}\right) = -\dfrac{7\pi}{4}$

EXAMPLE 3. Find the degree measure, to the nearest tenth of a degree, of the angle whose radian measure is 2.3.

SOLUTION

$$2.3 = (2.3)\left(\dfrac{180}{\pi}\right)^{\circ} = \left(\dfrac{414}{\pi}\right)^{\circ} \doteq 131.8^{\circ}$$

(Recall that the symbol $\doteq$ means "is approximately equal to.")

EXAMPLE 4. Find the radian measure, to the nearest hundredth of a radian, of the angle whose degree measure is 50°.

SOLUTION

$$50^{\circ} = (50)\left(\dfrac{\pi}{180}\right) = \dfrac{5\pi}{180} \doteq 0.873.$$

The reader should become familiar with the radian measure of angles of degree measures 30°, 45°, 60°, 90°, and the angles which have degree meas-

TRIGONOMETRIC RATIOS

Table 7.1

Measure in	
Degrees	Radians
0°	0
30°	$\dfrac{\pi}{6}$
45°	$\dfrac{\pi}{4}$
60°	$\dfrac{\pi}{3}$
90°	$\dfrac{\pi}{2}$
120°	$\dfrac{2\pi}{3}$
135°	$\dfrac{3\pi}{4}$
150°	$\dfrac{5\pi}{6}$
180°	π
210°	$\dfrac{7\pi}{6}$
225°	$\dfrac{5\pi}{4}$
240°	$\dfrac{4\pi}{3}$
270°	$\dfrac{3\pi}{2}$
300°	$\dfrac{5\pi}{3}$
315°	$\dfrac{7\pi}{4}$
330°	$\dfrac{11\pi}{6}$
360°	2π

ures which are integral multiples of 30°, 60°, 45°, and 90°. The radian measures of angles with these degree measures are shown in Table 7.1.

EXERCISES 7.1

Find the degree measure of the angle whose radian measure is as given.

1. $\dfrac{\pi}{5}$ 2. $\dfrac{3\pi}{4}$ 3. $\dfrac{5\pi}{3}$

4. $-\dfrac{5\pi}{4}$ 6. $\dfrac{5\pi}{12}$ 8. $\dfrac{3\pi}{2}$

5. 0   7. -3π 9. $-\dfrac{5\pi}{6}$

Find the degree measure, to the nearest tenth of a degree, of the angle whose radian measure is as given.

10. 1.25   12. 1.05   14. 1.45
11. 2.75   13. 9.23   15. -3.01

Find the radian measure, to the nearest hundredth of a radian, of the angle whose degree measure is as given (Exercises 16–24).

16. 750°   19. 48°   22. $-420°$
17. 12°   20. 36°   23. $-130°$
18. 320°   21. 85°   24. $-800°$

7.3 ✦ The Trigonometric Ratios

The coordinate axes divide the C─────────────────────────────────adrants which are numbered ────────────────────── ──── ──── ──────

Those points ─────────────────────────────────ates positive. Those points wl─────────────────────────────────negative. In Quadrant II, the─────────────────────────────────tes are positive. In Quadran────────────────────────────────rdinates are negative. Any po─────────────────────────────────r quadrants

Q-III   Q-IV

Figure 7.7

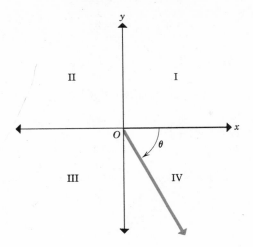

Figure 7.8

or on one of the axes. Recall that the axes are not part of any quadrant—they separate the quadrants.

When an angle is in standard position, its terminal side will either lie in one of the quadrants or on one of the axes. For example, the negative angle shown in Figure 7.8 has its terminal side in Quadrant IV and is said to be in Quadrant IV.

When two angles in standard position have the same terminal side, they are said to be **coterminal.** Thus, angles whose degree measures are 120°, −240°, and 480° are coterminal. Angles which are coterminal differ in degree measure by some integer multiple of 360° and in radian measure by some integer multiple of 2π.

The angles whose degree measures are 120°, −240°, and 480° are coterminal because

$$-240° = 120° - 360°$$
$$480° = 120° + 360°$$

The angles whose radian measures are $\dfrac{\pi}{4}$, $\dfrac{9\pi}{4}$, and $-\dfrac{7\pi}{4}$ are coterminal because

$$\frac{9\pi}{4} = \frac{\pi}{4} + 2\pi$$

$$-\frac{7\pi}{4} = \frac{\pi}{4} - 2\pi$$

When an angle is in standard position, its terminal side will form an angle with either the positive or negative part of the x-axis which is a positive angle of measure 90° or less or is an angle of measure 0°. This angle is called the **reference angle** of the given angle.

DEFINITION 7.1. The **reference angle** for any angle, θ, is the nonnegative angle of smallest measure between the terminal side of θ and the x-axis, where θ is in standard position.

By Definition 7.1 the reference angle of a given angle θ is the nonnegative angle of smallest measure between the terminal side of θ and the x-axis when θ is in standard position. Hence we find the degree measure of the reference angle of θ as follows (Figure 7.9).

(a) If θ is in Q-II, the degree measure of the reference angle is 180° − (the degree measure of θ).
(b) If θ is in Q-III, the dgree measure of the reference angle is (the degree measure of θ) − 180°.
(c) If θ is in Q-IV, the degree measure of the reference angle is 360° − (the degree measure of θ).

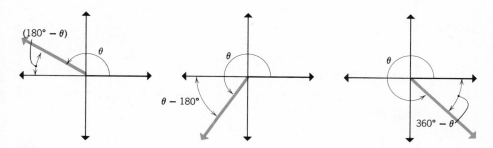

Figure 7.9

EXAMPLE 1. Find the degree measure of the reference angle of an angle whose degree measure is (a) 150°; (b) 225°; (c) 300°.

SOLUTION

(a) Since an angle of 150° is in Quadrant II, we find the degree measure of its reference angle by subtracting 150° from 180°:

$$180° - 150° = 30°$$

The degree measure of the reference angle is 30°.

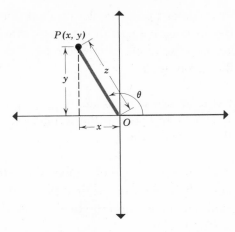

Figure 7.10

(b) An angle of degree measure 225° is in Q-III. The degree measure of its reference angle is

$$225° - 180° = 45°$$

(c) An angle of measure 300° is in Q-IV. The degree measure of its reference angle is

$$360° - 300° = 60°$$

Consider the angle θ in Figure 7.10 which has been generated by ray $\overrightarrow{OP}$. Let us mark off on the rotating ray line segment $\overline{OP}$ of length r. We let (x, y) be the coordinates of P, the point on the terminal side of the angle at a distance r from the origin. We shall call the line segment $\overline{OP}$, joining P to the origin, the **radius vector** of the point P. The length of $\overline{OP}$—which is nonnegative—is called the **polar distance** of P.

Thus, beginning with the angle θ in standard position, we can find three numbers, x, y, and r, the coordinates and the polar distance of the point P. We form six ratios with these numbers called the **trigonometric ratios** of the angle. These ratios are called **sine, cosine, tangent, cotangent, secant,** and **cosecant.**

DEFINITION 7.2. Let θ be any angle in standard position. Let P be the point with coordinates (x, y) on the terminal side of the angle θ with polar distance $r > 0$. The **trigonometric ratios** of θ are then defined as follows (the abbreviation of each name is given):

$$\sin \theta = \frac{y}{r}$$

$$\cos \theta = \frac{x}{r}$$

$$\tan \theta = \frac{y}{x}, \, x \neq 0$$

$$\cot \theta = \frac{x}{y}, \, y \neq 0$$

$$\sec \theta = \frac{r}{x}, \, x \neq 0$$

$$\csc \theta = \frac{r}{y}, \, y \neq 0$$

For any specific value of θ the values of the trigonometric ratios are uniquely determined, except for certain values of θ for which x or y is equal to 0—as indicated in the definition. The values of the trigonometric ratios depend only on θ and not on the polar distance, that is, the distance of P from the origin.

To show that for any particular value of θ the values of the trigonometric ratios are uniquely determined, we shall show that the value of $\sin \theta$ is independent of the choice of point P on the terminal side of θ. Let P and P' be two points on the terminal side of θ with coordinates (x, y) and (x', y'), respectively (Figure 7.11).

Using the coordinates of P', we have

$$\sin \theta = \frac{y'}{r'}$$

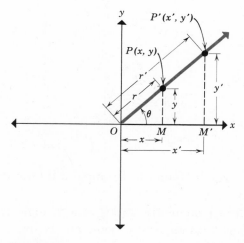

Figure 7.11

Using the coordinates of P, we have

$$\sin \theta = \frac{y}{r}$$

Since triangles $OP'M'$ and OPM are similar, it follows that

$$\frac{y'}{r'} = \frac{y}{r}$$

and the value of $\sin \theta$ is the same whether it is obtained by using P or by using P'.

In a similar fashion it can be shown that for a specified value of θ the other trigonometric ratios are uniquely determined.

Once the coordinates of a point (other than the origin) on the terminal side of the angle in standard position are known, the values of the trigonometric ratios of that angle can be found immediately.

EXAMPLE 2. Determine the values of the trigonometric ratios of θ if the point $(3, -4)$ lies on the terminal side of θ.

SOLUTION. The polar distance can be found directly from the distance formula and is

$$r = \sqrt{3^2 + (-4)^2} = \sqrt{9 + 16} = \sqrt{25} = 5$$

Definition 7.2 then gives

$$\sin \theta = \frac{y}{r} = \frac{-4}{5} = -\frac{4}{5}$$

$$\cos \theta = \frac{x}{r} = \frac{3}{5}$$

$$\tan \theta = \frac{y}{x} = \frac{-4}{3} = -\frac{4}{3}$$

$$\cot \theta = \frac{x}{y} = \frac{3}{-4} = -\frac{3}{4}$$

$$\sec \theta = \frac{r}{x} = \frac{5}{3}$$

$$\csc \theta = \frac{r}{y} = \frac{5}{-4} = -\frac{5}{4}$$

We note that the results found in Example 2 do not depend on the kind of measure used for θ.

The values of the trigonometric ratios of any angle depend only on the coordinates and polar distance of a point on its terminal side. Since the terminal sides of coterminal angles are the same, each of the six trigonometric ratios has the same value for coterminal angles.

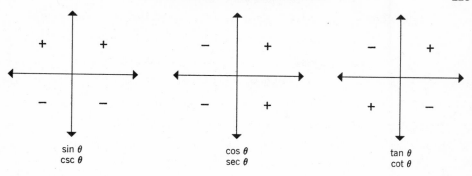

Figure 7.12

The sign of any of the trigonometric ratios will be determined by the quadrant which contains the terminal side of the angle placed in standard position. In the first quadrant both the x- and y-coordinates are positive, so that all of the trigonometric ratios of angles in the first quadrant are positive. In the second quadrant the x-coordinate is negative and the y-coordinate is positive so that the sine and cosecant of angles in the second quadrant are positive and all the other ratios are negative. In a similar manner we can determine the signs of the trigonometric ratios of angles in the four quadrants. These are shown in Figure 7.12.

We shall discuss the trigonometric ratios of angles whose terminal sides are on the axes later.

EXAMPLE 3. Find the remaining trigonometric ratios if $\sin \theta = -\frac{7}{25}$ and θ is in Q-III.

SOLUTION. Let $P(x, y)$ be a point on the terminal side of θ with polar distance r. Then, from the definition of $\sin \theta$, we have

$$\frac{y}{r} = -\frac{7}{25}$$

We may choose any positive value for r and then determine the value of y. However, we seek convenient numbers, so we choose $r = 25$. This makes $y = -7$. The distance formula tells us that

$$x^2 + y^2 = r^2$$

so that

$$x^2 + (-7)^2 = (25)^2$$
$$x^2 + 49 = 625$$
$$x^2 = 576$$
$$x = \pm 24$$

But, since θ is in Q-III, x must be negative so that $x = -24$. Thus

$$\cos \theta = \frac{-24}{25} = -\frac{24}{25}$$

$$\tan \theta = \frac{-7}{-24} = \frac{7}{24}$$

$$\cot \theta = \frac{-24}{-7} = \frac{24}{7}$$

$$\sec \theta = \frac{25}{-24} = -\frac{25}{24}$$

$$\csc \theta = \frac{25}{-7} = -\frac{25}{7}$$

EXERCISES 7.2

Find the values of the six trigonometric ratios of the angle θ if:

1. $(4, 3)$ is on the terminal side of θ.
2. $(-3, 4)$ is on the terminal side of θ.
3. $(24, -10)$ is on the terminal side of θ.
4. $(-1, 4)$ is on the terminal side of θ.
5. $(-12, -5)$ is on the terminal side of θ.
6. $(-8, -15)$ is on the terminal side of θ.
7. $(-2, 4)$ is on the terminal side of θ.
8. $(-1, 3)$ is on the terminal side of θ.
9. $(4, -2)$ is on the terminal side of θ.
10. $(-1, -1)$ is on the terminal side of θ.

Find the values of the remaining trigonometric ratios of θ.

11. $\sin \theta = \frac{4}{5}$, $\cos \theta$ is positive.
12. $\tan \theta = \frac{1}{2}$, $\sin \theta$ is negative.
13. $\cos \theta = -\frac{7}{12}$, $\tan \theta$ is positive.
14. $\csc \theta = -\frac{5}{3}$, θ is in Q-IV.
15. $\cot \theta = -\frac{15}{2}$, θ is in Q-II.
16. $\tan \theta = 5$, θ is in Q-III.
17. $\sin \theta = \frac{5}{13}$, θ is in Q-II.
18. $\cos \theta = \frac{-2}{7}$, θ is in Q-II.
19. $\tan \theta = \frac{9}{7}$, θ is in Q-III.
20. $\csc \theta = \frac{3}{2}$, θ is in Q-II.

Find the reference angles of angles with degree measures as given (Exercises 21–30).

21. $150°$ 23. $135°$
22. $240°$ 24. $300°$

25. 315°   33. 327°
26. −225°   34. 197° 20′
27. −330°   35. −312° 30′
28. −135°   36. −212° 40′
29. −120°   37. 267° 50′
30. −300°   38. −199° 30′
31. 170°   39. 116° 35′
32. 245°   40. 127° 35′

Find the reference angles of angles with radian measure as given (Exercises 41–50).

41. $\dfrac{3\pi}{4}$ 46. $-\dfrac{5\pi}{6}$

42. $\dfrac{7\pi}{6}$ 47. $-\dfrac{7\pi}{4}$

43. $\dfrac{5\pi}{3}$ 48. $-\dfrac{7\pi}{12}$

44. $\dfrac{11\pi}{6}$ 49. $\dfrac{11\pi}{12}$

45. $\dfrac{5\pi}{4}$ 50. $\dfrac{11\pi}{4}$

7.4 ◆ Trigonometric Ratios of Special Angles

There are methods by which we can find approximations for the values of the trigonometric ratios of an angle that are accurate to any desired degree of accuracy. However, for certain special angles, we can find the exact values of these ratios. In Figure 7.13 angles that measure 30°, 45°, and 60° are shown with a convenient point chosen on the terminal side of each angle. We find the coordinates of these points by using the properties of a right triangle. Thus, in Figures 7.13a and 7.13c the length of the side of the right

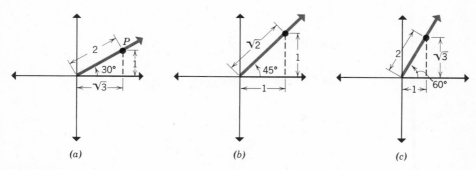

(a)   (b)   (c)

Figure 7.13

triangle which is opposite the angle of 30° is one-half the length of the hypotenuse. In Figure 7.13*a* we choose $r = 2$ and $y = 1$, so that $x = \sqrt{3}$. In Figure 7.13*c* we choose $x = 1$ and $r = 2$ to get $y = \sqrt{3}$. In Figure 7.13*b* we have an isosceles right triangle, so if we choose $x = y = 1$, then $r = \sqrt{2}$.

Using Definition 7.2, we find the values of the trigonometric ratios in Table 7.2.

Table 7.2

θ	$\sin \theta$	$\cos \theta$	$\tan \theta$	$\cot \theta$	$\sec \theta$	$\csc \theta$
30°	$\dfrac{1}{2}$	$\dfrac{\sqrt{3}}{2}$	$\dfrac{1}{\sqrt{3}} = \dfrac{\sqrt{3}}{3}$	$\sqrt{3}$	$\dfrac{2}{\sqrt{3}} = \dfrac{2\sqrt{3}}{3}$	2
45°	$\dfrac{1}{\sqrt{2}} = \dfrac{\sqrt{2}}{2}$	$\dfrac{1}{\sqrt{2}} = \dfrac{\sqrt{2}}{2}$	1	1	$\sqrt{2}$	$\sqrt{2}$
60°	$\dfrac{\sqrt{3}}{2}$	$\dfrac{1}{2}$	$\sqrt{3}$	$\dfrac{1}{\sqrt{3}} = \dfrac{\sqrt{3}}{3}$	2	$\dfrac{2}{\sqrt{3}} = \dfrac{2\sqrt{3}}{3}$

We can use a similar procedure to determine the trigonometric ratios of any angle for which 30°, 45°, or 60° is the reference angle. In doing so we must take into account the appropriate signs of the *x*- and *y*-coordinates in the four quadrants.

EXAMPLE 1. Find the values of the six trigonometric ratios of an angle of degree measure 240°.

SOLUTION. With the angle of degree measure 240°, we have a reference angle of degree measure 60° ($240° - 180° = 60°$). The point $(-1, -\sqrt{3})$ with $r = 2$ is chosen on the terminal side (see Figure 7.14). Then

$$\sin 240° = \frac{-\sqrt{3}}{2} = -\frac{\sqrt{3}}{2}$$

$$\cos 240° = \frac{-1}{2} = -\frac{1}{2}$$

$$\tan 240° = \frac{-\sqrt{3}}{-1} = \sqrt{3}$$

$$\cot 240° = \frac{-1}{-\sqrt{3}} = \frac{1}{\sqrt{3}} = \frac{\sqrt{3}}{3}$$

$$\sec 240° = \frac{2}{-1} = -2$$

$$\csc 240° = \frac{2}{-\sqrt{3}} = -\frac{2\sqrt{3}}{3}$$

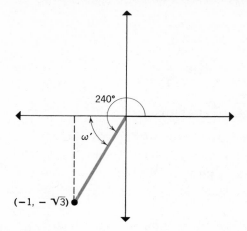

Figure 7.14

EXAMPLE 2. Find the values of the six trigonometric ratios of an angle whose degree measure is 135°.

SOLUTION. With the angle of degree measure 135° we have a reference angle of degree measure 45° ($180° - 135° = 45°$). The point $(-1, 1)$ with $r = \sqrt{2}$ is chosen on the terminal side (see Figure 7.15).

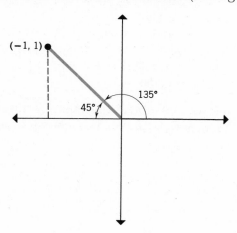

Figure 7.15

Then

$$\sin 135° = \frac{1}{\sqrt{2}} = \frac{\sqrt{2}}{2}$$

$$\cos 135° = \frac{-1}{\sqrt{2}} = -\frac{\sqrt{2}}{2}$$

$$\tan 135° = \frac{-1}{1} = -1$$
$$\cot 135° = -1$$
$$\sec 135° = -\sqrt{2}$$
$$\csc 135° = \sqrt{2}$$

7.5 ◆ Trigonometric Ratios of Axes Angles

We now find the exact values of the trigonometric ratios of angles whose terminal sides are on the axes. Such angles are called **axes angles.** For any axis angle in standard position, a point on the terminal side can always be chosen with $r = 1$. Then either x or y will be 0 and the other will be ± 1.

EXAMPLE 1. Find the values of the six trigonometric ratios which exist for an angle of degree measure 90°.

SOLUTION. The terminal side of an angle of degree measure 90° is the positive y-axis. We choose the point on this side for which $r = 1$, $x = 0$, and $y = 1$. The trigonometric ratios which involve x in the denominator do not exist and we have

$$\sin 90° = \tfrac{1}{1} = 1$$
$$\cos 90° = \tfrac{0}{1} = 0$$
$$\tan 90° \text{ does not exist}$$
$$\cot 90° = \tfrac{0}{1} = 0$$
$$\sec 90° \text{ does not exist}$$
$$\csc 90° = \tfrac{1}{1} = 1$$

In the same manner we may obtain the values in Table 7.3. A dash in the table indicates that the corresponding trigonometric ratio does not exist.

Table 7.3

	$\sin \theta$	$\cos \theta$	$\tan \theta$	$\cot \theta$	$\sec \theta$	$\csc \theta$
0°	0	1	0	—	1	—
90°	1	0	—	0	—	1
180°	0	-1	0	—	-1	0
270°	-1	0	—	0	—	-1

EXERCISES 7.3

Draw the angles with measures given in Exercises 1–20. Determine the values of their six trigonometric ratios.

1. $60°$

2. $45°$

3. $\dfrac{\pi}{6}$

4. $210°$

5. $315°$
6. $-30°$
7. $-315°$
8. $300°$

9. $330°$

10. $\dfrac{3\pi}{2}$

11. $\dfrac{\pi}{3}$

12. $\dfrac{5\pi}{6}$

13. $\dfrac{3\pi}{4}$

14. $\dfrac{7\pi}{4}$

15. $-300°$
16. $240°$
17. $-120°$
18. $150°$

19. $-\dfrac{4\pi}{3}$

20. $-135°$

Using the values of the trigonometric ratios of the indicated angles, verify the statements below (Exercises 21–30).

21. $\sin^2 30° + \cos^2 30° = 1$.
22. $\sin 120° = 2 \sin 60° \cos 60°$.
23. $\tan^2 45° + 1 = \sec^2 45°$.
24. $\cos 240° = \cos^2 120° - \sin^2 120°$.

25. $\dfrac{\sin 300°}{\cos 300°} = \tan 300°$.

26. $\csc 150° = \dfrac{1}{\sin 150°}$.

27. $1 - \cos^2 240° = \sin^2 240°$.
28. $\sin(-135°)\csc(-135°) = 1$.
29. $\tan(-150°)\cot(-150°) = 1$.
30. $\tfrac{1}{2}\sin 240° = \sin 120° \cos 120°$.

In each of the following replace the symbol $\sim$ with the symbols $<$, $=$, or $>$ to make true statements (Exercises 31–40).

31. $\cos 60° \sim \cos 30°$
32. $\sin 30° + \cos 60° \sim \sin 90°$
33. $\sin 45° \sim \cos 45°$
34. $\sin 45° + \cos 45° \sim \tan 45°$
35. $\sin 90° \sim \sin 0°$
36. $\sin 90° \sim \sin 180°$
37. $\sin 180° \sim \sin 270°$

38. $\cos 0° \sim \cos 90°$

39. $\tan 0° \sim \tan 45°$

40. $\sin^2 180° \sim 1 - \cos^2 180°$

7.6 ⋆ Tables of Trigonometric Ratios

The trigonometric ratios exist for all angles except for certain ratios for the axes angles. In Sections 7.4 and 7.5 we found exact values for these ratios for certain special angles. Table I in the appendix gives values, correct to four decimal places, of the trigonometric ratios of angles with degree measures between $0°$ and $90°$.

In this table angles of degree measure less than $45°$ are listed in the column at the left of the page. The measure of the angles increases as we read down. Angles of measure between $45°$ and $90°$ are listed in the column at the right of the page and the measure of the angle increases as we read up the column. The labelings at the tops of the columns apply to angles of degree measure less than $45°$ and the labelings at the bottom apply to angles of degree measure greater than $45°$. We see, then, that each entry in the table serves as the value of two different ratios for two different angles.

The arrangement of the table is based on an important property of the trigonometric ratios which we now discuss. Let θ and θ' be **complementary angles,** that is, angles with degree measures between $0°$ and $90°$ for which the sum of the degree measures is $90°$. In Figure 7.16 two such angles are shown. The points $P(x, y)$ and $P'(x', y')$ are chosen so that $OP = OP'$. It is easy to show by the use of congruent triangles that $x' = y$ and $y' = x$.

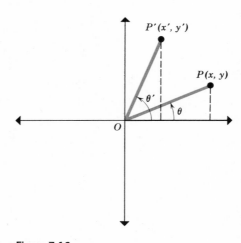

Figure 7.16

Hence, by Definition 7.2, we have

$$\sin \theta = \frac{y}{r} = \frac{x'}{r} = \cos \theta'$$

$$\cos \theta = \frac{x}{r} = \frac{y'}{r} = \sin \theta'$$

$$\tan \theta = \frac{y}{x} = \frac{x'}{y'} = \cot \theta'$$

$$\cot \theta = \frac{x}{y} = \frac{y'}{x'} = \tan \theta'$$

$$\sec \theta = \frac{r}{x} = \frac{r}{y'} = \csc \theta'$$

$$\csc \theta = \frac{r}{y} = \frac{r}{x'} = \sec \theta'$$

The six trigonometric ratios fall into three pairs of **coratios,** the sine and cosine, the tangent and cotangent, and the secant and cosecant.

The following examples show how to use Table I to find the values of trigonometric ratios of an angle whose measure is expressed in terms of degrees and an integer multiple of 10′.

EXAMPLE 1. Find the value of sin 22° 10′.

SOLUTION. In Table I we read down the column at the left, headed "Degrees," until we come to 22° 10′. Then we go horizontally to the right until we come to the column headed "Sin." There we find the entry 0.3773. Hence

$$\sin 22° \ 10' = 0.3773$$

EXAMPLE 2. Find the value of tan 60° 20′.

SOLUTION. In Table I we read up the column at the right, headed "Degrees," from the bottom of the page until we come to 60° 20′. Then we go horizontally to the left until we come to the column designated at the bottom of the page by "Tan." There we find the entry 1.756. Hence

$$\tan 60° \ 20' = 1.756$$

Table I may be used to find the measure of an angle whose degree measure is between 0° and 90° if one of its trigonometric ratios is known. The process is illustrated by the examples below.

EXAMPLE 3. If $\sin \theta = 0.2979$, find the degree measure of θ.

SOLUTION. We locate the number of 0.2979 in a Sin column in Table I and, by going horizontally to the left to the Degrees column, we find that it corresponds to an angle of degree measure 17° 20'. Hence

$$\theta = 17° \ 20'$$

EXAMPLE 4. If $\tan \theta = 2.145$, find the measure of θ.

SOLUTION. We locate the number 2.145 in a Tan column in Table I and, by going horizontally to the right to the Degrees column, we find that it corresponds to an angle of degree measure of 65°. Therefore

$$\theta = 65°$$

EXERCISES 7.4

Use Table I to find the value of each of the following (Exercises 1–30).

1. $\sin 32° \ 20'$	16. $\tan 33° \ 30'$
2. $\cos 16° \ 30'$	17. $\cos 66° \ 40'$
3. $\tan 25° \ 50'$	18. $\sec 7° \ 10'$
4. $\cot 40°$	19. $\csc 12° \ 20'$
5. $\sec 14° \ 10'$	20. $\tan 11° \ 50'$
6. $\sin 56° \ 40'$	21. $\sin 80° \ 10'$
7. $\cos 82° \ 20'$	22. $\cos 66° \ 20'$
8. $\tan 74° \ 30'$	23. $\cot 57° \ 40'$
9. $\csc 36° \ 50'$	24. $\sin 3° \ 10'$
10. $\sec 57° \ 30'$	25. $\cot 46° \ 10'$
11. $\cot 72° \ 40'$	26. $\csc 18° \ 40'$
12. $\cos 47° \ 50'$	27. $\csc 72° \ 50'$
13. $\sin 78° \ 10'$	28. $\cos 50° \ 10'$
14. $\sin 88° \ 10'$	29. $\sec 38° \ 40'$
15. $\tan 7° \ 20'$	30. $\sin 79° \ 50'$

Use Table I to find the angles with the given trigonometric ratios (Exercises 31–60).

31. $\sin \theta = 0.1305$	38. $\sin \theta = 0.6820$
32. $\tan \theta = 6.968$	39. $\csc \theta = 1.414$
33. $\cos \theta = 0.9621$	40. $\cot \theta = 0.6494$
34. $\cot \theta = 4.511$	41. $\sin \theta = 0.8192$
35. $\sec \theta = 1.086$	42. $\cos \theta = 0.5150$
36. $\cot \theta = 0.4245$	43. $\csc \theta = 1.572$
37. $\cos \theta = 0.4014$	44. $\cot \theta = 1.111$

45. sec $\theta = 1.051$

46. cos $\theta = 0.9724$

47. tan $\theta = 0.4040$

48. sin $\theta = 0.1622$

49. cot $\theta = 0.4986$

50. cot $\theta = 5.396$

51. cot $\theta = 3.047$

52. cos $\theta = 0.8208$

53. tan $\theta = 0.5169$

54. cot $\theta = 0.7954$

55. sec $\theta = 1.766$

56. tan $\theta = 1.006$

57. sin $\theta = 0.0087$

58. tan $\theta = 0.0582$

59. csc $\theta = 1.117$

60. cot $\theta = 0.3906$

7.7 ◆ Values of the Trigonometric Ratios of Other Angles

The use of Table I can be extended to angles of any degree measure. We know that the reference angle is always an angle of degree measure between $0°$ and $90°$. Furthermore, whenever x, y, and r are known for a point on the terminal side of an angle, then, if the reference angle is placed in standard position and a point on its terminal side is chosen with the same r, this point will have coordinates $(|x|, |y|)$. Hence the trigonometric ratios of an angle and its reference angle differ at most in sign.

EXAMPLE 1. Find the value of cos $172°$ $10'$.

SOLUTION. The reference angle of $172°$ $10'$ is $7°$ $50'$. From the table we find

$$\cos 7°\ 50' = 0.9907$$

Since an angle of $172°$ $10'$ is in Q-II, the value of the cosine is negative. Hence

$$\cos 172°\ 10' = -\cos 7°\ 50' = -0.9907$$

EXAMPLE 2. Find the value of cot $(-127°)$.

SOLUTION. The reference angle of $-127°$ is $53°$. From the table we find

$$\cot 53° = 0.7536$$

Since an angle of $-127°$ is in Quadrant III and the cotangent ratio is positive for an angle in Q-III, we have

$$\cot 53° = \cot (-127°) = 0.7536$$

EXAMPLE 3. Given sin $\theta = 0.6157$ and $90° < \theta < 180°$, find θ.

SOLUTION. From the table we find

$$\sin 38° = 0.6157$$

Angle θ is in Quadrant II and its reference angle is $38°$, therefore

$$\theta = 180° - 38° = 142°$$

EXAMPLE 4. Given cot $\theta = -1.098$ and $270° < \theta < 360°$, find θ.

SOLUTION. From the table we find

$$\cot 42° \ 20' = 1.098$$

Since θ is in the fourth quadrant,

$$\theta = 360° - 42° \ 20' = 317° \ 40'$$

EXERCISES 7.5

Use Table I to find the following.

1. sin 124° 10′	11. cot 280° 40′
2. cos 135° 20′	12. sec 290° 50′
3. tan 150° 30′	13. csc 340° 40′
4. cot 160° 50′	14. tan 337° 10′
5. sec 175° 30′	15. sin 350°
6. csc 166° 20′	16. cos 300° 40′
7. sin 185° 30′	17. sin 187° 40′
8. cos 200° 50′	18. tan 253° 10′
9. tan 245° 10′	19. csc 188° 30′
10. cos 250° 30′	20. cos 345° 40′

Use Table I to find the following angles.

21. $\sin \theta = 0.3584$, $90° < \theta < 180°$.
22. $\cos \theta = -0.5568$, $90° < \theta < 180°$.
23. $\cot \theta = -2.145$, $270° < \theta < 360°$.
24. $\csc \theta = -1.796$, $270° < \theta < 360°$.
25. $\tan \theta = 2.006$, $180° < \theta < 270°$.
26. $\sec \theta = -1.076$, $180° < \theta < 270°$.
27. $\sin \theta = -0.9346$, $180° < \theta < 270°$.
28. $\csc \theta = -1.161$, $270° < \theta < 360°$.
29. $\cos \theta = 0.6777$, $270° < \theta < 360°$.
30. $\tan \theta = -1.111$, $270° < \theta < 360°$.

7.8 ◆ Interpolation

We have seen that for an angle whose measure is expressed in terms of degrees and an integral multiple of ten minutes, the values of the trigonometric ratios can be found directly in Table I. If the number of minutes in the measure of an angle is not a multiple of ten, it is necessary to use **linear interpolation** to find the trigonometric ratios. The following examples illustrate the method which is similar to that used in the case of logarithms.

EXAMPLE 1. Find the value of sin 15° 32'.

SOLUTION. The angle whose measure is 15° 32' is not listed in the table, but it lies between an angle whose measure is 15° 30' and one whose measure is 15° 40'. We assume that sin 15° 32' occupies the same relative position between sin 15° ȯ0' and sin 15° 40' as does 15° 32' between the measures of 15° 30' and 15° 40'. The work done in getting the proper value is shown below:

$$10' \begin{bmatrix} 2' \begin{bmatrix} \sin 15°\ 30' = 0.2672 \\ \sin 15°\ 32' = ? \end{bmatrix} y \\ \sin 15°\ 40' = 0.2700 \end{bmatrix} 0.0028$$

To locate sin 15° 32' in the proper relative position we form the proportion

$$\frac{y}{0.0028} = \frac{2}{10}$$

$$y = \frac{2(0.0028)}{10} = 0.00056 \doteq 0.0006$$

We take for the value of the correction the number of significant figures in the table, which is four, so we replace 0.00056 by 0.0006. The correction is added to the sine of the angle that has a measure smaller than our angle. This gives

$$\sin 15°\ 32' = 0.2672 + 0.0006 = 0.2678$$

EXAMPLE 2. Find the value of cos 72° 48'.

SOLUTION

$$10' \begin{bmatrix} 8' \begin{bmatrix} \cos 72°\ 40' = 0.2979 \\ \cos 72°\ 48' = ? \end{bmatrix} y \\ \cos 72°\ 50' = 0.2952 \end{bmatrix} 0.0027$$

We form the proportion

$$\frac{y}{0.0027} = \frac{8}{10}$$

$$y = \frac{8(0.0027)}{10} = 0.00216 \doteq 0.0022$$

We take y to be 0.0022 and this time we *subtract* the correction from the cosine of the angle that has a measure smaller than our angle. Why? This gives

$$\cos 72°\ 48' = 0.2979 - 0.0022 = 0.2957$$

It is important to observe that for the ratios whose values increase as the measure of the angle increases from 0° to 90°, that is, the sine, tangent, and secant, the correction is added. For the cosine, cotangent, and cosecant, whose values decrease as the measure of the angle increases from 0° to 90°, the correction is subtracted. In each case the correction is added to or subtracted from the ratio of the angle with the smaller measure.

EXAMPLE 3. If $\sin \theta = 0.1465$, find the degree measure of θ.

SOLUTION. We attempt to find 0.1465 in a sin column of Table I. This we are unable to do, but we do locate two numbers between which our number lies. A process of interpolation is applied to these entries to determine the measure of θ:

$$10'\left[x'\left[\begin{array}{l}\sin 8°\ 20' = 0.1449 \\ \sin \theta \quad\ \ = 0.1465\end{array}\right]0.0016 \atop \sin 8°\ 30' = 0.1478\right]0.0029$$

We form the proportion

$$\frac{x}{10} = \frac{16}{29}$$

$$x = \frac{16(10)}{29} = 5.5 \doteq 6$$

Adding 6' to 8° 20', we have

$$\theta = 8°\ 26'$$

EXAMPLE 5. If $\cot \theta = 1.123$, find the degree measure of θ.

SOLUTION

$$10'\left[x'\left[\begin{array}{l}\cot 41°\ 40' = 1.124 \\ \cot \theta \quad\ \ = 1.123\end{array}\right]0.001 \atop \cot 41°\ 50' = 1.117\right]0.007$$

We form the proportion

$$\frac{x}{10} = \frac{1}{7}$$

$$x = \frac{1(10)}{7} \doteq 1$$

Adding 1' to 41° 40' we have

$$\theta = 41° 41'$$

EXERCISES 7.6

Use Table I to find the following (Exercises 1–28).

1. sin 20° 15'
2. cos 10° 32'
3. tan 36° 37'
4. cot 42° 51'
5. sec 56° 27'
6. csc 38° 11'
7. sin 39° 57'
8. cot 78° 46'
9. sin 86° 54'
10. cos 37° 17'
11. tan 49° 49'
12. csc 76° 56'
13. sin 79° 41'
14. sec 32° 48'
15. cot 12° 12'
16. tan 48° 26'
17. sin 69° 7'
18. cos 14° 14'
19. tan 32° 37'
20. cot 89° 1'
21. csc 18° 38'
22. sin 77° 11'
23. cos 59° 25'
24. tan 39° 8'
25. cos 44° 19'
26. sec 13° 12'
27. tan 9° 8'
28. cos 47° 42'

Find the measure of θ to the nearest minute (Exercises 29–48).

29. $\sin \theta = 0.1462$
30. $\cos \theta = 0.9869$
31. $\tan \theta = 0.4111$
32. $\cot \theta = 1.102$
33. $\sec \theta = 1.110$
34. $\cot \theta = 0.6784$
35. $\csc \theta = 1.229$
36. $\tan \theta = 1.788$
37. $\cot \theta = 7.604$
38. $\cos \theta = 0.5617$
39. $\csc \theta = 1.433$
40. $\cot \theta = 1.345$
41. $\tan \theta = 0.5217$
42. $\sin \theta = 0.6199$
43. $\cos \theta = 0.7110$
44. $\tan \theta = 6.107$
45. $\csc \theta = 7.111$
46. $\cos \theta = 0.3095$
47. $\sin \theta = 0.4380$
48. $\tan \theta = 3.806$

CHAPTER REVIEW

1. Convert to radian measure:
   (a) 15°   (b) 310°   (c) 642°   (d) $-83°$
2. Convert to degree measure:

   (a) $\dfrac{3\pi}{2}$ (b) $-\dfrac{16\pi}{5}$ (c) 54π (d) 90

Find the values of the six trigonometric ratios for the following (Exercises 3–8).

3. $\theta = \dfrac{2\pi}{3}$ 6. $\cos \theta = -\frac{1}{2}$, θ is in Q-III

4. $\theta = 315°$ 7. $\sin \theta = -\frac{1}{2}\sqrt{2}$, θ is in Q-IV

5. $\theta = -\dfrac{\pi}{6}$ 8. $\cot \theta = -1$, θ is in Q-II

Evaluate (Exercises 9–20).

9. $\cos 135°$ 15. $\cot 137° \, 40'$

10. $\sin \dfrac{11\pi}{2}$ 16. $\sec 311° \, 20'$

11. $\cos 37° \, 10'$ 17. $\csc 453° \, 10'$

12. $\tan 41° \, 30'$ 18. $\cos 46° \, 12'$

13. θ if $\cos \theta = 0.5150$ and $0 < \theta < \dfrac{\pi}{2}$ 19. $\sin 81° \, 2'$

14. θ, if $\sin \theta° = 0.6884$ and $0° < \theta° < -90°$ 20. $\cot 99° \, 43'$

CHAPTER 8

Numerical Trigonometry

8.1 ◆ Right Triangles

If the point (x, y) is on the terminal side of an angle θ in the first quadrant, this point determines a right triangle whose sides have lengths x and y and whose hypotenuse has length $\sqrt{x^2 + y^2}$. Therefore we can see that the ratios of the lengths of the sides of a right triangle are trigonometric ratios (Figure 8.1).

$$\sin \theta = \frac{\text{length of side opposite } \theta}{\text{length of hypotenuse}}$$

$$\cos \theta = \frac{\text{length of side adjacent to } \theta}{\text{length of hypotenuse}}$$

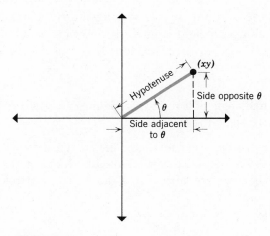

Figure 8.1

$$\tan \theta = \frac{\text{length of side opposite } \theta}{\text{length of side adjacent to } \theta}$$

$$\cot \theta = \frac{\text{length of side adjacent to } \theta}{\text{length of side opposite to } \theta}$$

$$\sec \theta = \frac{\text{length of hypotenuse}}{\text{length of side adjacent to } \theta}$$

$$\csc \theta = \frac{\text{length of hypotenuse}}{\text{length of side opposite to } \theta}$$

With these definitions it is not necessary that the right triangle be situated in such a way that θ is in standard position. The ratios defined above are equally valid for a right triangle in any position (Figure 8.2).

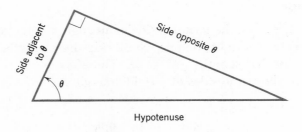

Figure 8.2

If we are given some parts of a right triangle, that is, the measures of some angles and the lengths of some sides, and are asked to find the remaining parts, we say that we are **solving** the right triangle. From plane geometry we recall that a right triangle is determined when, in addition to the right angle, the measures of any two parts, of which at least one is a side, are known.

EXAMPLE 1. If one angle of a right triangle measures 32° and the hypotenuse has length 24, solve the triangle.

SOLUTION. First we make a sketch labeling the measures of the given parts. Generally the hypotenuse is labeled c, the legs a and b, and the angles opposite a, b, and c are labeled A, B, and C respectively (Figure 8.3).

In this case

$$B = 90° - A = 90° - 32° = 58°$$

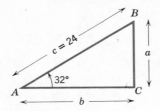

Figure 8.3

Next we have

$$\sin 32° = \frac{a}{c} = \frac{a}{24}$$

or

$$24 \sin 32° = a$$

and

$$\cos 32° = \frac{b}{c} = \frac{b}{24}$$

or

$$24 \cos 32° = b$$

From Table I we find $\sin 32° = 0.5299$ and $\cos 32° = 0.8480$. Hence,

$$a = (24)(0.5299) = 12.7176 \doteq 12.7$$
$$b = (24)(0.8480) = 20.3520 \doteq 20.4$$

The solution is

$$A = 58°, B = 32°, a = 12.7, b = 20.4, c = 24$$

EXAMPLE 2. In the right triangle ABC, $a = 6$ and $b = 8$. Solve the triangle.

SOLUTION. We make a sketch labeling the measures of the given parts (Figure 8.4).

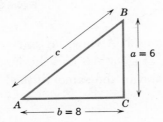

Figure 8.4

In this case

$$\tan A = \frac{a}{b} = \frac{6}{8} = 0.7500.$$

Using Table I, we find

$$A = 36° 52'$$

Then $B = 90° - A = 90° - 36° 52' = 53° 8'$. Using the Pythagorean relation, we have

$$c^2 = a^2 + b^2$$
$$= 6^2 + 8^2$$
$$= 36 + 64 = 100$$

and $c = 10$. The solution is $B = 53° 8'$, $A = 36° 52'$, $c = 10$, $a = 6$, $b = 8$.

Applications of trigonometry arise in which the determination of an angle or of a distance can be made to depend on the solution of one or more right triangles or of finding parts of a right triangle given certain other parts.

One type of application involves problems stated in terms of angle of depression and angle of elevation.

The **angle of depression** of a point B as seen by an observer at O is the positive angle POB formed by the line of sight $\overrightarrow{OB}$ and the horizontal line $\overrightarrow{OP}$ through the eye of the observer and in the same vertical plane as B (Figure 8.5).

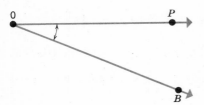

Figure 8.5

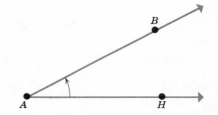

Figure 8.6

In Figure 8.6, an observer at A views an object at B. The positive angle HAB formed by the line of sight $\overrightarrow{AB}$ and the horizontal line $\overrightarrow{AH}$ through the eye of the observer and in the same vertical plane as B is called the **angle of elevation.**

EXAMPLE 3. A tower 40 feet high stands on the bank of a river. From the top of the tower the angle of depression of a point on the opposite bank is 14°. How wide is the river?

SOLUTION. The situation is depicted in Figure 8.7 with BC as the desired distance.

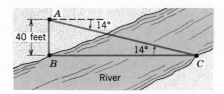

Figure 8.7

From geometry we know that angle ACB has measure of $14°$. We have from right triangle ABC

$$\cot 14° = \frac{BC}{40}$$

and

$$\begin{aligned} BC &= 40 \cot 14° \\ &= 40(4.011) \\ &= 160.440 \end{aligned}$$

The river is about 160 feet wide.

EXAMPLE 4. A vertical stake 18.2 inches high casts a horizontal shadow 50 inches long. What time is it if the sun is directly overhead at 12:25 P.M. and will set at 6:25 P.M.?

SOLUTION. The angle of elevation of the sun (Figure 8.8) is found by

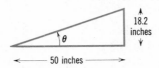

Figure 8.8

$$\begin{aligned} \tan \theta &= \frac{18.2}{50} \\ &= 0.3640 \\ \theta &= 20° \end{aligned}$$

It takes the earth 6 hours to rotate $90°$. Since the rotation is uniform, each degree of elevation of the sun will correspond to $\frac{6}{90}$ of an hour, or 4 minutes. Consequently a rotation through $20°$ will require 80 minutes, since

$$(20)(4) = 80$$

But 80 minutes is 1 hour and 20 minutes. Hence the time is 1 hour and 20 minutes before sunset or 5:05 P.M.

EXERCISES 8.1

Solve the following right triangles (Exercises 1–5).

1. $A = 30°, b = 12$ 4. $a = 18, b = 26$
2. $B = 54°, c = 87$ 5. $b = 500, a = 393$
3. $A = 38° 50', b = 311$
6. An isosceles triangle has sides of length 40, 40, and 60. Find the measures of its angles.
7. A railroad track makes an angle of 7° with the horizontal. How many feet does a train rise while traveling 100 miles along the track?
8. A vertical stake 20.0 inches high casts a horizontal shadow 12.5 inches long. What time is it if the sun rose at 6:00 A.M. and will be directly overhead at noon?
9. From a lighthouse 80 feet above the water the angles of depression of two buoys in the same straight line with the lighthouse are 15° and 20°. What is the distance between the buoys?
10. From the top of a lighthouse 110 feet above sea level at high tide the angle of depression of a buoy is 53° 50' at high tide and 56° 20' at low tide. How high is the tide?
11. An airplane passes directly overhead at an altitude of 8 miles. Six minutes later its angle of elevation is measured as 10° 11'. What is the ground speed of the plane?
12. The angle of depression of a launch 205 feet from the bottom of Niagara Falls as seen from Goat Island at the top is 38°. How high is Niagara Falls?
13. If a man 6 feet tall casts a shadow 5 feet in length, what is the angle of elevation of the sun?
14. A 10-foot ladder, with its foot anchored in an alley, will reach 8 feet up the building on one side of the alley and 7 feet up a building on the other. How wide is the alley?
15. From a point on the ground 450 feet from the foot of the Washington Monument, the angle of elevation of the top of the monument is 51°. How high is the monument?
16. A ladder 28 feet long is leaning against a house on a level lot. Its foot is 8 feet from the house. How high above the ground is the top of the ladder? What angle does it make with the ground?
17. A highway runs up a hill at a constant angle of 5°. What is the difference in elevation of two points on the highway 1000 feet part measured along the highway?

18. The length of a tightly stretched kite string is 500 feet. It makes an angle of 37° with the horizontal. How high is the kite?

19. A column 55 feet high stands on a pedestal 25 feet high. From a certain point on the ground in the same horizontal plane with the foot of the pedestal the pedestal subtends an angle of 12°. What angle does the column subtend at this point? (An angle subtended by an object at the eye is the angle between the lines of sight to the extremities A and B of the object, as angle AOB in the figure below.)

20. From a point A in the horizontal plane with the foot F of a tower, the angle of elevation of its top is 18°. From a point B, 150 feet nearer to the foot of the tower and on the line $\overleftrightarrow{AF}$, the angle of elevation of its top is 25°. Find the height of the tower.

8.2 ◆ Logarithms of the Trigonometric Ratios

The work in numerical trigonometry can often be simplified by the use of logarithms. For example, it may be necessary in the solution of a triangle to compute the value of $a = 126 \sin 46° \, 10'$. From Table I (in the appendix) we find $\sin 46° \, 10' = 0.7214$. Then, for $\log_{10} a$ we have, using the laws of logarithms and Table II,

$$
\begin{aligned}
\log_{10} a &= \log_{10} 126 + \log_{10} 0.7214 \\
&= 2.1004 + (9.8581 - 10) \\
&= 1.9585
\end{aligned}
$$

and $a = 90.88$

We see in the above computation, references to two tables were required to find $\log_{10} \sin 46° \, 10'$. Table III gives the common logarithms of the trigonometric ratios so that only one reference table is required.

The sine and cosine of any angle whose degree measure is less than 90° is less than 1. The same is true of the tangent of an angle between 0° and 45° and the cotangent of an angle between 45° and 90°. Consequently, the logarithms of such ratios are negative. To conserve space in the table, the

"-10" is understood with each entry. For example, the table entry for $\log_{10} \sin 40° 20'$ is 9.8111, but we understand that

$$\log_{10} \sin 40° 20' = 9.8111 - 10$$

We show how to use Table III by means of examples.

EXAMPLE 1. Find the value of $\log_{10} \sin 40° 24'$.

SOLUTION. The desired value is between $\log_{10} \sin 40° 20'$ and $\log_{10} \sin 40° 30'$. The interpolation is shown below.

$$10'\left[4'\begin{bmatrix}\log_{10} \sin 40° 20' = 9.8111 - 10 \\ \log_{10} \sin 40° 24' = x\end{bmatrix}h \middle| 0.0014 \right.$$
$$\left. \quad\quad \log_{10} \sin 40° 30' = 9.8125 - 10\right]$$

$$\frac{4}{10} = \frac{h}{0.0014}$$

$$h = \frac{4(0.0014)}{10} = \frac{0.0056}{10} = 0.00056 \doteq 0.0006$$

$$\log_{10} \sin 40° 24' = (9.8111 - 10) + 0.0006$$
$$= 9.8117 - 10$$

EXAMPLE 2. Find θ if $\log_{10} \cos \theta = 9.8560 - 10$.

SOLUTION. Using the table, we see that θ must be between $44°$ and $44° 10'$. The interpolation is shown below:

$$10'\left[x'\begin{bmatrix}\log_{10} \cos 44° 00' = 9.8569 - 10 \\ \log_{10} \cos \quad \theta \quad = 9.8560 - 10\end{bmatrix}0.0009 \middle| 0.0012 \right.$$
$$\left. \quad\quad \log_{10} \cos 44° 10' = 9.8557 - 10\right]$$

$$\frac{x}{10} = \frac{9}{12} \quad \text{or} \quad x = \frac{90}{12} \doteq 8$$

Then $\theta = 44° 8'$.

EXERCISES 8.2

Find the values of each of the following.

1. $\log_{10} \cos 48°$
2. $\log_{10} \sin 37°$
3. $\log_{10} \tan 56°$
4. $\log_{10} \cot 74°$
5. $\log_{10} \tan 27°$

6. $\log_{10} \sin 36° 50'$
7. $\log_{10} \cos 73° 42'$
8. $\log_{10} \tan 57° 33'$
9. $\log_{10} \cot 43° 27'$
10. $\log_{10} \cot 54° 12'$

Find a value of θ corresponding to each of the following.

11. $\log_{10} \sin \theta = 9.7438 - 10$ 16. $\log_{10} \sin \theta = 9.9841 - 10$
12. $\log_{10} \cos \theta = 9.6340 - 10$ 17. $\log_{10} \tan \theta = 9.9332 - 10$
13. $\log_{10} \tan \theta = 0.1124$ 18. $\log_{10} \cot \theta = 0.8314$
14. $\log_{10} \cot \theta = 9.837 - 10$ 19. $\log_{10} \sin \theta = 9.1101 - 10$
15. $\log_{10} \cos \theta = 9.8733 - 10$ 20. $\log_{10} \cos \theta = 9.4641 - 10$

8.3 ◆ The Law of Sines

An oblique triangle is a triangle that does not contain a right angle. In the discussion of oblique triangles that follows we shall use a standard lettering to name the triangles. The vertices will be labeled A, B, and C; the corresponding angles will be named similarly, that is, A, B, and C; the side of the triangle opposite any vertex will be labeled with the corresponding lowercase letter.

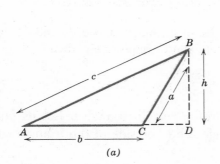 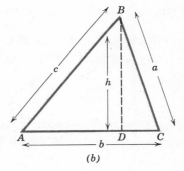

(a)   (b)

Figure 8.9

Let ABC be either oblique triangle shown in Figure 8.9. The altitude h is drawn from vertex B to meet side $\overline{AC}$ in point D, with D as the foot of the altitude. From right triangle ABD in Figure 8.9a we have

$$\sin A = \frac{h}{c} \qquad \text{or} \qquad h = c \sin A$$

and

$$\sin (180° - C) = \frac{h}{a}$$

But $\sin (180° - C) = \sin C$, and we have

$$\sin C = \frac{h}{a} \qquad \text{or} \qquad h = a \sin C$$

By the transitive property of equality we have

$$c \sin A = a \sin C$$

Dividing both members of this equation by ac, we obtain

(1)
$$\frac{\sin A}{a} = \frac{\sin C}{c}$$

In a similar manner we obtain

(2)
$$\frac{\sin A}{a} = \frac{\sin B}{b}$$

A combination of Equations 1 and 2 gives

$$\frac{\sin A}{a} = \frac{\sin B}{b} = \frac{\sin C}{c}$$

This result is called the **Law of Sines.** For triangle ABC in Figure 8.9b Equations 1 and 2 follow immediately by using right triangle ABD.

THEOREM 8.1. If A, B, and C are angles of a triangle and a, b, and c are lengths of the sides opposite A, B, and C, respectively, then

$$\frac{\sin A}{a} = \frac{\sin B}{b} = \frac{\sin C}{c}$$

The Law of Sines can be used to solve certain triangles. Let us first consider an example in which two angles and one side of the triangle are given.

EXAMPLE 1. Solve the triangle for which $A = 60°$, $B = 45°$, and $b = 12$.

SOLUTION. First we make a sketch labeling the measures of the given parts (Figure 8.10).

By Theorem 8.1 we have

$$\frac{\sin 60°}{a} = \frac{\sin 45°}{12}$$

from which we obtain

$$a = \frac{12 \sin 60°}{\sin 45°}$$

Using logarithms, we have

$$\log_{10} a = \log_{10} 12 + \log_{10} \sin 60° - \log_{10} \sin 45°$$
$$= 1.0792 + (9.9375 - 10) - (9.8495 - 10)$$
$$= 1.1672$$

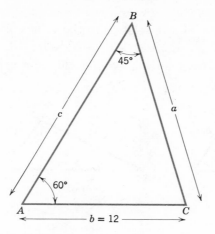

Figure 8.10

and $$a = 14.69 \doteq 14.7$$

Next we observe that

$$A = 60°, B = 45°, \text{ and}$$
$$C = 180° - A - B$$
$$= 180° - 60° - 45° = 75°$$

Again, using Theorem 8.1, we have

$$\frac{\sin 45°}{12} = \frac{\sin 75°}{c}$$

from which we obtain

$$c = \frac{12 \sin 75°}{\sin 45°}$$

Using logarithms, we have

$$\log_{10} c = \log_{10} 12 + \log_{10} \sin 75° - \log_{10} \sin 45°$$
$$= 1.0792 + (9.9849 - 10) - (9.8495 - 10)$$
$$= 1.2146$$

and $$c = 16.39 \doteq 16.4.$$

The solution is $A = 60°$, $B = 45°$, $C = 75°$, $a = 14.7$, $b = 12$ and $c = 16.4$.

If the lengths of two sides of a triangle, say, a and b, and the measure of an angle opposite one of these sides, say, A, is given, we may encounter the possibility of no triangle, one triangle, or two triangles having these parts, depending on the value of a in relation to the values of b and A.

In Figure 8.11 we choose $A < 90°$ and we hold b and A constant and observe the possible situations as a assumes different values.

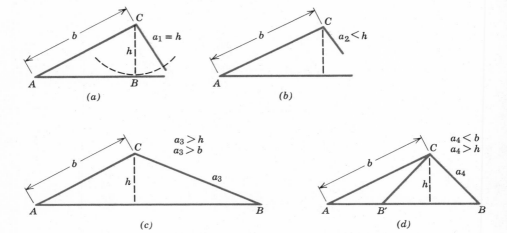

Figure 8.11

We see from Figure 8.11 that the possibility of no triangle, one triangle, or two triangles depends on the relationship of a to h, the altitude, and b. Also, we observe that

$$(3) \qquad\qquad \sin A = \frac{h}{b} \qquad \text{or} \qquad b \sin A = h$$

Now, using Figure 8.11 and Equation 3, we see that if:

I. $a = b \sin A$, we have a unique solution which is a right triangle (Figure 8.11a).

II. $a < b \sin A$, there is no triangle possible and hence no solution (Figure 8.11b).

III. $a > b \sin A$ and $b \le a$, we have a unique solution. In fact, if $b = a$ we have an isosceles triangle (Figure 8.11c).

IV. $a > b \sin A$ and $a < b$, we have two possible triangles and therefore two solutions. The two possible triangles are ABC and $AB'C$ in Figure 8.11d.

If $A > 90°$, then any specified measures of the sides and the angles permit only two possibilities:

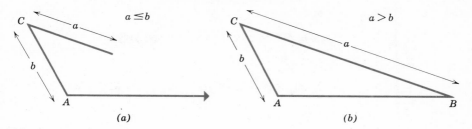

Figure 8.12

I. $a \leq b$, no triangle (Figure 8.12a).

II. $a > b$, one triangle (Figure 8.12b).

EXAMPLE 2. Solve the triangle for which $a = 3$, $b = 4$, and $A = 30°$. Find all angle measurements to the nearest 10′.

SOLUTION. First, we sketch a figure and observe that we have given two sides and an angle opposite one of them (Figure 8.13). Hence we see that this gives rise to the possibility of no triangle, one triangle or two triangles. We first check for the number of possible solutions and observe that $A < 90°$ and $a > b \sin A$, $a < b$. Therefore, there are two solutions

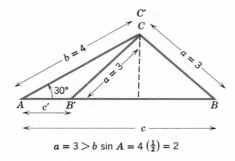

$$a = 3 > b \sin A = 4 \left(\tfrac{1}{2}\right) = 2$$

Figure 8.13

Using Theorem 8.1, we obtain

$$\frac{\sin 30°}{3} = \frac{\sin B}{4}$$

which is equivalent to

$$\sin B = \tfrac{4}{3} \sin 30°$$
$$= \left(\tfrac{4}{3}\right)\left(\tfrac{1}{2}\right) = 0.6667$$

Using Table I, we find

$$B = 41° 50' \qquad \text{(correct to nearest 10')}$$

If $B \doteq 41° 50'$, then

$$C = 180° - 30° - 41° 50' = 108° 10' \qquad \text{(correct to nearest 10')}$$

To find c we again use Theorem 8.1 which gives

$$\frac{\sin 108° 10'}{c} = \frac{\sin 30°}{3}$$

or

$$\begin{aligned} c &= \frac{3 \sin 108° 10'}{\sin 30°} \\ &= \frac{3(0.9502)}{0.5000} \\ &= 5.7012 \doteq 5.7 \end{aligned}$$

The computed parts of triangle ABC are $B = 41° 50'$, $C = 108° 10'$, and $c = 5.7$.

We shall now solve triangle $AB'C'$. Since triangle BCB' is isosceles, angle $BB'C'$ has the same measure as angle B. Hence $B' = 180° - B = 180° - 41° 50' = 138° 10'$. Also, angle

$$C' = 180° - 138° 10' - 30° = 11° 50'$$

Using Theorem 8.1, we obtain

$$\frac{\sin 30°}{3} = \frac{\sin 11° 50'}{c'}$$

or

$$\begin{aligned} c' &= \frac{3 \sin 11° 50'}{\sin 30°} \\ &= \frac{3(0.2051)}{0.5000} \\ &= 1.2306 \doteq 1.2 \end{aligned}$$

Thus the two solutions are

$$B = 41° 50', \ C = 108° 10', \ A = 30°, \ a = 3, \ b = 4, \ c = 5.7$$

and

$$B' = 138° 10', \ C = 11° 50', \ A = 30°, \ a = 3, \ b = 4, \ c' = 1.2$$

EXERCISES 8.3

Solve the triangle ABC for each of the following.

1. $A = 30°$, $B = 80°$, $a = 15$.

2. $B = 75°$, $C = 45°$, $a = 47.5$.
3. $a = 4.22$, $A = 72°, 20'$, $B = 45° \ 30'$.
4. $A = 47°$, $B = 65°$, $b = 68$.
5. $B = 18°$, $A = 36°$, $b = 146$.
6. $A = 37°$, $b = 20$, $a = 11$.
7. $A = 75°$, $a = 95$, $b = 75$.
8. $a = 3{,}333$, $b = 4{,}567$, $A = 41° \ 52'$.
9. $A = 29° \ 24'$, $a = 32.16$, $b = 34.77$.
10. $A = 60° \ 16'$, $a = 4.201$, $b = 3.981$.
11. A surveyor wishes to find the distance across a river from point A to point B. He finds that the distance from A to a point C on the same side of the river is 687.4 feet and angles BAC and BCA are $49° \ 53'$ and $58° \ 16'$, respectively. Find the distance AB.
12. Two sides of a triangle are 10 feet and 20 feet. The angle opposite the shorter side is $29°$. Find the angle opposite the longer side.
13. The diagonals of a parallelogram intersect at an angle of $52° \ 10'$. One diagonal is 3325 feet long and one side is 2995 feet long. Find the length of the other diagonal. (There are two solutions.)
14. One angle of a rhombus is $120°$ and the length of the longer diagonal is 180 feet. Find the length of a side and the length of the shorter diagonal.
15. Watertown is 75 miles from Blubberg and Sleepyville is 60 miles from Blubberg. From Watertown it is observed that an angle between the line joining Watertown and Blubberg and the line joining Watertown and Sleepyville is $29°$. How far is Watertown from Sleepyville?
16. At a certain time the lines from the Earth to the sun and to Mars are observed to make an angle of $117°$. Using the mean distances 9.3×10^7 of the Earth to the sun and 1.41×10^8 miles of Mars to the sun, find the distance between the Earth and Mars at that time.

8.4 ◆ The Law of Cosines

We now prove the Law of Cosines.

THEOREM 8.2. If A, B, and C are angles of a triangle and if a, b, and c are lengths of the sides opposite A, B, and C, respectively, then

$$a^2 = b^2 + c^2 - 2bc \cos A$$
$$b^2 = a^2 + c^2 - 2ac \cos B$$
$$c^2 = a^2 + b^2 - 2ab \cos C$$

PROOF

Case I. All the angles of the triangle have measures less than $90°$.

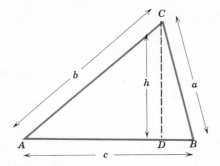

Figure 8.14

We draw an altitude h from the vertex C intersecting side $\overline{AB}$ at point D (Figure 8.14). Then, in right triangle DBC,

$$a^2 = h^2 + (DB)^2$$
$$= h^2 + (c - AD)^2$$
$$= \underbrace{h^2 + (AD)^2}_{b^2} + c^2 - 2c(AD)$$
$$= \qquad b^2 + c^2 - 2c(AD)$$

But

$$\text{Cos } A = \frac{AD}{b} \qquad \text{or} \qquad AD = b \cos A$$

Hence we have

$$a^2 = b^2 + c^2 - 2bc \cos A$$

Case II. One angle of the triangle has measure greater than $90°$.

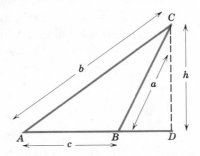

Figure 8.15

We draw an altitude from vertex C which intersects side $\overline{AB}$ at D. Then

$$a^2 = h^2 + (BD)^2$$
$$= h^2 + (AD - c)^2$$
$$= \underbrace{h^2 + (AD)^2}_{b^2} + c^2 - 2c(AD)$$
$$= \qquad b^2 + c^2 - 2c(AD)$$

But

$$\text{Cos } A = \frac{AD}{b} \qquad \text{or} \qquad b \cos A = AD$$

Hence we have

$$a^2 = b^2 + c^2 - 2bc \cos A$$

By relettering the triangle we may at once write the other forms of the Law of Cosines,

$$b^2 = a^2 + c^2 - 2ac \cos B$$
$$c^2 = a^2 + b^2 - 2ab \cos C$$

Theorem 8.2 may be used to find the solutions of certain triangles.

EXAMPLE 1. Solve the triangle for which $A = 150°$, $b = 6$, $c = 8$. Find the measures of the angles correct to the nearest 10′.

SOLUTION. Make a sketch labeling the given measures (Figure 8.16).

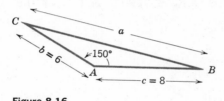

Figure 8.16

We can find a by using the Law of Cosines:

$$\begin{aligned}
a^2 &= b^2 + c^2 - 2bc \cos A \\
&= 6^2 + 8^2 - (2)(6)(8)(\cos 150°) \\
&= 36 + 64 - (96)(-0.8660) \\
&= 36 + 64 + 83.1360 \\
&= 183.1360
\end{aligned}$$

and

$$a \doteq 13.5$$

Next we find B. We can use either the Law of Sines or the Law of Cosines to find B. Let us use the Law of Cosines:

$$b^2 = c^2 + a^2 - 2ac \cos B$$

or

$$\begin{aligned}
\text{Cos } B &= \frac{c^2 + a^2 - b^2}{2ac} \\
&= \frac{8^2 + (13.5)^2 - 6^2}{(2)(13.5)(8)}
\end{aligned}$$

$$= \frac{64 + 182.25 - 36}{216}$$

$$= \frac{210.25}{216}$$

$$\doteq 0.9734$$

Using Table I in the appendix we find

$$B \doteq 13° \ 10'$$

Then $C = 180° - 150° - 13° \ 10'$
$$= 16° \ 50'$$

The solution is

$$C = 16° \ 50', \ B = 13° \ 10', \ a = 12.5, \ b = 6, \ c = 8, \ A = 150°$$

EXAMPLE 2. Solve the triangle for which $a = 8$, $b = 5$, and $c = 4$. Find the angle measures correct to the nearest 10'.

SOLUTION. First we draw a sketch and label the measures given (Figure 8.17).

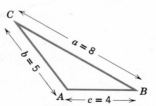

Figure 8.17

We use the Law of Cosines to find A:

$$a^2 = b^2 + c^2 - 2bc \cos A$$

or

$$\text{Cos } A = \frac{b^2 + c^2 - a^2}{2bc}$$

$$= \frac{5^2 + 4^2 - 8^2}{2 \cdot 5 \cdot 4}$$

$$= -\frac{23}{40}$$

$$\doteq -0.5750$$

Since Cos A is negative, A must be an obtuse angle. Using Table I we find

$$A = 125° \ 10'$$

To find C we again use the Law of Cosines:

$$c^2 = a^2 + b^2 - 2bc \cos C$$

or

$$\cos C = \frac{a^2 + b^2 - c^2}{2ab}$$

$$= \frac{8^2 + 5^2 - 4^2}{2 \cdot 8 \cdot 5}$$

$$= \frac{73}{80}$$

$$= 0.9125$$

which gives

$$C = 24° \ 10'$$

Then

$$B = 180° - 125° \ 10' - 24° \ 10' = 30° \ 40'$$

The solution is

$$A = 125° \ 10', B = 30° \ 40', \text{ and } C = 24° \ 10'$$
$$a = 8, b = 5, c = 4$$

Summary of the Solution of Oblique Triangles

We see from the examples in Sections 8.3 and 8.4 that the Law of Cosines is used to solve oblique triangles in the case in which three sides are given and in the case in which two sides and an included angle are given. These cases are often labeled SSS and SAS, respectively.

The Law of Sines is used to solve oblique triangles in the case in which two sides and an angle opposite one of the given sides are given and in the case in which two angles and a side are given. These cases are often labeled SSA and SAA, respectively.

The case in which two sides and an angle opposite one of the given sides are given is called the ambiguous case because in this case the triangle is not always uniquely determined. With the given parts we may be able to construct two triangles, only one triangle, or no triangle at all.

EXERCISES 8.4

Solve triangle ABC for each of the following. Find all angle measures correct to the nearest $10'$ (Exercises 1–10).

1. $a = 8, b = 5, c = 4$
2. $a = 300, b = 250, C = 58° 40'$
3. $b = 50, c = 110, A = 150°$
4. $a = 43, b = 37, c = 55$
5. $c = 21, b = 16, A = 76°$
6. $a = 20, b = 30, c = 40$
7. $A = 105°, a = 75, b = 50$
8. $b = 9, c = 13, A = 115° 10'$
9. $b = 65, c = 310, A = 65° 10'$
10. $b = 43.70, c = 41.70, A = 61° 10'$

11. Two points, R and S, are on opposite sides of an impassable swamp. A point T is chosen which is 1513 feet from R and 1925 feet from S. Angle STR is measured to be $43° 56'$. Find RS, the width of the swamp.
12. Two sides of a parallelogram are 70 and 90 inches and one of the diagonals is 30 inches long. Find the angles of the parallelogram.
13. An airplane has a heading of $200°$ with airspeed 205 miles per hour, but a 25 mile per hour wind is blowing from the east. Find the airspeed of the plane.
14. A railroad track crosses a highway at an angle of $60°$. A train is 30 yards from the intersection when an automobile is 50 yards from the crossing. Find the distance from the train to the car.
15. Two streets meet at an angle of $57° 30'$. A triangular lot has frontage of 110.9 and 120.2 feet, respectively, on the two streets. Find the angles that the rear side of the lot makes with the two streets.
16. The adjacent sides of a parallelogram are 10.62 and 14.73 inches. The longer diagonal is 22.04 inches. Find the length of the shorter diagonal.

8.5 ◆ Vectors

To specify the velocity of motion in a plane we must give both the direction and the magnitude of the velocity. Quantities that are determined by both magnitude and direction are called vector quantities and are represented by directed line segments called **vectors.** Units of length on the vector indicate the units of magnitude of the vector quantity, and the direction of the vector indicates the direction of the vector quantity.

For example, with a scale of 50 pounds to a unit, the vector **OA** in Figure 8.18a might represent a force of 150 pounds in a direction making an angle of $30°$ with the positive x-axis. With a scale of 100 miles per hour to the unit, the vector **OP** in Figure 8.16b represents a velocity of 400 miles per hour at an angle of $45°$ with the positive y-axis (this angle, called the **heading,** is measured clockwise from north to the vector).

We shall represent a vector by a boldface letter such as **v,** and its length will be designated by $\|\mathbf{v}\|$.

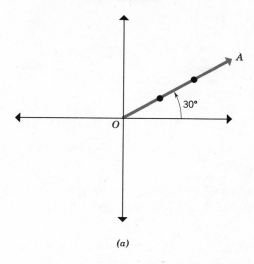

(a)

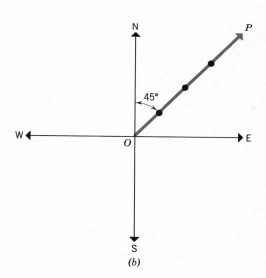

(b)

Figure 8.18

If two vectors v_1 and v_2 have the same initial point, O, the vector **sum** or the **resultant** of v_1 and v_2 is defined as the diagonal $v_1 + v_2$ of the parallelogram constructed with v_1 and v_2 as sides (Figure 8.19).

Often a problem requires the breaking down of a vector rather than the addition of two vectors. Whenever a vector is expressed as a resultant of two vectors, these two vectors are called **components** of the vector. The

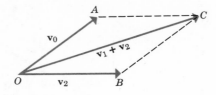

Figure 8.19

breaking down of a vector into its components is called a **resolution** of the vector into components.

EXAMPLE 1. A force of 100 pounds is acting in a direction of 30° with the ground. Find the horizontal and vertical components of the force.

SOLUTION. Let $\mathbf{F}_x$ and $\mathbf{F}_y$ represent the horizontal and vertical components, respectively, of the force. From the right triangle formed, we obtain

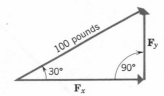

Figure 8.20

$$\sin 30° = \frac{\|\mathbf{F}_y\|}{100}$$

or
$$\|\mathbf{F}_y\| = 100 \sin 30°$$
$$= (100)(\tfrac{1}{2}) = 50$$

The vertical component of the force is 50 pounds. Also

$$\cos 30° = \frac{\|\mathbf{F}_x\|}{100}$$

or
$$\|\mathbf{F}_x\| = 100 \cos 30°$$
$$= (100)(0.8660)$$
$$= 86.60$$

The horizontal component of the force is 86.60 pounds.

EXAMPLE 2. An airplane has a heading of 180° and an airspeed (AS) of 300 miles per hour. A wind of 25 miles per hour is blowing in the direction of 270°. Find the course correct to the nearest 10′ and the groundspeed (GS) of the plane.

SOLUTION. In navigation, directions are given by angles measured from true north in a clockwise direction from 0° to 360°. These angles are regarded as positive. The airspeed of 200 miles per hour means that should the plane fly through still air it would go 200 miles per hour. The 25 miles per hour wind is blowing due west. The vector representing the actual motion of the plane in relation to the ground is the resultant of the airspeed vector, magnitude 200 miles per hour and direction 180° (due south), and the wind vector, magnitude 25 miles per hour and direction 270° (due west). The magnitude of the resultant gives the ground speed of the plane, that is, the speed of the plane in relation to the ground, and its direction gives the course of the plane. The angle D in the figure which is the difference between the heading and the course is called the **drift angle.**

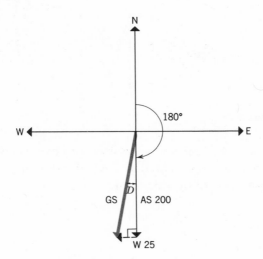

Figure 8.21

From the figure we observe

$$\tan D = \tfrac{25}{200}$$
$$= 0.1250$$

from which we have

$$D \doteq 7° \ 10′ \qquad \text{(correct to the nearest 10′)}$$

The course is $180° + 7° \ 10′ \doteq 187° \ 10′$.

Also, $\sec 7° \ 10' = \dfrac{\text{GS}}{200}$

and $\sec 7° \ 10' = 1.008$

from which we obtain

$$\begin{aligned} \text{GS} &= (200)(\sec 7° \ 10') \\ &= (200)(1.008) = 201.6 \end{aligned}$$

The groundspeed is approximately 202 miles per hour.

EXAMPLE 3. What is the minimum force required to prevent a 400-pound oil drum from rolling down a plane that makes an angle of 15° with the horizontal?

SOLUTION. The weight or force of 400 pounds represented by $\overrightarrow{OR}$ acts

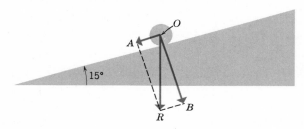

Figure 8.22

vertically downward and can be resolved into two components, one parallel to the plane represented by $\overrightarrow{OA}$ and the other perpendicular to $\overrightarrow{OA}$ and represented by $\overrightarrow{OB}$. Then

$$\sin \angle ORA = \frac{\|OA\|}{\|OR\|}$$

or

$$\begin{aligned} \|OA\| &= \|OR\| \sin \angle ORA \\ &= (400)(\sin 15°) \\ &= (400)(0.2588) \\ &= 103.52 \doteq 104 \end{aligned}$$

The force required to prevent the drum from rolling is 104 pounds (to the nearest pound).

EXERCISES 8.5

1. A child is pushing a toy boat by means of a stick which makes an angle of

22° 30′ with the horizontal. If the push along the stick is 8.25 pounds, what force tends to move the boat?

2. An automobile weighing 2500 pounds is parked on a hill which makes an angle of 10° with the horizontal. What force tends to pull the car down the hill?

3. An automobile traveling 55 miles per hour encounters rain that is falling straight down. Find the vertical speed of the raindrops if the streaks they form on the side windows make an angle of 41° with the vertical.

4. A cable that can withstand a tension of 10,000 pounds is used to pull automobiles up a ramp to the second floor of a storage garage. If the inclination of the ramp is 36° 16′, find the weight of the heaviest truck that can be pulled by the cable.

5. An airplane with cruising speed 178 miles per hour in still air is headed due north. If a west wind of 27.5 miles per hour is blowing, find the direction correct to the nearest 10′ and groundspeed of the plane.

6. A barrel weighing 150 pounds rests on an inclined plane. If a force of 40 pounds is needed to keep the barrel from rolling down, what angle does the plane make with the horizontal?

7. A boat is moving in the direction 36° 20′ at the rate of 15 miles per hour. A man walks across the deck at right angles to the boat's travel at the rate of 5.6 miles per hour. In what direction is the man traveling with respect to the surface of the water and what is the speed in that direction?

8. A billiard table whose dimensions are 36 inches by 72 inches has a ball resting against the midpoint of one end. Where should the ball be made to strike a side so as to be deflected to the opposite corner?

9. A boat capable of a speed of 528 feet per minute in still water is at the west bank of a river that flows south at 115 feet per minute. In what direction must the boat be steered if it is to land at a pier due east of the starting point? How long will the trip take if the river is 600 feet wide at this point?

10. In opening a vertical window from the top by means of a window pole, a man pulls with a force of 50 pounds. To the nearest pound, what part of the man's force lowers the window if the pole makes an angle of 25° with the window?

11. Find the force necessary to pull a loaded trailer weighing 1500 pounds up a grade of 32°.

12. A ship sails north at 18 miles per hour and a man walks across the deck toward the east at 4 miles per hour. What are his speed and direction with reference to the earth?

CHAPTER REVIEW

1. Given the right triangle ABC with angle $A = 56°$, angle C the right angle, and $c = 80$, solve the triangle.

2. Given the oblique triangle ABC with $b = 40$, $c = 50$, and $A = 36°$, find a.
3. Given the oblique triangle ABC with $a = 7$, $b = 8$, and $c = 13$, find C.
4. Given the triangle ABC with $a = 3333$, $b = 4567$, and $A = 41° 52'$, solve the triangle.
5. Given the triangle ABC with $a = 846.15$, $b = 900.48$, and $A = 70° 12.3'$, solve the triangle.
6. A helicopter pilot and his copilot are flying at an altitude of 300 feet directly over a tower. The pilot measures the angle of depression of a tower due east to be $30°$ while the copilot measures the angle of depression of a tower due west to be $40°$. How far apart are the two towers?
7. The resultants of forces of 50 pounds and 20 pounds is a force of 60 pounds. Find the angles the resultant makes with each of its components.
8. A ship heads due east from a dock at a speed of 18 miles per hour. After traveling 30 miles it turns due south and continues at the same speed. Find the distance and bearing from the dock after 5 hours.
9. In an hour how far north will an airplane fly if its airspeed and heading are 600 miles per hour at $40°$, and if winds of 120 miles per hour are blowing in (not from) the direction $330°$?
10. If a ball is thrown due east at 15 yards per second from a car that is traveling due north at the rate of 20 yards per second, find the velocity of the ball and the direction of its path.

CHAPTER 9

Graphs of the Trigonometric Functions

9.1 ◆ The Trigonometric Functions

In Chapter 6 we defined the trigonometric functions: cosine, sine, tangent, secant, cosecant, and cotangent. For all of the trigonometric functions, the domain is the set of all real numbers, θ, for which the corresponding trigonometric ratio for an angle of θ radians exists. Thus, each trigonometric function is a set of ordered pairs of numbers with a number θ in the domain and the corresponding trigonometric ratio of an angle of θ radians in the range. These are tabulated in Table 9.1.

Table 9.1

Function	Ordered Pair	Domain
Cosine	$(\theta, \cos \theta)$	R, the set of real numbers
Sine	$(\theta, \sin \theta)$	R, the set of real numbers
Tangent	$(\theta, \tan \theta)$	$\{\theta \mid \theta \neq \frac{\pi}{2} + n\pi, n = 0, \pm 1, \pm 2, \ldots\}$
Secant	$(\theta, \sec \theta)$	$\{\theta \mid \theta \neq \frac{\pi}{2} + n\pi, n = 0, \pm 1, \pm 2, \ldots\}$
Cosecant	$(\theta, \csc \theta)$	$\{\theta \mid \theta \neq n\pi, n = 0, \pm 1, \pm 2, \ldots\}$
Cotangent	$(\theta, \cot \theta)$	$\{\theta \mid \theta \neq n\pi, n = 0, \pm 1, \pm 2, \ldots\}$

9.2 ◆ Periodicity of the Trigonometric Functions

Since coterminal angles have the same trigonometric ratios, we know that any given trigonometric ratio of an angle θ is exactly equal to the same ratio of the angle $\theta + 2n\pi$, where n is any integer, positive or negative. For example,

$$\sin \frac{\pi}{12} = \sin \frac{25\pi}{12} = \sin \frac{49\pi}{12} = \sin \left(- \frac{23\pi}{12} \right)$$

and so forth. Thus, $\sin \theta$ takes on all its possible values as θ varies from 0 to 2π. It then repeats these values as θ varies from 2π to 4π, from 4π to 6π, and so forth. Since

$$\sin \theta = \sin (\theta + 2n\pi) \qquad n = 0, \pm 1, \pm 2, \ldots$$

we see that the sine function, that is, the set of all ordered pairs of the form $(\theta, \sin \theta)$, is a periodic function with period 2π.

Since it is also true that

$$\cos \theta = \cos (\theta + 2n\pi) \qquad n = 0, \pm 1, \pm 2, \ldots$$

the cosine function is also a periodic function with period 2π.

We also saw in Chapter 6 that the tangent function had period π. Thus

$$\tan \theta = \tan (\theta + n\pi) \qquad n = 0, \pm 1, \pm 2, \ldots$$

With the exception of those values of θ for which the given trigonometric ratios do not exist, we have

$$\sec \theta = \frac{1}{\cos \theta}, \; \csc \theta = \frac{1}{\sin \theta}, \; \cot \theta = \frac{1}{\tan \theta}$$

Thus the cosecant and secant functions must have the same period, 2π, as the sine and cosine functions and the cotangent function must have the same period, π, as the tangent function.

EXAMPLE 1. Find the period of the function specified by $y = \cos 2x$.

SOLUTION. Since the cosine function has period 2π, the values of $\cos z$ will pass through a complete period when $0 \leq z \leq 2\pi$. Thus the values of the function specified by $y = \cos 2x$ will pass through a complete period when

$$0 \leq 2x \leq 2\pi$$

which is equivalent to

$$0 \leq x \leq \pi$$

Hence the period of this function is π.

EXAMPLE 2. Find the period of the function specified by $y = \tan \dfrac{x}{4}$.

SOLUTION. Since the tangent function has period π, the values of $\tan z$ will pass through a complete period when $0 \leq z \leq \pi$. Thus the values of the function specified by $y = \tan \dfrac{x}{4}$ will pass through a complete period when

$$0 \le \frac{x}{4} \le \pi$$

which is equivalent to

$$0 \le x \le 4\pi$$

Hence the period of this function is 4π.

The above examples may be generalized in the theorem below which we accept without proof.

THEOREM 9.1. If a function f is periodic with period p and if the function g is specified by the equation $g(x) = f(ax)$, where $a \ne 0$, then g is a periodic function with period $\dfrac{p}{|a|}$.

EXAMPLE 3. Find the period of the function specified by $y = \cos (x - 1)$.

SOLUTION. If, in a function $f(x)$, the variable x is replaced by $x - k$, the effect on the graph of $f(x)$ is a translation of the curve an amount k parallel to the x-axis and in a positive direction. For, if (x_1, y_1) are the coordinates of a point of the graph of $y = f(x)$, then the point $(x_1 + k, y_1)$ that results from translating (x_1, y_1) an amount k in the positive x direction, will be on the graph of $y = f(x - k)$. If $k > 0$, the translation is to the right, and if $k < 0$, the translation is to the left.

We see, then, that the period of the function specified by $y = \cos (x - 1)$ is exactly the same as the period of the function specified by $y = \cos x$. In fact, the graphs of the two functions will be exactly the same but the graph of $y = \cos(x - 1)$ will be translated one unit in the positive direction.

The period of the given function is 2π.

We now state Theorem 9.2 without proof.

THEOREM 9.2. If b is a positive real number, the functions specified by $y = \sin (bx + c)$, $y = \cos (bx + c)$, $y = \sec (bx + c)$ and $y = \csc (bx + c)$ will have the period $\dfrac{2\pi}{b}$. The functions specified by $y = \tan (bx + c)$ and $y = \cot (bx + c)$ will both have period $\dfrac{\pi}{b}$.

The graphs of the functions specified by the given equations may be obtained by translating the graphs of $y = \sin bx$, $y = \cos bx$, $y = \tan bx$, $y = \cot bx$, $y = \sec bx$ and $y = \csc bx$, respectively, an amount $\dfrac{c}{b}$ in the negative x-direction.

EXAMPLE 4. Find the period of the function specified by $y = \sin\left(3x + \dfrac{\pi}{4}\right)$.

SOLUTION. Applying Theorem 9.2 with $b = 3$ and $c = \dfrac{\pi}{4}$ we find the period of the given function is

$$\frac{2\pi}{b} = \frac{2\pi}{3}$$

The graph of the given function is obtained by translating the graph of $y = \sin 3x$ $\dfrac{\frac{\pi}{4}}{3} = \dfrac{\pi}{12}$ units in the negative x-direction.

EXERCISES 9.1

In Exercises 1–12 find the values of each of the following.

1. $\cos \dfrac{7\pi}{4}$

2. $\sin \dfrac{5\pi}{4}$

3. $\tan \dfrac{2\pi}{3}$

4. $\sec \left(-\dfrac{1}{4}\pi\right)$

5. $\csc \dfrac{13\pi}{6}$

6. $\cot \left(-\dfrac{5\pi}{4}\right)$

7. $\sin \left(\dfrac{13\pi}{6}\right)$

8. $\cos \left(\dfrac{13\pi}{3}\right)$

9. $\tan \left(-\dfrac{39\pi}{4}\right)$

10. $\sec \left(\dfrac{52\pi}{3}\right)$

11. $\csc \left(-\dfrac{37\pi}{4}\right)$

12. $\cot \left(\dfrac{11\pi}{3}\right)$

Find the periods of the functions specified by each of the following (Exercises 13–20).

13. $y = 2 \sin x$

14. $y = \sin 2x$

15. $y = \cos (x - \pi)$

16. $y = \cos 3(x - \pi)$

17. $y = \tan 3x$

18. $y = \cot \dfrac{x}{4}$

19. $y = \sec (5x + 1)$

20. $y = \csc \dfrac{x + 1}{3}$

9.3 ◆ Behavior of the Trigonometric Functions

We now consider how the values of the trigonometric functions vary with the values of θ. Since we know from the preceding section that all six of the trigonometric functions are periodic, it will be sufficient to study their behaviors over one complete period.

In what follows it is understood that whenever reference is made to an angle θ, we mean an angle of radian measure θ. If we draw a unit circle with center at the origin and choose a point P of this circle that is on the terminal side of the angle θ in standard position, the definition of the functions cosine and sine tells us that P has coordinates $(\cos \theta, \sin \theta)$. See Figure 9.1.

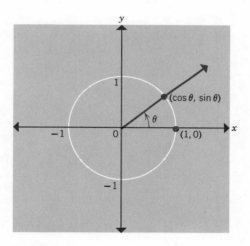

Figure 9.1

Using the special angles, those of degree measure 30°, 45°, and 60° $\left(\text{radian measure } \dfrac{\pi}{6}, \dfrac{\pi}{4}, \text{ and } \dfrac{\pi}{3}\right)$, and the fact that one complete period of the cosine function is 2π, we find values for the cosine, sine, and tangent functions in Table 9.2.

The table shows the values of the cosine, sine, and tangent functions for the special angles and for the axes angles. Since

$$\tan \theta = \frac{\sin \theta}{\cos \theta}$$

we see that for θ slightly less than $\dfrac{\pi}{2}$, $\sin \theta$ is positive and near 1, while

Table 9.2

θ	$\cos \theta$	$\sin \theta$	$\tan \theta$
0	1	0	0
$\dfrac{\pi}{6}$	0.866	0.5	0.577
$\dfrac{\pi}{4}$	0.707	0.707	1
$\dfrac{\pi}{3}$	0.5	0.866	1.732
$\dfrac{\pi}{2}$	0	1	—
$\dfrac{2\pi}{3}$	-0.5	0.866	-1.732
$\dfrac{3\pi}{4}$	-0.707	0.707	-1
$\dfrac{5}{6}\pi$	-0.866	0.5	-0.577
π	-1	0	0
$\dfrac{7}{6}\pi$	-0.866	-0.5	0.577
$\dfrac{5}{4}\pi$	-0.707	-0.707	1
$\dfrac{4}{3}\pi$	-0.5	-0.866	1.732
$\dfrac{3}{2}\pi$	0	-1	—
$\dfrac{7}{4}\pi$	0.5	-0.866	-1.732
$\dfrac{5}{3}\pi$	0.707	-0.707	-1
$\dfrac{11}{6}\pi$	0.866	-0.5	-0.577
2π	1	0	0

$\cos \theta$ is positive and near 0. Thus, for such values of θ, $\tan \theta$ is positive and very large. We express this in symbols by

$$\tan \left(\frac{\pi}{2}\right)^{-} = +\infty$$

For values of θ slightly larger than $\dfrac{\pi}{2}$, $\sin \theta$ is positive and near 1, while $\cos \theta$ is negative and near 0. Thus, for such values of θ, $\tan \theta$ is negative with very large absolute value. We express this in symbols by

$$\tan \left(\frac{\pi}{2}\right)^{+} = -\infty$$

The behavior of the six trigonometric functions are given in Table 9.3. We use an upward arrow ($\nearrow$) to indicate "increasing to" and a downward arrow ($\searrow$) to indicate "decreasing to."

Table 9.3

θ	$Q = \mathrm{I}$ $0 \nearrow \frac{\pi}{2}$	$Q = \mathrm{II}$ $\frac{\pi}{2} \nearrow \pi$	$Q = \mathrm{III}$ $\pi \nearrow \frac{3}{2}\pi$	$Q = \mathrm{IV}$ $\frac{3}{2}\pi \nearrow 2\pi$
$\cos\theta$	$1 \searrow 0$	$0 \searrow -1$	$-1 \nearrow 0$	$0 \nearrow 1$
$\sin\theta$	$0 \nearrow 1$	$1 \searrow 0$	$0 \searrow -1$	$-1 \nearrow 1$
$\tan\theta$	$0 \nearrow +\infty$	$-\infty \nearrow 0$	$0 \nearrow +\infty$	$-\infty \searrow 0$
$\sec\theta$	$1 \nearrow +\infty$	$-\infty \nearrow -1$	$-1 \searrow -\infty$	$+\infty \searrow 1$
$\csc\theta$	$+\infty \searrow 1$	$1 \nearrow +\infty$	$-\infty \nearrow -1$	$-1 \searrow -\infty$
$\cot\theta$	$+\infty \searrow 0$	$0 \searrow -\infty$	$+\infty \searrow 0$	$0 \searrow -\infty$

EXERCISES 9.2

For what values of θ, in the interval described by $0 \leq \theta \leq 2\pi$, is:

1. $\cos\theta$ increasing?
2. $\sin\theta$ increasing?
3. $\sin\theta$ and $\cos\theta$ both increasing?
4. $\tan\theta$ increasing?
5. $\csc\theta \leq 1$?
6. $\sec\theta \leq 1$?
7. $\sec\theta \geq 1$?
8. $\csc\theta \geq 1$?
9. $\tan\theta \geq 0$?
10. $\cot\theta \leq 0$?
11. $\sec\theta \leq -1$?
12. $\csc\theta \leq -1$?
13. $\cot\theta$ decreasing?
14. $\sec\theta$ decreasing?
15. $\csc\theta$ decreasing?

9.4 ◆ The Graphs of the Trigonometric Functions

Now let us consider the function specified by the equation

$$(1) \qquad\qquad y = \sin x$$

The graph of this function consists of the set of all points with coordinates (x, y) which satisfy Equation 1. Here x and y are real numbers, and in this function x is a given real number value and the corresponding value of y is found by obtaining the sine of an angle of radian measure x. The domain of the function is the set R of real numbers.

Using Table I we obtain the values for x and y given in Table 9.4.

The points (x, y) with $y = \sin x$ from Table 9.3 are plotted and joined by a smooth curve to give the graph in Figure 9.2. Because the sine function

Table 9.4

x	$y = \sin x$
0	0
$\dfrac{\pi}{6}$	0.50
$\dfrac{\pi}{4}$	0.71
$\dfrac{\pi}{3}$	0.87
$\dfrac{\pi}{2}$	1
$\dfrac{2\pi}{3}$	0.87
$\dfrac{3\pi}{4}$	0.71
$\dfrac{5\pi}{6}$	0.50
π	0
$\dfrac{7\pi}{6}$	-0.50
$\dfrac{5\pi}{4}$	-0.71
$\dfrac{4\pi}{3}$	-0.87
$\dfrac{3\pi}{2}$	-1
$\dfrac{5\pi}{3}$	-0.87
$\dfrac{7\pi}{4}$	-0.71
$\dfrac{11\pi}{6}$	-0.50
2π	0

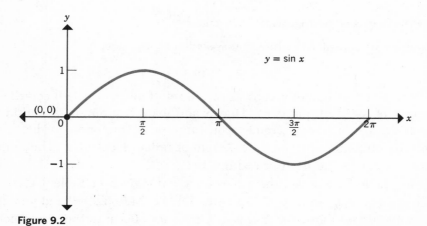

Figure 9.2

is periodic with period 2π, the part of the graph in Figure 9.2 may be repeated indefinitely to the right and to the left to give the complete graph.

The reader should practice drawing the portion of the graph shown in Figure 9.2 until he can make a hasty sketch of it from memory. This graph is called a **sine curve** or a **sinusoid.** The maximum value of the function is called the **amplitude** of the sine curve. The period of the function is called the **wavelength** of the sine curve.

A similar procedure with the function specified by

$$y = \cos x$$

results in the graph shown in Figure 9.3.

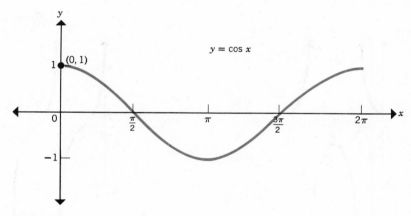

Figure 9.3

By using the same method as we used to graph the sine function, we can draw graphs of the other trigonometric functions. These graphs are shown in Figure 9.4 (see page 274).

EXAMPLE 1. Construct one period of the graph of the function specified by $y = 3 \sin x$.

SOLUTION. Using the values of the sine function for special angles, we have Table 9.5 (see page 275).

Plotting the ordered pairs $(x, 3 \sin x)$ and connecting them by a smooth curve gives the graph shown in Figure 9.5 (see page 276).

EXAMPLE 2. Construct one period of the graph of the function specified by

$$y = \cos 2x$$

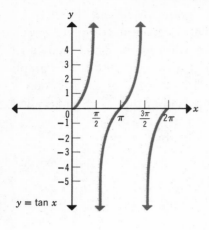

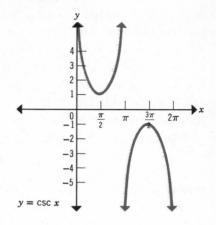

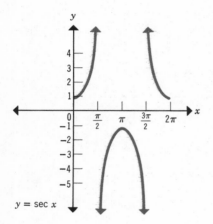

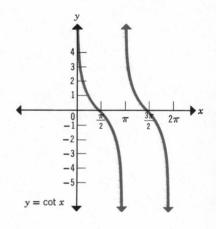

Figure 9.4

SOLUTION. Since the cosine function has period 2π we shall have one complete period for this function when

$$0 \leq 2x \leq 2\pi$$

which is equivalent to

$$0 \leq x \leq \pi$$

A table of values will be:

x	0	$\dfrac{\pi}{4}$	$\dfrac{\pi}{2}$	$\dfrac{3}{4}\pi$	π
$2x$	0	$\dfrac{\pi}{2}$	π	$\dfrac{3}{2}\pi$	2π
$\cos 2x$	1	0	-1	0	1

Table 9.5

x	$\sin x$	$3 \sin x$
0	0	0
$\dfrac{\pi}{6}$	0.5	1.5
$\dfrac{\pi}{4}$	0.707	2.121
$\dfrac{\pi}{3}$	0.866	2.598
$\dfrac{\pi}{2}$	1	3
$\dfrac{2\pi}{3}$	0.866	2.598
$\dfrac{3\pi}{4}$	0.707	2.121
$\dfrac{5\pi}{6}$	0.5	1.5
π	0	0
$\dfrac{7\pi}{6}$	-0.5	-1.5
$\dfrac{5\pi}{4}$	-0.707	-2.121
$\dfrac{4\pi}{3}$	-0.866	-2.598
$\dfrac{2\pi}{3}$	-1	-3
$\dfrac{5\pi}{3}$	-0.866	-2.598
$\dfrac{7\pi}{4}$	-0.707	-2.121
$\dfrac{11\pi}{6}$	-0.5	-1.5
2π	0	0

Plotting the ordered pairs $(x, \cos 2x)$ and connecting them with a smooth curve gives the graph shown in Figure 9.6 (see page 276).

EXERCISES 9.3

Construct the graph of the function specified by each equation below in the intervals given.

1. $y = \sin x$; $-2\pi \leq x \leq 2\pi$
2. $y = \cos x$; $-2\pi \leq x \leq 2\pi$
3. $y = \sec x$; $-2\pi \leq x \leq 2\pi$
4. $y = \cot x$; $-2\pi \leq x \leq 2\pi$

5. $y = \tan x$; $-2\pi \leq x \leq 2\pi$
6. $y = \csc x$; $-2\pi \leq x \leq 2\pi$
7. $y = \sin (-x)$; $-2\pi \leq x \leq 2\pi$
8. $y = \cos (-x)$; $-2\pi \leq x \leq 2\pi$

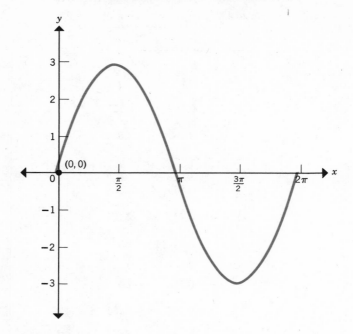

Figure 9.5

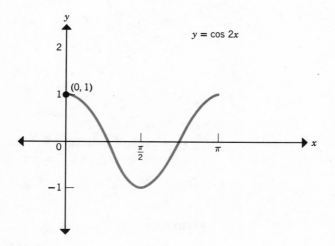

Figure 9.6

9. $y = \tan(-x)$; $-2\pi \leq x \leq 2\pi$
10. $y = 2 \sin x$; $-2\pi \leq x \leq 2\pi$
11. $y = \sin 2x$; $-\pi \leq x \leq \pi$
12. $y = \cos \dfrac{1}{2} x$; $-\dfrac{3\pi}{4} \leq x \leq \dfrac{3\pi}{4}$
13. $y = 3 \sin 3x$; $-\pi \leq x \leq \pi$
14. $y = \tan 2x$; $-\pi \leq x \leq \pi$

16. $y = \csc 3x$; $-\pi \leq x \leq \pi$
17. $y = \sec 2x$; $-\pi \leq x \leq \pi$
18. $y = 3 \cot \dfrac{x}{4}$; $-2\pi \leq x \leq 2\pi$
19. $y = \cos\left(x + \dfrac{\pi}{4}\right)$; $-2\pi \leq x \leq \pi$
20. $y = \cot(-2x)$; $-\pi \leq x \leq \pi$

CHAPTER REVIEW

1. Find the values of each of the following.

(a) $\sin \dfrac{\pi}{4}$
(b) $\cos \dfrac{11\pi}{6}$
(c) $\cot \dfrac{2\pi}{3}$
(d) $\csc \dfrac{43\pi}{4}$
(e) $\csc\left(-\dfrac{5\pi}{6}\right)$
(f) $\sec \dfrac{13\pi}{6}$
(g) $\cot\left(-\dfrac{5\pi}{4}\right)$
(h) $\csc\left(-\dfrac{5\pi}{3}\right)$

2. Find the periods of the functions specified by the equations given below.

(a) $y = 2 \cos x$
(b) $y = \tan 3x$
(c) $y = \cot \frac{1}{4}x$
(d) $y = \sin \frac{1}{2}x$
(e) $y = \cos(x + \pi)$
(f) $y = \sin(x - \pi)$
(g) $y = \tan(x + 1)$
(h) $y = \csc \dfrac{x + 1}{2}$

3. Sketch the graph of the function specified by $y = 2 \cos x$; $-\pi \leq x \leq \pi$
4. Sketch the graph of the function specified by $y = 3 \sin \frac{1}{2}x$; $-2\pi \leq x \leq 2\pi$
5. Sketch the graph of the function specified by $y = \tan 3x$; $-\dfrac{\pi}{2} < x < \dfrac{\pi}{2}$
6. Sketch the graph of the function specified by $y = 2 \cot \frac{1}{2}x$; $0 < x < 2\pi$
7. Sketch the graph of the function specified by $y = \sec 3x$; $-\pi \leq x \leq \pi$
8. Sketch the graph of the function specified by $y = 2 \csc \frac{1}{4}x$; $-4\pi \leq x \leq 4\pi$
9. Sketch the graph of the function specified by $y = 4 \tan 5x$; $-\dfrac{3\pi}{10} \leq x \leq \dfrac{3\pi}{10}$
10. Sketch the graph of the function specified by $y = \cot(-x)$; $-\pi \leq x \leq \pi$
11. Sketch the graph of the function specified by $y = 2 \sin(-2x)$; $-\pi \leq x \leq \pi$
12. Sketch the graph of the function specified by $y = 3 \cot(-\frac{1}{4}x)$; $-2\pi \leq x \leq 2\pi$
13. Sketch the graph of the function specified by $y = \frac{1}{2} \tan 2x$; $0 \leq x \leq \pi$
14. Sketch the graph of the function specified by $y = 2 \sec(-3x)$; $0 \leq x \leq \dfrac{2\pi}{3}$
15. Sketch the graph of the function specified by $y = \cos(-\frac{1}{4}x)$; $0 \leq x \leq 4\pi$

CHAPTER 10

Inverse Trigonometric Functions

10.1 ◆ Inverse Functions

In Section 5.2 we defined the inverse relation, S^{-1}, of a relation S, as the relation obtained when the components of each ordered pair of S are interchanged. Let us consider the relations f and g:

$$f = \{(1, 2), (2, 4), (3, 6), (4, 8)\},$$
$$g = \{(1, 3), (2, 5), (3, 5), (4, 7)\}.$$

The inverse relations of f and g are:

$$f^{-1} = \{(2, 1), (4, 2), (6, 3), (8, 4)\}$$
$$g^{-1} = \{(3, 1), (5, 2), (5, 3), (7, 4)\}$$

We observe that both relations f and g are functions but that only f^{-1} is a function, called the **inverse function** of f. We recall that although *every function has an inverse relation, not every function has an inverse function.*

DEFINITION 10.1. If f is a function with the property that no two of its ordered pairs have the same second component, then the **inverse function** of f, denoted by f^{-1}, is the set of ordered pairs obtained by interchanging the order of the components of every ordered pair of f.

EXAMPLE 1. Determine whether the function

$$f = \{(3, 1), (4, 5), (6, 7), (9, 8)\}$$

has an inverse function.

SOLUTION. Interchanging the order of the components of the ordered pairs of f yields

$$\{(1, 3), (5, 1), (7, 6), (8, 9)\}$$

Since no two of these ordered pairs have the same first component and different second components, this set of ordered pairs is a function, which is the inverse function of f. Thus, f has an inverse function.

EXAMPLE 2. Determine whether the function

$$g = \{(x, y) \mid y = x^2 + 1\}$$

has an inverse function.

SOLUTION. Solving the equation $y = x^2 + 1$ explicitly for x, we have

$$x^2 = y - 1$$

which is equivalent to

$$x = \sqrt{y - 1} \qquad \text{or} \qquad x = -\sqrt{y - 1}$$

Since every value of y $(y > 1)$ gives two distinct values for x, the function g does *not* have an inverse function. We could also determine this by noting that the ordered pairs $(-1, 2)$ and $(1, 2)$ are elements of the function g. Thus the inverse relation of g will contain the ordered pairs $(2, -1)$ and $(2, 1)$, so that this inverse relation cannot be a function.

EXERCISES 10.1

For each function f, state whether or not it has an inverse function f^{-1} (Exercises 1–5).

1. $f = \{(1, 3), (3, 7), (5, 8), (9, 11), (11, 17)\}$.
2. $f = \{(4, 7), (5, 7), (9, 3), (8, -1), (9, 0), (1, 1)\}$.
3. $f = \{(-1, 0), (0, 0), (1, 0), (2, 0)\}$.
4. $f = \{(-3, 9), (-2, 4), (-1, 1), (0, 0)\}$.
5. $f = \{(0, 0), (1, 1), (2, 4), (-2, 4), (3, 3)\}$.

For the functions specified by the equations below, state whether or not each has an inverse function.

6. $y = x + 3$ 9. $y = x^2 + 6$
7. $y = 2x - 7$ 10. $y = x^3$
8. $y = x^2 - 3$

10.2 ◆ Inverse Trigonometric Relations

For each trigonometric function we can construct an inverse relation. Thus, for the sine function,

$$\text{sine} = \{(t, \sin t)\}$$

the inverse relation is the set of ordered pairs,

$$\{(\sin t, t)\}$$

Since we like to express the second component in terms of the first, we need a way of writing a general solution of the equation

$$w = \sin t$$

expressing t in terms of w. For this we write

$$t = \text{arc sin } w$$

which is read "t is equal to the arc sine of w." Thus we have the relation

$$\text{arc sine} = \{(x, \text{arc sin } x)\}$$

Since

$$y = \text{arc sin } x$$

is equivalent to

$$x = \sin y$$

we see that y can be any real number, while x is a value of the sine function. Thus the domain of this relation is $\{x \mid -1 \leq x \leq 1\}$. Because of the periodicity of the sine function, there are infinitely many values of arc sin x for any given value of x. For example, arc sin $\dfrac{1}{2}$ has values $\dfrac{\pi}{6}$, $\dfrac{5\pi}{6}$, $\dfrac{13\pi}{6}$, $-\dfrac{7\pi}{6}$, and so forth. We say that arc sin x is **multiple-valued.**

We obtain information concerning the equation $y = $ arc sin x from our knowledge of the sine function. We can interpret arc sin x as "angle whose sine is x" by using radian measure for the angle. The equivalent equations for each of the trigonometric functions are shown in Table 10.1.

Table 10.1

$x = \sin y$	$y = \text{arc sin } x$
$x = \cos y$	$y = \text{arc cos } x$
$x = \tan y$	$y = \text{arc tan } x$
$x = \csc y$	$y = \text{arc csc } x$
$x = \sec y$	$y = \text{arc sec } x$
$x = \cot y$	$y = \text{arc cot } x$

We now draw the graph of the inverse trigonometric relation specified by the equation

$$y = \arc \sin x$$

To do this we need merely draw the graph of

$$x = \sin y$$

and interchange the order of the components of the ordered pairs of the sine function, giving the graph in Figure 10.1.

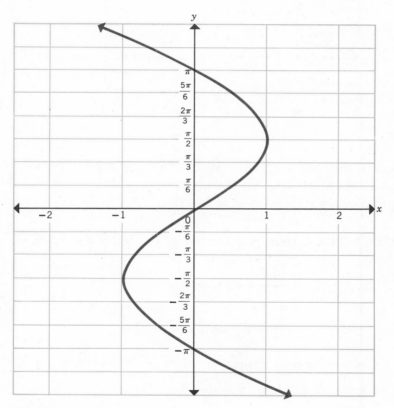

Figure 10.1

Observing the graph, we see that the inverse relation of the sine function is *not* a function. Hence we see that the sine function does not have an inverse function.

The graphs of the other inverse trigonometric relations may be obtained in a similar manner and are shown in Figure 10.2.

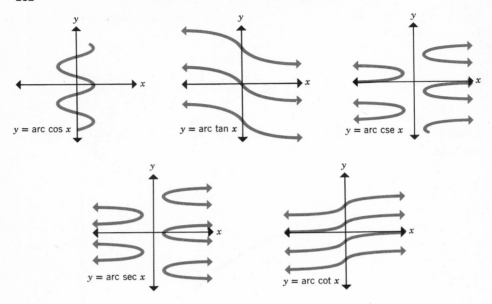

Figure 10.2

Observing the graphs in Figure 10.2, we see that none of the inverse relations of the trigonometric functions is a function. Thus, none of the trigonometric functions has an inverse function.

EXAMPLE 1. Find the least positive value of y such that $y = \text{arc sin} \dfrac{\sqrt{3}}{2}$.

SOLUTION. $y = \text{arc sin} \dfrac{\sqrt{3}}{2}$ is equivalent to

$$\sin y = \frac{\sqrt{3}}{2}$$

Examining the graph of the sine function shows that the least positive
value of y for which $\sin\ y = \dfrac{\sqrt{3}}{2}$ is $\dfrac{\pi}{3}$.

EXAMPLE 2. Solve the equation $y = 2 \sin \dfrac{x}{2}$ for x in terms of y.

SOLUTION. $y = 3 \sin \dfrac{x}{2}$ is equivalent to

$$\frac{y}{3} = \sin \frac{x}{2}$$

$$\frac{x}{2} = \text{arc sin } \frac{y}{3}$$

$$x = 2 \text{ arc sin } \frac{y}{3}$$

EXERCISES 10.2

Find the least nonnegative value of each of the following (Exercises 1–10).

1. arc sin $\frac{1}{2}$
2. arc tan 1
3. arc cos $\left(-\frac{1}{2}\right)$
4. arc csc 1
5. arc tan 0
6. arc csc 2
7. arc tan (-1)
8. arc cos $\left(-\frac{\sqrt{3}}{2}\right)$
9. arc sin $\left(\frac{-1}{\sqrt{2}}\right)$
10. arc cos $\left(-\frac{1}{\sqrt{2}}\right)$

11. arc sec 2
12. arc csc $\sqrt{2}$
13. arc cot (-1)
14. arc sin (-1)
15. arc cos 0
16. arc sec (-1)
17. arc sin $\left(-\frac{1}{2}\right)$
18. arc tan $\sqrt{3}$
19. arc cos (-1)
20. arc cot $\sqrt{3}$

Solve each of the following for x in terms of y (Exercises 21–25).

21. $y = \tan x$
22. $y = 2 \sin 3x$
23. $2y = \cot 3x$
24. $y + 2 = \cos 5x$
25. $y = 2 \text{ arc sin } x$

Sketch the graph of each of the following (Exercises 26–30).

26. $y = \text{arc cot } x$
27. $y = \text{arc sin } 2x$
28. $y = 2 \text{ arc cos } x$
29. $y = \text{arc sin } (x + 1)$
30. $y = 3 \text{ arc tan } 2x$

10.3 ⁕ Principal Values of the Inverse Relations

The fact that in the equation

$$y = \text{arc sin } x$$

y is multiple-valued tells us that the relation specified by this equation is *not* a function. If we restrict the allowable values of y in this equation so that to each value of x there will be one and only one value of y, then we will have constructed an inverse function of a particular subset of the sine function. We call this the **principal value** of the inverse trigonometric relation, or the **inverse trigonometric function.** We designate the function by capitalizing the initial letter of its name.

In a similar manner we define principal values of the other inverse trigonometric relations. The domains and ranges for these inverse functions are given in Table 10.2.

Table 10.2

Function	Domain	Range
$y = \text{Arc sin } x$	$-1 \leq x \leq 1$	$-\dfrac{\pi}{2} \leq y \leq \dfrac{\pi}{2}$
$y = \text{Arc cos } x$	$-1 \leq x \leq 1$	$0 \leq y \leq \pi$
$y = \text{Arc tan } x$	All real numbers	$-\dfrac{\pi}{2} < y < \dfrac{\pi}{2}$
$y = \text{Arc cot } x$	All real numbers	$0 < y < \pi$
$y = \text{Arc sec } x$	$x \geq 1 \quad \text{or} \quad x \leq -1$	$0 \leq y \leq \pi; y \neq \dfrac{\pi}{2}$
$y = \text{Arc csc } x$	$x \geq 1 \quad \text{or} \quad x \leq -1$	$-\dfrac{\pi}{2} \leq y \leq \dfrac{\pi}{2}; y \neq 0$

With these restrictions, the functions that are the principal values of the inverse trigonometric relations are:

$$\text{Arc sine} = \left\{ (x, y) \mid y = \text{Arc sin } x, \ -\frac{\pi}{2} \leq y \leq \frac{\pi}{2} \right\}$$

$$\text{Arc cosine} = \{ (x, y) \mid y = \text{Arc cos } x, \ 0 \leq y \leq \pi \}$$

$$\text{Arc tangent} = \left\{ (x, y) \mid y = \text{Arc tan } x, \ -\frac{\pi}{2} < y < \frac{\pi}{2} \right\}$$

$$\text{Arc cotangent} = \{ (x, y) \mid y = \text{Arc cot } x, \ 0 < y < \pi \}$$

$$\text{Arc secant} = \left\{ (x, y) \mid y = \text{Arc sec } x, \ 0 \leq y \leq \pi, \ y \neq \frac{\pi}{2} \right\}$$

$$\text{Arc cosecant} = \left\{ (x, y) \mid y = \text{Arc csc } x, \ -\frac{\pi}{2} \leq y \leq \frac{\pi}{2}, \ y \neq 0 \right\}$$

EXAMPLE 1. Find Arc cot $(-\sqrt{3})$

SOLUTION. Since

$$\cot 150° = \cot \frac{5\pi}{6} = -\sqrt{3}$$

and $$0 < \frac{5\pi}{6} < \pi$$

$$\text{Arc cot } (-\sqrt{3}) = \frac{5\pi}{6}$$

EXAMPLE 2. Find Arc sin 0.6820.

SOLUTION. Turning to Table I, we find in the column headed "Sin" the number 0.6820. Since this corresponds to $\sin 43° = \sin 0.7505$ and $-\frac{\pi}{2} \le 0.7505 \le \frac{\pi}{2}$, we have

$$\text{Arc sin } 0.6820 = 0.7505$$

EXAMPLE 3. Find sin (Arc sin x).

SOLUTION. Let

$$\text{Arc sin } x = \theta$$

Then

$$\sin (\text{Arc sin } x) = \sin \theta$$

But, by definition,

$$\text{Arc sin } x = \theta$$

is equivalent to

$$\sin \theta = x$$

Hence

$$\sin (\text{Arc sin } x) = x$$

EXAMPLE 4. Find sin (Arc cot $-\frac{12}{5}$).

SOLUTION. We interpret symbols such as sin (Arc cot $-\frac{12}{5}$) by using the definitions of the trigonometric ratios. This problem asks for the sine of an angle between 0 and π whose cotangent is $-\frac{12}{5}$:

$$\cot \theta = \frac{x}{y} = -\frac{12}{5}$$

Choosing $x = -12$, $y = 5$, we find

$$r = \sqrt{(-12)^2 + 5^2}$$
$$= \sqrt{144 + 25} = 13$$

Then

$$\sin (\text{Arc cot } -\tfrac{12}{5}) = \tfrac{5}{13}$$

EXERCISES 10.3

Find the values of the following. You may wish to use Table I (Exercises 1–10).

1. Arc sin $\frac{1}{2}$

2. Arc tan 1

3. Arc cos (-1)

4. Arc cos $-\dfrac{\sqrt{2}}{2}$

5. Arc sin $\left(-\dfrac{\sqrt{3}}{2}\right)$

6. Arc cos $(-\frac{1}{2})$

7. Arc tan $\left(-\dfrac{\sqrt{3}}{3}\right)$

8. Arc cot 2.747

9. Arc sec (-2.203)

10. Arc csc (-2.130)

Find the values of each of the following. (Exercises 11–20).

11. sin (Arc sin x)
12. tan (Arc tan x)
13. cos (Arc cos x)
14. sec (Arc sec x)
15. tan (Arc cos $\frac{1}{2}$)

16. sec (Arc tan x)
17. sec (Arc sin x)
18. cot (Arc cos x)
19. csc (Arc sin x)
20. sin (Arc sec x)

CHAPTER REVIEW

1. For each function, f, state whether or not it has an inverse function f^{-1}.
   (a) $f = \{(1, 2), (3, 5), (4, 6), (9, 10), (17, 11)\}$.
   (b) $f = \{(-2, 6), (-1, 5), (0, 0), (1, 5), (2, 6)\}$.
   (c) $f = \{(-2, 4), (-1, 1), (0, 0), (1, 1), (2, 4)\}$.
   (d) $f = \{(-1, 3), (-2, 0), (-1, 6), (0, 7), (1, 1), (2, 3)\}$.

2. For the functions specified by the equations below, state whether or not each has an inverse function.
   (a) $y = 3x + 5$ (c) $y = 3x^2 + 6$
   (b) $y = x^2 - 2$ (d) $y = 3x$

3. Find the least nonnegative value of each of the following.
   (a) arc cos $(-\frac{1}{2})$ (f) arc csc (-1)
   (b) arc tan 0   (g) arc cot 1
   (c) arc sin $(-\frac{1}{2})$ (h) arc sin $\dfrac{\sqrt{2}}{2}$
   (d) arc sin 1   (i) arc sec $\sqrt{2}$
   (e) arc cos $\dfrac{\sqrt{3}}{2}$ (j) arc sin (-1)

4. Sketch the graph of each of the following.
   (a) $y = $ arc cos x (b) $y = $ arc sin $(x + 1)$

5. Find the values of the following.

(a) Arc sin 0   (d) Arc sin $\dfrac{\sqrt{3}}{2}$

(b) Arc sec $\sqrt{2}$ (e) Arc sin $\left(-\dfrac{\sqrt{2}}{2}\right)$

(c) Arc cos $\dfrac{\sqrt{3}}{2}$

6. Find the values of the following.
   (a) cos (Arc tan x)   (c) sec (Arc tan x)
   (b) cos (Arc sin x)   (d) csc (Arc csc x)
7. Find the value of each of the following.
   (a) cos [Arc sin $(-\frac{2}{3})$]   (c) cos (Arc cos $2x$)
   (b) sin [Arc tan (-3)]   (d) cos $(\pi - \text{Arc sin } \frac{5}{6})$
8. Find the value of each of the following.
   (a) sin (Arc cos x + Arc cos y)   (b) cos (Arc sin x + Arc sin y)
9. Prove: Arc cos $(-\frac{1}{2}) = 2$ Arc cos $\frac{1}{2}$.

10. Prove: Arc sin x + Arc cos $x = \dfrac{\pi}{2}$, $0 \le x \le 1$.

Identify the following as true or false and give reasons for your answers.

11. Arc csc $x = \dfrac{1}{\text{Arc sin } x}$.

12. sec (Arc cos x) $= \dfrac{1}{x}$.

13. Arc sin $(-x) = -\text{Arc sin } x$.
14. Arc cos $(-x) = -\text{Arc cos } x$.

15. Arc cos $x = \dfrac{\pi}{2} - \text{Arc sin } x$.

CHAPTER 11

Determinants

11.1 ◆ Second Order Determinants

In mathematics it is frequently convenient to consider rectangular arrays of numbers such as

$$\begin{pmatrix} a_{11} & a_{12} & a_{13} & a_{14} \\ a_{21} & a_{22} & a_{23} & a_{24} \\ a_{31} & a_{32} & a_{33} & a_{34} \\ a_{41} & a_{42} & a_{43} & a_{44} \end{pmatrix}$$

Such a rectangular array of numbers is called a **matrix**. The numbers $a_{11}, a_{12}, \ldots, a_{44}$ are called the **elements** of the matrix. The first subscript of each element denotes the horizontal line in which the number occurs. These horizontal lines are called the **rows** of the matrix. The second subscript denotes the vertical line in which the number occurs. The vertical lines of numbers are called the **columns** of the matrix. Thus a_{34} is the element in the third row and the fourth column. In general, if a matrix has m rows and n columns, it is called an m by n matrix (written $n \times n$). If $m = n$, as in the matrix above, the matrix is called a **square matrix.** A square matrix of n rows and n columns is said to be of **order n.**

With each square matrix, we associate a number called the **determinant** of the matrix. For a matrix of order two

$$\begin{pmatrix} a_{11} & a_{12} \\ a_{21} & a_{22} \end{pmatrix}$$

the determinant is defined as $a_{11}a_{22} - a_{21}a_{12}$. This determinant is symbolized by enclosing the square array of numbers in the matrix by vertical lines,

$$\begin{vmatrix} a_{11} & a_{12} \\ a_{21} & a_{22} \end{vmatrix}$$

and is called a **second order determinant.** The value of this determinant can be remembered by referring to the following diagram:

$$\begin{vmatrix} a_{11} & a_{12} \\ a_{21} & a_{22} \end{vmatrix} = a_{11}a_{22} - a_{21}a_{12}$$

Notice that the first term in the value of the determinant is the product of the elements in the principal diagonal as indicated by the downward-sloping arrow and the second term is the opposite of the product of the elements in the secondary diagonal as indicated by the upward-sloping arrow.

EXAMPLE 1. Find the value of $\begin{vmatrix} 1 & 3 \\ 2 & 4 \end{vmatrix}$.

SOLUTON

$$\begin{vmatrix} 1 & 3 \\ 2 & 4 \end{vmatrix} = (1)(4) - (2)(3) = 4 - 6 = -2$$

EXAMPLE 2. Evaluate $\begin{vmatrix} -1 & -3 \\ 3 & 2 \end{vmatrix}$.

SOLUTION

$$\begin{vmatrix} -1 & -3 \\ 3 & 2 \end{vmatrix} = (-1)(2) - (3)(-3) = -2 + 9 = 7$$

EXERCISES 11.1

Evaluate the following determinants (Exercises 1–24).

1. $\begin{vmatrix} 2 & 3 \\ 4 & 5 \end{vmatrix}$

4. $\begin{vmatrix} -1 & 1 \\ 0 & -2 \end{vmatrix}$

2. $\begin{vmatrix} -1 & 2 \\ 0 & 1 \end{vmatrix}$

5. $\begin{vmatrix} 1 & -3 \\ 2 & 4 \end{vmatrix}$

3. $\begin{vmatrix} -1 & 2 \\ 3 & 0 \end{vmatrix}$

6. $\begin{vmatrix} 1 & 3 \\ -2 & -4 \end{vmatrix}$

7. $\begin{vmatrix} 1 & -3 \\ -2 & 4 \end{vmatrix}$ 16. $\begin{vmatrix} 1 & 2x \\ 3x & -1 \end{vmatrix}$

8. $\begin{vmatrix} -2 & 3 \\ 1 & -4 \end{vmatrix}$ 17. $\begin{vmatrix} x^2 & x \\ 1 & 2 \end{vmatrix}$

9. $\begin{vmatrix} -2 & -3 \\ 1 & -4 \end{vmatrix}$ 18. $\begin{vmatrix} x^2 & 1 \\ x & 2 \end{vmatrix}$

10. $\begin{vmatrix} -2 & 3 \\ -1 & -4 \end{vmatrix}$ 19. $\begin{vmatrix} x^2 & 2 \\ 1 & 3x \end{vmatrix}$

11. $\begin{vmatrix} 2 & -3 \\ -4 & -1 \end{vmatrix}$ 20. $\begin{vmatrix} 2 & x^2 \\ 3x & 1 \end{vmatrix}$

12. $\begin{vmatrix} -2 & -4 \\ -1 & 3 \end{vmatrix}$ 21. $\begin{vmatrix} 3x & 4x \\ 5x & x \end{vmatrix}$

13. $\begin{vmatrix} x & 2 \\ 3x & 1 \end{vmatrix}$ 22. $\begin{vmatrix} \cos \theta & \sin \theta \\ \sin \theta & \cos \theta \end{vmatrix}$

14. $\begin{vmatrix} x & 2x \\ 1 & 2 \end{vmatrix}$ 23. $\begin{vmatrix} \cos \theta & \sin \theta \\ -\sin \theta & \cos \theta \end{vmatrix}$

15. $\begin{vmatrix} 4 & x \\ 3 & 2x \end{vmatrix}$ 24. $\begin{vmatrix} \sin \theta & -\cos \theta \\ \cos \theta & \sin \theta \end{vmatrix}$

25. Find the solution set of $\begin{vmatrix} 2 & 3 \\ 4 & x \end{vmatrix} = 0$.

26. Find the solution set of $\begin{vmatrix} x-2 & 10 \\ x-2 & 4 \end{vmatrix} = 4(x-2)$.

11.2 ◆ Third Order Determinants

Now let us consider the third order matrix

(1) $$\begin{pmatrix} a_{11} & a_{12} & a_{13} \\ a_{21} & a_{22} & a_{23} \\ a_{31} & a_{32} & a_{33} \end{pmatrix}$$

The determinant of this matrix is written

$$\begin{vmatrix} a_{11} & a_{12} & a_{13} \\ a_{21} & a_{22} & a_{23} \\ a_{31} & a_{32} & a_{33} \end{vmatrix}$$

For each element of matrix 1 there exists a matrix of order two obtained by deleting the row and column in which this element lies. The second order determinant associated with this matrix is called the **minor** of the element under consideration. For example, the minor of a_{11}, denoted by A_{11}, of matrix 1 is

$$\begin{vmatrix} a_{22} & a_{23} \\ a_{32} & a_{33} \end{vmatrix}$$

the minor of element a_{12}, denoted by A_{12}, is

$$\begin{vmatrix} a_{21} & a_{23} \\ a_{31} & a_{33} \end{vmatrix} ; \text{ and so forth.}$$

The value of a third order determinant is defined by the following equation:

$$\begin{vmatrix} a_{11} & a_{12} & a_{13} \\ a_{21} & a_{22} & a_{23} \\ a_{31} & a_{32} & a_{33} \end{vmatrix} = a_{11}\begin{vmatrix} a_{22} & a_{23} \\ a_{32} & a_{33} \end{vmatrix} - a_{21}\begin{vmatrix} a_{12} & a_{13} \\ a_{32} & a_{33} \end{vmatrix} + a_{31}\begin{vmatrix} a_{12} & a_{13} \\ a_{22} & a_{23} \end{vmatrix}$$

which is called the **expansion** of the third order determinant in terms of the minors of the elements of the first column.

In the above definition, the elements of the first column and their minors were used. It can be shown that the minors of the elements of any column or row can be used in the expansion if the proper sign is placed before each term. Thus

$$a_{11}A_{11} - a_{21}A_{21} + a_{31}A_{31} = -a_{12}A_{12} + a_{22}A_{22} - a_{32}A_{32}$$
$$= a_{13}A_{13} - a_{23}A_{23} + a_{33}A_{33}$$
$$= a_{11}A_{11} - a_{12}A_{12} + a_{13}A_{13}$$
$$= -a_{21}A_{21} + a_{22}A_{22} - a_{23}A_{23}$$
$$= a_{31}A_{31} - a_{32}A_{32} + a_{33}A_{33}$$

The product $a_{ij}A_{ij}$ has the sign $(-1)^{i+j}$ attached to it.

EXAMPLE 1. Evaluate

$$\begin{vmatrix} 2 & 3 & 4 \\ 2 & -1 & 3 \\ 4 & -5 & 6 \end{vmatrix}$$

SOLUTION. We shall evaluate the given determinant in terms of the minors of the elements of the first row:

$$\begin{vmatrix} 2 & 3 & 4 \\ 2 & -1 & 3 \\ 4 & -5 & 6 \end{vmatrix} = 2 \begin{vmatrix} -1 & 3 \\ -5 & 6 \end{vmatrix} - 3 \begin{vmatrix} 2 & 3 \\ 4 & 6 \end{vmatrix} + 4 \begin{vmatrix} 2 & -1 \\ 4 & -5 \end{vmatrix}$$

$$= (2)(-6 + 15) - (3)(12 - 12) + 4(-10 + 4)$$
$$= (2)(9) - (3)(0) + (4)(-6)$$
$$= 18 - 0 - 24 = -6$$

EXAMPLE 2. Evaluate

$$\begin{vmatrix} \sin x & \cos x & \tan x \\ 0 & 0 & 1 \\ -\cos x & \sin x & \cot x \end{vmatrix}.$$

SOLUTION. We shall evaluate the given determinant in terms of the minors of the elements of the first row:

$$\begin{vmatrix} \sin x & \cos x & \tan x \\ 0 & 0 & 1 \\ -\cos x & \sin x & \cot x \end{vmatrix} = \sin x \begin{vmatrix} 0 & 1 \\ \sin x & \cot x \end{vmatrix} - \cos x \begin{vmatrix} 0 & 1 \\ -\cos x & \cot x \end{vmatrix}$$

$$+ \tan x \begin{vmatrix} 0 & 0 \\ -\cos x & \sin x \end{vmatrix}$$

$$= (\sin x)(-\sin x) - (\cos x)(\cos x) + (\tan x)(0)$$

$$= -\sin^2 x - \cos^2 x$$
$$= -(\sin^2 x + \cos^2 x)$$
$$= -1$$

An easier way to evaluate this determinant is to use the minors of the elements of the second row. Why? Then

$$\begin{vmatrix} \sin x & \cos x & \tan x \\ 0 & 0 & 1 \\ -\cos x & \sin x & \cot x \end{vmatrix} = -0 \begin{vmatrix} \cos x & \tan x \\ \sin x & \cot x \end{vmatrix} + 0 \begin{vmatrix} \sin x & \tan x \\ -\cos x & \cot x \end{vmatrix} - 1 \begin{vmatrix} \sin x & \cos x \\ -\cos x & \sin x \end{vmatrix}$$

$$= 0 + 0 - (\sin^2 x + \cos^2 x)$$
$$= -1$$

EXERCISES 11.2

Evaluate (Exercises 1–12).

1. $\begin{vmatrix} 2 & 3 & 7 \\ 5 & 9 & 8 \\ 0 & 3 & 0 \end{vmatrix}$

7. $\begin{vmatrix} 0 & 6 & -2 \\ -2 & -6 & 7 \\ 1 & 5 & 1 \end{vmatrix}$

2. $\begin{vmatrix} 6 & 7 & 5 \\ 0 & 8 & 6 \\ 4 & 9 & 3 \end{vmatrix}$

8. $\begin{vmatrix} 3 & 1 & 2 \\ 1 & -5 & -1 \\ 2 & 3 & 2 \end{vmatrix}$

3. $\begin{vmatrix} 6 & 4 & 3 \\ 1 & -2 & 1 \\ 1 & 2 & 3 \end{vmatrix}$

9. $\begin{vmatrix} 0 & 1 & 3 \\ 1 & 5 & 0 \\ 2 & 0 & 3 \end{vmatrix}$

4. $\begin{vmatrix} 2 & 4 & 3 \\ 1 & 0 & -2 \\ -3 & 1 & 1 \end{vmatrix}$

10. $\begin{vmatrix} 4 & 8 & -3 \\ 2 & -5 & 2 \\ 24 & 2 & 1 \end{vmatrix}$

5. $\begin{vmatrix} -1 & 5 & 2 \\ 4 & -1 & -2 \\ 7 & 6 & 3 \end{vmatrix}$

11. $\begin{vmatrix} 3 & -5 & 0 \\ 5 & 0 & -8 \\ 1 & 1 & 1 \end{vmatrix}$

6. $\begin{vmatrix} 0 & 6 & 0 \\ 1 & 2 & -5 \\ -2 & -1 & 4 \end{vmatrix}$

12. $\begin{vmatrix} 1 & 3 & 10 \\ 0 & -2 & 4 \\ 2 & 0 & 5 \end{vmatrix}$

13. Express $\begin{vmatrix} a_1 & b_1 & c_1 \\ a_2 & b_2 & c_2 \\ a_3 & b_3 & c_3 \end{vmatrix}$ in terms of the elements only.

14. Find the solution set of

$$\begin{vmatrix} -2 - x & 3 \\ 3 & -5 - x \end{vmatrix} = 0$$

15. Find the solution set of

$$\begin{vmatrix} 1-x & 2 & 0 \\ 0 & -2 & 3 \\ 2 & -1-x & -2 \end{vmatrix} = 0$$

16. Evaluate Vandermonde's determinant of order three:

$$\begin{vmatrix} 1 & 1 & 1 \\ x & y & z \\ x^2 & y^2 & z^2 \end{vmatrix}$$

17. Evaluate the skew-symmetric determinant:

$$\begin{vmatrix} 0 & a & b \\ -a & 0 & c \\ -b & -c & 0 \end{vmatrix}$$

11.3 ◆ Determinants of Order n

We now wish to define the general determinant of order n, which we designate as

$$\begin{vmatrix} a_{11} & a_{12} & \cdots & a_{1n} \\ a_{21} & a_{22} & \cdots & a_{2n} \\ \vdots & & & \vdots \\ a_{n1} & a_{n2} & \cdots & a_{nn} \end{vmatrix}$$

This determinant has n rows and n columns and n^2 elements. The element a_{ij} is the number in the ith row and the jth column. If the ith row and the jth column are deleted from the determinant, the resulting determinant of order $n-1$ is designated by A_{ij} and is called the **minor** of element a_{ij}.

We define the value of the determinant of order n as

$$\begin{vmatrix} a_{11} & a_{12} & \cdots & a_{1n} \\ a_{21} & a_{22} & \cdots & a_{2n} \\ \vdots & & & \vdots \\ a_{n1} & a_{n2} & \cdots & a_{nn} \end{vmatrix}$$

$$= a_{11}A_{11} - a_{21}A_{21} + a_{31}A_{31} - a_{41}A_{41} + \cdots + (-1)^{n+1}a_{n1}A_{n1}$$

This is called the **expansion** of the determinant of order n in terms of the minors of the elements of the first column. In the above definition the expansion of the nth order determinant was given in terms of the minors of the elements of the first column of the determinant. In general, it can be shown that in the expansion of a determinant of order n, the elements of any row or column together with their minors may be used with the proper sign attached to each product. The product $a_{ij}A_{ij}$ has the sign $(-1)^{i+j}$ assigned to it. The minor A_{ij} together with the proper sign is called the **cofactor** of a_{ij} and is usually denoted by C_{ij}. Thus $C_{ij} = (-1)^{i+j}A_{ij}$.

EXAMPLE 1. Evaluate

$$\begin{vmatrix} -1 & 2 & 3 & 4 \\ 0 & 1 & -1 & 2 \\ 3 & -2 & -1 & -3 \\ 0 & 1 & 2 & 1 \end{vmatrix}$$

SOLUTION. Using the elements of the fourth row and their minors in the expansion, we have

$$\begin{vmatrix} -1 & 2 & 3 & 4 \\ 0 & 1 & -1 & 2 \\ 3 & -2 & -1 & -3 \\ 0 & 1 & 2 & 1 \end{vmatrix} = -0 \begin{vmatrix} 2 & 3 & 4 \\ 1 & -1 & 2 \\ -2 & -1 & -3 \end{vmatrix} + 1 \begin{vmatrix} -1 & 3 & 4 \\ 0 & -1 & 2 \\ 3 & -1 & -3 \end{vmatrix}$$

$$- 2 \begin{vmatrix} -1 & 2 & 4 \\ 0 & 1 & 2 \\ 3 & -2 & -3 \end{vmatrix} + 1 \begin{vmatrix} -1 & 2 & 3 \\ 0 & 1 & -1 \\ 3 & -2 & -1 \end{vmatrix}$$

$$= 0 + 1 \left\{ -0 \begin{vmatrix} 3 & 4 \\ -1 & -3 \end{vmatrix} + (-1) \begin{vmatrix} -1 & 4 \\ 3 & -3 \end{vmatrix} - 2 \begin{vmatrix} -1 & 3 \\ 3 & -1 \end{vmatrix} \right\}$$

$$- 2 \left\{ -0 \begin{vmatrix} 2 & 4 \\ -2 & -3 \end{vmatrix} + 1 \begin{vmatrix} -1 & 4 \\ 3 & -3 \end{vmatrix} - 2 \begin{vmatrix} -1 & 2 \\ 3 & -2 \end{vmatrix} \right\}$$

$$+ 1 \left\{ -0 \begin{vmatrix} 2 & 3 \\ -2 & -1 \end{vmatrix} + 1 \begin{vmatrix} -1 & 3 \\ 3 & -1 \end{vmatrix} - (-1) \begin{vmatrix} -1 & 2 \\ 3 & -2 \end{vmatrix} \right\}$$

$$= 0 + 1\{0 + (-1)(-9) - 2(-8)\} - 2\{0 + 1(-9) - 2(-4)\}$$
$$\quad + 1\{0 + 1(-8) - (-1)(-4)\}$$
$$= 0 + 1(0 + 9 + 16) - 2(0 - 9 + 8) + 1(0 - 8 - 4)$$
$$= 0 + 25 + 2 - 12$$
$$= 15$$

EXAMPLE 2. Evaluate

$$\begin{vmatrix} 0 & 1 & 1 & 1 \\ 1 & 0 & a & b \\ 1 & a & 0 & c \\ 1 & b & c & 0 \end{vmatrix}$$

SOLUTION. We shall evaluate the determinant using the minors of the elements of the first row.

$$\begin{vmatrix} 0 & 1 & 1 & 1 \\ 1 & 0 & a & b \\ 1 & a & 0 & c \\ 1 & b & c & 0 \end{vmatrix} = 0\begin{vmatrix} 0 & a & b \\ a & 0 & c \\ b & c & 0 \end{vmatrix} - 1\begin{vmatrix} 1 & a & b \\ 1 & 0 & c \\ 1 & c & 0 \end{vmatrix} + 1\begin{vmatrix} 1 & 0 & b \\ 1 & a & c \\ 1 & b & 0 \end{vmatrix} - 1\begin{vmatrix} 1 & 0 & a \\ 1 & a & 0 \\ 1 & b & c \end{vmatrix}$$

$$= 0 - 1\left\{1\begin{vmatrix} 0 & c \\ c & 0 \end{vmatrix} - a\begin{vmatrix} 1 & c \\ 1 & 0 \end{vmatrix} + b\begin{vmatrix} 1 & 0 \\ 1 & c \end{vmatrix}\right\}$$

$$\quad + 1\left\{1\begin{vmatrix} a & c \\ b & 0 \end{vmatrix} - 0\begin{vmatrix} 1 & c \\ 1 & 0 \end{vmatrix} + b\begin{vmatrix} 1 & a \\ 1 & b \end{vmatrix}\right\}$$

$$\quad - 1\left\{1\begin{vmatrix} a & 0 \\ b & c \end{vmatrix} - 0\begin{vmatrix} 1 & 0 \\ 1 & c \end{vmatrix} + a\begin{vmatrix} 1 & a \\ 1 & b \end{vmatrix}\right\}$$

$$= 0 - 1[1(-c^2) - a(0 - c) + bc]$$
$$\quad + 1[(1)(-bc) - 0(-c) + b(b - a)] - 1[(1)(ac) - 0(c)$$
$$\quad + a(b - a)]$$
$$= 0 - (-c^2 + ac + bc) + (-bc + b^2 - ab) - (ac + ab - a^2)$$
$$= a^2 + b^2 + c^2 - 2ac - 2bc - 2ab$$

EXERCISES 11.3

Evaluate the following.

1. $\begin{vmatrix} 5 & 1 & 3 & -1 \\ 1 & 4 & -2 & 3 \\ 1 & 1 & -3 & 2 \\ 2 & 0 & 1 & 5 \end{vmatrix}$

2. $\begin{vmatrix} 2 & 9 & 1 & 7 \\ 8 & 5 & -1 & 3 \\ 6 & 4 & 1 & -2 \\ 0 & 0 & 1 & 3 \end{vmatrix}$

3. $\begin{vmatrix} 2 & 0 & 4 & 6 \\ 2 & 4 & 1 & 3 \\ 1 & 3 & 0 & 2 \\ 4 & 8 & 2 & 6 \end{vmatrix}$

4. $\begin{vmatrix} 6 & 1 & 2 & 3 \\ -4 & -3 & 4 & -5 \\ 12 & 2 & 4 & 6 \\ 0 & 6 & -7 & 1 \end{vmatrix}$

5. $\begin{vmatrix} \sin\theta & \sin 2\theta & \sin 3\theta \\ \sin^2\theta & \sin^2 2\theta & \sin^2 3\theta \\ \sin 2\theta & \sin 4\theta & \sin 6\theta \end{vmatrix}$

6. $\begin{vmatrix} 1 & 1 & 1 & 1 \\ 1 & 1+x & 1 & 1 \\ 1 & 1 & 1+y & 1 \\ 1 & 1 & 1 & 1+z \end{vmatrix}$

7. $\begin{vmatrix} \sin x \cos y & \sin x \sin y & \cos x \\ \cos x \cos y & \cos x \sin y & -\sin x \\ -\sin x \sin y & \sin x \cos y & 0 \end{vmatrix}$

8. $\begin{vmatrix} 3 & 0 & 1 & 2 \\ -1 & 1 & 4 & 3 \\ 2 & 2 & 0 & 1 \\ 7 & 3 & 5 & 0 \end{vmatrix}$

9. $\begin{vmatrix} 1 & a & a^2 & a^3 \\ 1 & b & b^2 & b^3 \\ 1 & c & c^2 & c^3 \\ 1 & d & d^2 & d^3 \end{vmatrix}$

10. $\begin{vmatrix} 0 & a & b & c & d \\ -a & 0 & e & f & g \\ -b & -e & 0 & h & j \\ -c & -f & -h & 0 & k \\ -d & -g & -j & -k & 0 \end{vmatrix}$

11. $\begin{vmatrix} 2 & 3 & 1 & 4 & 5 \\ 1 & 2 & 5 & -3 & 6 \\ 2 & 3 & 1 & 4 & 5 \\ 5 & 1 & 3 & 2 & 7 \\ 3 & 5 & 3 & 1 & 4 \end{vmatrix}$

12. $\begin{vmatrix} 1 & 1 & 1 & 1 \\ 1 & 1-x & 1 & 1 \\ 1 & 1 & 1-y & 1 \\ 1 & 1 & 1 & 1-z \end{vmatrix}$

13. $\begin{vmatrix} a & b & a+b \\ b & a+b & a \\ a+b & a & b \end{vmatrix}$

14. $\begin{vmatrix} b+c & a & a \\ b & c+a & b \\ c & c & a+b \end{vmatrix}$

15. $\begin{vmatrix} 1 & x & y & z+w \\ 1 & y & z & x+w \\ 1 & z & w & x+y \\ 1 & w & x & y+z \end{vmatrix}$

16. $\begin{vmatrix} 10 & 4 & 3 & 8 \\ \frac{2}{3} & -2 & \frac{1}{3} & -\frac{2}{3} \\ 0 & -2 & \frac{5}{2} & 3 \\ \frac{5}{6} & \frac{1}{3} & \frac{1}{4} & \frac{2}{3} \end{vmatrix}$

17. $\begin{vmatrix} a+2b & a+4b & a+6b & a+8b \\ a+3b & a+5b & a+7b & a+9b \\ a+4b & a+6b & a+8b & a+10b \\ a+5b & a+7b & a+9b & a+11b \end{vmatrix}$

18. $\begin{vmatrix} a & b+c & d & e \\ b & c+d & a & e \\ c & a+d & b & e \\ d & a+b & c & e \end{vmatrix}$

11.4 ◆ Properties of Determinants

We shall now state some elementary properties of determinants. These properties simplify the work of evaluating a determinant.

PROPERTY D-1. The value of a determinant is unchanged if all the corresponding rows and columns are interchanged.

By Property D-1 we know that the value is unchanged if the first row becomes the first column of a new determinant, the second row becomes the second column of the new determinant, and so forth. Thus

$$\begin{vmatrix} a_{11} & a_{12} & a_{13} \\ a_{21} & a_{22} & a_{23} \\ a_{31} & a_{32} & a_{33} \end{vmatrix} = \begin{vmatrix} a_{11} & a_{21} & a_{31} \\ a_{12} & a_{22} & a_{32} \\ a_{13} & a_{23} & a_{33} \end{vmatrix}$$

We shall accept Property D-1 without proof, but the reader should verify the property for the third order determinant.

As a consequence of Property D-1 the words "row" and "column" can always be interchanged in a statement of a property of a determinant.

PROPERTY D-2. If all the elements of a column (row) are multiplied by the same number k, the value of the determinant is multiplied by k.

PROOF. Let the elements of the jth column of determinant D be multiplied by k. Let us call the resulting determinant D'. Then

$$D = \begin{vmatrix} a_{11} & a_{12} & \cdots & a_{1j} & \cdots & a_{1n} \\ a_{21} & a_{22} & \cdots & a_{2j} & \cdots & a_{2n} \\ \vdots & & & & & \vdots \\ a_{n1} & a_{n2} & \cdots & a_{nj} & \cdots & a_{nn} \end{vmatrix}$$

and

$$D' = \begin{vmatrix} a_{11} & a_{12} & \cdots & ka_{1j} & \cdots & a_{2n} \\ a_{21} & a_{22} & \cdots & ka_{2j} & \cdots & a_{2n} \\ \vdots & & & & & \vdots \\ a_{n1} & a_{n2} & \cdots & ka_{nj} & \cdots & a_{nn} \end{vmatrix}$$

Evaluating D' using the minors of the elements of the jth row, we obtain:

$$\begin{aligned} D' &= (-1)^{1+j}ka_{1j}A_{1j} + (-1)^{2+j}ka_{2j}A_{2j} + \cdots + (-1)^{n+j}ka_{nj}A_{nj} \\ &= k[(-1)^{1+j}a_{1j}A_{1j} + (-1)^{2+j}a^{2j}A_{2j} + \cdots + (-1)^{n+j}a_{nj}A_{nj}] \\ &= kD \end{aligned}$$

PROPERTY D-3. If any two columns (rows) of a determinant are interchanged, the sign of the determinant is changed.

We shall accept this property without proof. The reader should verify the property for the third order determinant.

PROPERTY D-4. If every element of any column (row) is zero, the value of the determinant is zero.

PROOF. The proof of this property is obvious; consider the expansion using the minors of the elements of the column (row) whose elements are all zero.

PROPERTY D-5. If two columns (rows) of a determinant are identical, the value of the determinant is zero.

PROOF. Let D be the value of the determinant with the two identical columns. Let us interchange the two identical columns. The value of the resulting determinant is $-D$ by Property D-3. But since the two interchanged columns are identical, the two determinants are identical and hence have the same value. Thus

$$D = -D$$

from which we obtain $D = 0$

PROPERTY D-6. Any number that is a factor of each element of any column (row) is actually a factor of the determinant.

PROOF. This property follows directly from Property D-2.

PROPERTY D-7. If each element of any column (row) of a determinant is multiplied by the same number k and added to the corresponding elements of another column (row), the value of the determinant is unchanged.

PROOF. Let us consider the two determinants D and D'.

$$D = \begin{vmatrix} a_{11} & a_{12} & \cdots & a_{1n} \\ a_{21} & a_{22} & \cdots & a_{2n} \\ \vdots & & & \vdots \\ a_{n1} & a_{n2} & \cdots & a_{nn} \end{vmatrix}$$

$$D' = \begin{vmatrix} a_{11} & a_{12} & \cdots & (a_{ij} + ka_{1s}) & \cdots & a_{1n} \\ a_{21} & a_{22} & \cdots & (a_{2j} + ka_{2s}) & \cdots & a_{2n} \\ \vdots & & & & & \vdots \\ a_{n1} & a_{n2} & \cdots & (a_{nj} + ka_{ns}) & \cdots & a_{nn} \end{vmatrix}$$

Let us expand D' using the minors of the elements in the jth column. Then

$$\begin{aligned} D' &= (-1)^{1+j}(a_{ij} + ka_{1s})A_{1j} + (-1)^{2+j}(a_{2j} + ka_{2s})A_{2j} + \cdots \\ &\qquad\qquad\qquad\qquad + (-1)^{n+j}(a_{nj} + ka_{ns})A_{nj} \\ &= [(-1)^{1+j}a_{1j}A_{1j} + (-1)^{2+j}a_{2j}A_{2j} + \cdots + (-1)^{n+j}a_{nj}A_{nj}] \\ &\quad + k[(-1)^{1+j}a_{1s}A_{1j} + (-1)^{2+j}a_{2s}A_{2j} + \cdots + (-1)^{n+j}a_{ns}A_{nj}] \\ &= D + k[(-1)^{1+j}a_{1s}A_{1j} + (-1)^{2+j}a_{2s}A_{2j} + \cdots + (-1)^{n+j}a_{ns}A_{nj}] \end{aligned}$$

The sum

$$(-1)^{1+j}a_{1s}A_{1j} + (-1)^{2+j}a_{2s}A_{2j} + \cdots + (-1)^{n+j}a_{ns}A_{nj}$$

is the expansion of a determinant which differs from D only in that the elements of the jth column are $a_{1s}, a_{2s}, \ldots, a_{ns}$ making this column identical with the sth column; the value of this determinant is zero by Property D-5. Hence, $D = D'$ regardless of the value of k.

EXAMPLE 1. Using the properties of determinants, evaluate

(1)
$$\begin{vmatrix} 1 & 3 & 1 & 4 \\ 3 & -3 & -5 & 2 \\ -3 & 1 & 8 & -7 \\ 2 & 4 & 3 & 2 \end{vmatrix}$$

SOLUTION. Using Property D-7, we make the first row consist of the elements 1, 0, 0, 0 as follows.

1. Multiply each element of column one by -3 and add to the corresponding elements of column two.
2. Multiply each element of column one by -1 and add to the corresponding elements of column three.
3. Multiply each element of column one by -4 and add to the corresponding elements of column four.

The result, which is equal to the given determinant by D-7 is

(2)
$$\begin{vmatrix} 1 & 0 & 0 & 0 \\ 3 & -12 & -8 & -10 \\ -3 & 10 & 11 & 5 \\ 2 & -2 & 1 & -6 \end{vmatrix}$$

All the elements in the first column of determinant 2, except $a_{11} = 1$, can be replaced by 0 without affecting the value of the determinant for these elements do not enter into the expansion using the elements of the first row and their minors. We now have the given determinant equal to

(3)
$$\begin{vmatrix} 1 & 0 & 0 & 0 \\ 0 & -12 & -8 & -10 \\ 0 & 10 & 11 & 5 \\ 0 & -2 & 1 & -6 \end{vmatrix}$$

Determinant 3 is equal to

(1)
$$\begin{vmatrix} -12 & -8 & -10 \\ 10 & 11 & 5 \\ -2 & 1 & -6 \end{vmatrix} = (-2) \begin{vmatrix} 6 & -8 & -10 \\ -5 & 11 & 5 \\ 1 & 1 & -6 \end{vmatrix} \qquad \text{by D-6}$$

$$
= (4) \begin{vmatrix} -3 & 4 & 5 \\ -5 & 11 & 5 \\ 1 & 1 & -6 \end{vmatrix} \qquad \text{by D-6}
$$

$$
= -(4) \begin{vmatrix} 1 & 1 & -6 \\ -5 & 11 & 5 \\ -3 & 4 & 5 \end{vmatrix} \qquad \text{by D-3}
$$

$$
= (-4) \begin{vmatrix} 1 & 0 & 0 \\ -5 & 16 & -25 \\ -3 & 7 & -13 \end{vmatrix}
$$

by D-7, where the elements of the first column are multiplied by -1 and added to the corresponding elements of the second column and the elements of the first column are multiplied by 6 and added to the corresponding elements of the third column.

$$
= (-4) \begin{vmatrix} 1 & 0 & 0 \\ 0 & 16 & -25 \\ 0 & 7 & -13 \end{vmatrix} \qquad \text{(Why?)}
$$

$$
= (-4)(1) \begin{vmatrix} 16 & -25 \\ 7 & -13 \end{vmatrix} \qquad \text{(Why?)}
$$

$$
= (-4)(-208 + 175)
$$
$$
= (-4)(-33)
$$
$$
= 132
$$

EXAMPLE 2. Show by use of the properties of determinants that

$$
\begin{vmatrix} 1 & 1 & 1 \\ a & b & c \\ a^2 & b^2 & c^2 \end{vmatrix} = (a - b)(b - c)(c - a)
$$

SOLUTION

$$
\begin{vmatrix} 1 & 1 & 1 \\ a & b & c \\ a^2 & b^2 & c^2 \end{vmatrix} = \begin{vmatrix} 1 & 0 & 1 \\ a & b - c & c \\ a^2 & b^2 - c^2 & c^2 \end{vmatrix} \qquad \text{(Why?)}
$$

$$= \begin{vmatrix} 1 & 0 & 0 \\ a & b-c & c-a \\ a^2 & b^2-c^2 & c^2-a^2 \end{vmatrix} \quad \text{(Why?)}$$

$$= \begin{vmatrix} 1 & 0 & 0 \\ 0 & b-c & c-a \\ 0 & b^2-c^2 & c^2-a^2 \end{vmatrix} \quad \text{(Why?)}$$

$$= (1)\begin{vmatrix} b-c & c-a \\ b^2-c^2 & c^2-a^2 \end{vmatrix}$$

$$= (b-c)(c-a)\begin{vmatrix} 1 & 1 \\ b+c & c+a \end{vmatrix} \quad \text{(Why?)}$$

$$= (b-c)(c-a)(c+a-b-c)$$
$$= (b-c)(c-a)(a-b)$$

EXERCISES 11.4

Use the properties of determinants to show that the following are true (Exercises 1–10).

1. $\begin{vmatrix} 2 & 6 & 10 & 4 \\ 1 & 6 & 5 & 2 \\ 4 & 0 & 8 & 7 \\ 5 & 9 & 3 & 1 \end{vmatrix} = 6\begin{vmatrix} 1 & 1 & 5 & 2 \\ 1 & 2 & 5 & 2 \\ 4 & 0 & 8 & 7 \\ 5 & 3 & 3 & 1 \end{vmatrix}$.

2. $\begin{vmatrix} 3 & 4 & -1 & 2 \\ 4 & -2 & 7 & -1 \\ 7 & 6 & 1 & 3 \\ 0 & -10 & 11 & -5 \end{vmatrix} = 0$.

3. $\begin{vmatrix} 3a & 3b & 3c \\ 4a & 4b & 4c \\ d & e & f \end{vmatrix} = 0$.

4. $\begin{vmatrix} a_1 & b_1 & c_1 \\ a_2 & b_2 & c_2 \\ a_3 & b_3 & c_3 \end{vmatrix} = \begin{vmatrix} a_2 & c_2 & b_2 \\ a_1 & c_1 & b_1 \\ a_3 & c_3 & b_3 \end{vmatrix}.$

5. $\begin{vmatrix} 5k & p & 6k \\ 5m & t & 6m \\ 5n & 5 & 6n \end{vmatrix} = 0.$

6. $\begin{vmatrix} 1 & 1 & 1 \\ a & b & c \\ a^3 & b^3 & c^3 \end{vmatrix} = (a - b)(b - c)(c - a)(a + b + c).$

7. $\begin{vmatrix} c & a & d & b \\ a & c & d & b \\ a & c & b & d \\ c & a & b & d \end{vmatrix} = 0.$

8. $\begin{vmatrix} 1 & a & b & c + d \\ 1 & b & c & a + d \\ 1 & c & d & a + b \\ 1 & d & a & b + c \end{vmatrix} = 0.$

9. $\begin{vmatrix} b + c & c + a & a + b \\ b_1 + c_1 & c_1 + a_1 & a_1 + b_1 \\ b_2 + c_2 & c_2 + a_2 & a_2 + b_2 \end{vmatrix} = 2 \begin{vmatrix} a & b & c \\ a_1 & b_1 & c_1 \\ a_2 & b_2 & c_2 \end{vmatrix}.$

Evaluate (Exercises 10–14).

10. $\begin{vmatrix} 5 & -3 & 0 & 4 & 0 \\ 1 & 2 & 0 & 3 & 0 \\ 3 & -1 & 1 & -2 & 6 \\ 2 & 1 & 0 & 5 & 7 \\ 0 & 4 & -6 & 1 & 2 \end{vmatrix}.$

11.
$$\begin{vmatrix} a & b & c & d \\ a^2 & b^2 & c^2 & d^2 \\ a^3 & b^3 & c^3 & d^3 \\ a^4 & b^4 & c^4 & d^4 \end{vmatrix}.$$

12.
$$\begin{vmatrix} 1 + a_1b_1 & 1 + a_2b_1 & 1 + a_3b_1 & 1 + a_4b_1 \\ 1 + a_1b_2 & 1 + a_2b_2 & 1 + a_3b_2 & 1 + a_4b_2 \\ 1 + a_1b_3 & 1 + a_2b_3 & 1 + a_3b_3 & 1 + a_4b_3 \\ 1 + a_1b_4 & 1 + a_2b_4 & 1 + a_3b_4 & 1 + a_4b_4 \end{vmatrix}.$$

13.
$$\begin{vmatrix} 1 & 1 & 1 & 1 & 1 \\ 1 & 0 & 0 & 0 & x_1 \\ x_2 & 1 & 0 & 0 & x_2 \\ x_3 & x_3 & 1 & 0 & x_3 \\ x_4 & x_4 & x_4 & 1 & x_4 \end{vmatrix}.$$

14.
$$\begin{vmatrix} a & b & b & b \\ a & b & a & a \\ b & b & a & b \\ a & a & a & b \end{vmatrix}.$$

11.5 ◆ Solution of Systems of Equations Using Determinants

Let us consider the following system of two linear equations in two variables.

(1)
$$a_1x + b_1y = c_1$$
(2)
$$a_2x + b_2y = c_2$$

Let us multiply equation 1 by b_2 and Equation 2 by $-b_1$, and then add the two resulting equations member by member. Then

$$
\begin{aligned}
a_1b_2x + b_1b_2y &= b_2c_1 \\
-a_2b_1x - b_1b_2y &= -b_1c_2 \\
\hline
(a_1b_2 - a_2b_1)x &= b_2c_1 - b_1c_2
\end{aligned}
$$

(3)

Solving Equation 3 for x, we have

$$x = \frac{b_2 c_1 - b_1 c_2}{a_1 b_2 - a_2 b_1}, \; a_1 b_2 - a_2 b_1 \neq 0$$

In a similar fashion we find

$$y = \frac{a_1 c_2 - a_2 c_1}{a_1 b_2 - a_2 b_1}, \; a_1 b_2 - a_2 b_1 \neq 0$$

Note that

$$a_1 b_2 - a_2 b_1 = \begin{vmatrix} a_1 & b_1 \\ a_2 & b_2 \end{vmatrix}, \; b_2 c_1 - b_1 c_2 = \begin{vmatrix} c_1 & b_1 \\ c_2 & b_2 \end{vmatrix}$$

and

$$a_1 c_2 - a_2 c_1 = \begin{vmatrix} a_1 & c_1 \\ a_2 & c_2 \end{vmatrix}$$

Hence we may write:

$$x = \frac{\begin{vmatrix} c_1 & b_1 \\ c_2 & b_2 \end{vmatrix}}{\begin{vmatrix} a_1 & b_1 \\ a_2 & b_2 \end{vmatrix}}, \quad y = \frac{\begin{vmatrix} a_1 & c_1 \\ a_2 & c_2 \end{vmatrix}}{\begin{vmatrix} a_1 & b_1 \\ a_2 & b_2 \end{vmatrix}}, \; \begin{vmatrix} a_1 & b_1 \\ a_2 & b_2 \end{vmatrix} \neq 0$$

The solution set of the system of two linear equations in two variables is

$$\left\{ \left(\frac{\begin{vmatrix} c_1 & b_1 \\ c_2 & b_2 \end{vmatrix}}{\begin{vmatrix} a_1 & b_1 \\ a_2 & b_2 \end{vmatrix}}, \; \frac{\begin{vmatrix} a_1 & c_1 \\ a_2 & c_2 \end{vmatrix}}{\begin{vmatrix} a_1 & b_1 \\ a_2 & b_2 \end{vmatrix}} \right), \; \begin{vmatrix} a_1 & b_1 \\ a_2 & b_2 \end{vmatrix} \neq 0 \right\}$$

Now let us consider the following system of three equations in three variables.

$$(4) \qquad\qquad\qquad a_1 x + b_1 y + c_1 z = d_1$$
$$(5) \qquad\qquad\qquad a_2 x + b_2 y + c_2 z = d_2$$
$$(6) \qquad\qquad\qquad a_3 x + b_3 y + c_3 z = d_3$$

If we multiply the members of Equation 5 by c_3 and those of Equation 6 by $-c_2$ and add the corresponding members of the two resulting equations, we obtain

(7)
$$\begin{vmatrix} a_2 & c_2 \\ a_3 & c_3 \end{vmatrix} x + \begin{vmatrix} b_2 & c_2 \\ b_3 & c_3 \end{vmatrix} y = \begin{vmatrix} d_2 & c_2 \\ d_3 & c_3 \end{vmatrix}$$

In a similar manner, if we multiply the members of Equation 4 by c_3 and those of Equation 6 by $-c_1$ and add the corresponding members of the two resulting equations, we obtain

(8)
$$\begin{vmatrix} a_1 & c_1 \\ a_3 & c_3 \end{vmatrix} x + \begin{vmatrix} b_1 & c_1 \\ b_3 & c_3 \end{vmatrix} y = \begin{vmatrix} d_1 & c_1 \\ d_3 & c_3 \end{vmatrix}$$

Finally, if we multiply the members of Equation 4 by c_2 and those of Equation 5 by $-c_1$ and add the corresponding members of the resulting equations, we obtain

(9)
$$\begin{vmatrix} a_1 & c_1 \\ a_2 & c_2 \end{vmatrix} x + \begin{vmatrix} b_1 & c_1 \\ b_2 & c_1 \end{vmatrix} y = \begin{vmatrix} d_1 & c_1 \\ d_2 & c_2 \end{vmatrix}$$

Let us now multiply the members of Equation 7 by a_1 and those of Equation 8 by $-a_2$, and those of Equation 9 by a_3 and add corresponding members of the resulting equations. We obtain

(10)
$$\left[a_1 \begin{vmatrix} b_2 & c_2 \\ b_3 & c_3 \end{vmatrix} - a_2 \begin{vmatrix} b_1 & c_1 \\ b_3 & c_3 \end{vmatrix} + a_3 \begin{vmatrix} b_1 & c_1 \\ b_2 & c_2 \end{vmatrix} \right] y$$
$$= a_1 \begin{vmatrix} d_2 & c_2 \\ d_3 & c_3 \end{vmatrix} - a_2 \begin{vmatrix} d_1 & c_1 \\ d_3 & c_3 \end{vmatrix} + a_3 \begin{vmatrix} d_1 & c_1 \\ d_2 & c_2 \end{vmatrix}$$

Solving Equation 10 for y, we have

$$y = \frac{a_1 \begin{vmatrix} d_2 & c_2 \\ d_3 & c_3 \end{vmatrix} - a_2 \begin{vmatrix} d_1 & c_1 \\ d_3 & c_3 \end{vmatrix} + a_3 \begin{vmatrix} d_1 & c_1 \\ d_2 & c_2 \end{vmatrix}}{a_1 \begin{vmatrix} b_2 & c_2 \\ b_3 & c_3 \end{vmatrix} - a_2 \begin{vmatrix} b_1 & c_1 \\ b_3 & c_3 \end{vmatrix} + a_3 \begin{vmatrix} b_1 & c_1 \\ b_2 & c_2 \end{vmatrix}}$$

$$= \frac{\begin{vmatrix} a_1 & d_1 & c_1 \\ a_2 & d_2 & c_2 \\ a_3 & d_3 & c_3 \end{vmatrix}}{\begin{vmatrix} a_1 & b_1 & c_1 \\ a_2 & b_2 & c_2 \\ a_3 & b_3 & c_3 \end{vmatrix}}, \quad \begin{vmatrix} a_1 & b_1 & c_1 \\ a_2 & b_2 & c_2 \\ a_3 & b_3 & c_3 \end{vmatrix} \neq 0$$

In a similar manner we obtain

$$
x = \frac{\begin{vmatrix} d_1 & b_1 & c_1 \\ d_2 & b_2 & c_2 \\ d_3 & b_3 & c_3 \\ \end{vmatrix}}{\begin{vmatrix} a_1 & b_1 & c_1 \\ a_2 & b_2 & c_2 \\ b_2 & b_3 & c_3 \\ \end{vmatrix}}, \qquad z = \frac{\begin{vmatrix} a_1 & b_1 & d_1 \\ a_2 & b_2 & d_2 \\ a_3 & b_3 & d_3 \\ \end{vmatrix}}{\begin{vmatrix} a_1 & b_1 & c_1 \\ a_2 & b_2 & c_2 \\ a_3 & b_3 & c_3 \\ \end{vmatrix}}
$$

Now let us consider the following system of n linear equations in n variables, $x_1, x_2, \ldots, x_n$:

$$
a_{11}x_1 + a_{12}x_2 + \cdots + a_{1n}x_n = k_1
$$
$$
a_{21}x_1 + a_{22}x_2 + \cdots + a_{2n}x_n = k_2
$$
$$
\vdots
$$
$$
a_{n1}x_1 + a_{n2}x_2 + \cdots + a_{nn}x_n = k_n
$$

The solution of this system may be expressed in terms of determinants. Let

$$
D = \begin{vmatrix} a_{11} & a_{12} & \cdots & a_{1n} \\ a_{21} & a_{22} & \cdots & a_{2n} \\ & & \vdots & \\ a_{n1} & a_{n2} & \cdots & a_{nn} \\ \end{vmatrix}
$$

The determinant D is called the **determinant of the coefficients of the variables** of the above system.

We can now prove Theorem 11.1, which is called **Cramer's Rule.**

THEOREM 11.1. If D is the determinant of the coefficients of the variables in a system of n linear equations in n variables, the product of D and any one of the variables is equal to the determinant D_i obtained from D by substituting the constant terms in the system in place of the coefficients of that variable and leaving the other elements unchanged.

PROOF. Since

$$
D = \begin{vmatrix} a_{11} & a_{12} & \cdots & a_{1n} \\ a_{21} & a_{22} & \cdots & a_{2n} \\ & & \vdots & \\ a_{n1} & a_{n2} & \cdots & a_{nn} \\ \end{vmatrix}
$$

by Property D-2,

$$Dx_1 = \begin{vmatrix} a_{11}x_1 & a_{12} & \cdots & a_{1n} \\ a_{21}x_1 & a_{22} & \cdots & a_{2n} \\ & \vdots & & \\ a_{n1}x_1 & a_{n2} & \cdots & a_{nn} \end{vmatrix}$$

Let us now multiply each element of the second column by x_2, each element of the third column by $x_3, \ldots$, and each element of the nth column by x_n, and add all these products to the corresponding elements of the first column. We have, by Property D-7,

$$Dx_1 = \begin{vmatrix} a_{11}x_1 + a_{12}x_2 + \cdots + a_{1n}x_2 & a_{12} & a_{13} & \cdots & a_{1n} \\ a_{21}x_2 + a_{22}x_2 + \cdots + a_{2n}x_n & a_{22} & a_{n23} & \cdots & a_{2n} \\ & & \vdots & & \\ a_{n1}x_1 + a_{n2}x_2 + \cdots + a_{nn}x_n & a_{n2} & a_{n3} & \cdots & a_{nn} \end{vmatrix}$$

$$= \begin{vmatrix} k_1 & a_{12} & a_{13} & \cdots & a_{1n} \\ k_2 & a_{22} & a_{23} & \cdots & a_{2n} \\ & & \vdots & & \\ k_n & a_{n2} & a_{n3} & \cdots & a_{n7} \end{vmatrix}$$

$$= D_1$$

In the same manner we obtain

$$Dx_2 = D_2$$
$$Dx_3 = D_3$$
$$\vdots$$
$$Dx_n = D_n$$

where D_i is obtained by substituting $k_1, k_2, \ldots, k_n$ for the elements $a_{1i}, a_{2i}, a_{3i}, \ldots, a_{ni}$ of the ith column, respectively.

If $D \neq O$, the unique solution of our system is obtained from the above. And

$$x_1 = \frac{D_1}{D}$$

$$x_2 = \frac{D_2}{D}$$

$$\vdots$$

$$x_n = \frac{D_n}{D}$$

If $D = 0$, the system has no solution if any of the determinants D_i, $i = 1, 2, \ldots, n$, is different from zero. In order to prove this, let us assume that $D = 0$ and that one of the other determinants, say, D_r, is different from zero and that there is a solution $x_1 = r_1$, $x_2 = r_2, \ldots, x_n = r_n$. Then, from Theorem 11.1, we have

$$x_r \cdot D = D_r \qquad \text{or} \qquad x_r \cdot 0 = D_r$$

Since $D_r \neq 0$, this is impossible and the assumption that there is a solution under these conditions is false.

If $D = 0$ and $D_1 = D_2 = \cdots = D_n = 0$, the system may or may not have solutions. We shall not discuss this case because it is beyond the scope of this book.

It is clear that

$$x_1 = \frac{D_1}{D_1}, \; x_2 = \frac{D_2}{D_1} \ldots, \; x_n = \frac{D_n}{D}$$

satisfy the system above and therefore form a solution. For example, the first equation is satisfied, since

$$k_1 D - a_{11} D_1 - a_{12} D_2 - \cdots - a_{1n} D_n$$

is the expansion of

$$\begin{vmatrix} k_1 & a_{11} & a_{12} & \cdots & a_{1n} \\ k_1 & a_{11} & a_{12} & \cdots & a_{1n} \\ k_2 & a_{21} & a_{22} & \cdots & a_{2n} \\ & & \vdots & & \\ k_n & a_{n1} & a_{n2} & \cdots & a_{nn} \end{vmatrix}$$

by the minors of the elements of the first row. But this determinant has value zero, since the first two rows are identical. The other equations of the system are similarly satisfied.

EXAMPLE 1. Using determinants, find the solution set of the system

$$3x + 2y = 4$$
$$5x - y = 11$$

SOLUTION. Using Cramer's Rule, we obtain

$$x = \frac{\begin{vmatrix} 4 & 2 \\ 11 & -1 \end{vmatrix}}{\begin{vmatrix} 3 & 2 \\ 5 & -1 \end{vmatrix}} = \frac{(4)(-1) - (11)(2)}{(3)(-1) - (5)(2)} = \frac{-26}{-13} = 2$$

$$y = \frac{\begin{vmatrix} 3 & 4 \\ 5 & 11 \end{vmatrix}}{\begin{vmatrix} 3 & 2 \\ 5 & -1 \end{vmatrix}} = \frac{(3)(11) - (5)(4)}{(3)(-1) - (5)(2)} = \frac{13}{-13} = -1$$

The solution set is $\{(2, -1)\}$.

EXAMPLE 2. Using determinants, find the solution set of

$$\begin{aligned} 2x + 3y - z &= 3 \\ x + y - 3z &= -4 \\ -x - y + 5z &= 8 \end{aligned}$$

SOLUTION. Using Cramer's Rule, we obtain

$$x = \frac{\begin{vmatrix} 3 & 3 & -1 \\ -4 & 1 & -3 \\ 8 & -1 & 5 \end{vmatrix}}{\begin{vmatrix} 2 & 3 & -1 \\ 1 & 1 & -3 \\ -1 & -1 & 5 \end{vmatrix}} = \frac{3 \begin{vmatrix} 1 & -3 \\ -1 & 5 \end{vmatrix} - 3 \begin{vmatrix} -4 & -3 \\ 8 & 5 \end{vmatrix} + (-1) \begin{vmatrix} -4 & 1 \\ 8 & -1 \end{vmatrix}}{2 \begin{vmatrix} 1 & -3 \\ -1 & 5 \end{vmatrix} - (1) \begin{vmatrix} 3 & -1 \\ -1 & 5 \end{vmatrix} + (-1) \begin{vmatrix} 3 & -1 \\ 1 & -3 \end{vmatrix}}$$

$$= \frac{3(2) - 3(4) + (-1)(-4)}{2(2) - 14 + (-1)(-8)}$$

$$= \frac{-2}{-2} = 1$$

$$y = \frac{\begin{vmatrix} 2 & 3 & -1 \\ 1 & -4 & -3 \\ -1 & 8 & 5 \end{vmatrix}}{\begin{vmatrix} 2 & 3 & 1 \\ 1 & 1 & -3 \\ -1 & -1 & 5 \end{vmatrix}}$$

$$= \frac{2 \begin{vmatrix} -4 & -3 \\ 8 & 5 \end{vmatrix} - 1 \begin{vmatrix} 3 & -1 \\ 8 & 5 \end{vmatrix} + (-1) \begin{vmatrix} 3 & -1 \\ -4 & -3 \end{vmatrix}}{-2}$$

$$= \frac{2(4) - (23) + (-1)(-13)}{-2} = \frac{-2}{-2} = 1$$

$$z = \frac{\begin{vmatrix} 2 & 3 & 3 \\ 1 & 1 & -4 \\ -1 & -1 & 8 \end{vmatrix}}{\begin{vmatrix} 2 & 3 & 1 \\ 1 & 1 & -3 \\ -1 & -1 & 5 \end{vmatrix}}$$

$$= \frac{2\begin{vmatrix} 1 & -4 \\ -1 & 8 \end{vmatrix} - 1\begin{vmatrix} 3 & 3 \\ -1 & 8 \end{vmatrix} + (-1)\begin{vmatrix} 3 & 3 \\ 1 & -4 \end{vmatrix}}{-2}$$

$$= \frac{2(4) - 27 + (-1)(-15)}{-2} = \frac{-4}{-2} = 2$$

The solution set is $\{(1, 1, 2)\}$.

EXERCISES   11.5

Find the solution sets of the following systems by using determinants.

1. $5x - 2y = 9$
   $4x + 5y = -6$

2. $7x + 3y = 5$
   $2x + 5y = -2$

3. $21x - 10y = 75$
   $3x + 7y = 36$

4. $8x + y = 34$
   $x - 8y = 53$

5. $x \cos \theta - y \sin \theta = 0$
   $x \sin \theta + y \cos \theta = 1$

6. $2x + 3y + 8z = 9$
   $6x + 12y + 4z = 4$
   $4x + 6y + 4z = 6$

7. $4x - 3y + 2z = 9$
   $2x + 5y - 3z = 4$
   $5x + 6y - 2z = 18$

8. $2x - z = 1$
   $2x + 4y - z = 1$
   $x - 8y - 3z = -2$

9. $2x + 3y = -2$
   $5x + z = 15$
   $4y - 6z = 44$

10. $x + 2y + 3z = 0$
    $2x - y + z = 0$
    $5x + 3y - 2z = 0$

11. $-5x - 11y + 13z = 18$
    $6x + 7y - 10z = -3$
    $13x - 5y + 17z = 54$

12. $4x - 3y + 9z = 27$
    $x - 5y - 7z = 6$
    $2x + y + 5z = 9$

13. $x + y + z = a$
    $x + (1 + a)y + z = 2a$
    $x + y + (1 + a)z = 0$

14. $x - y + z + w = 3$
$x + y - z + w = 2$
$-x + y + z + w = 4$
$x + y + z - w = 1$

15. $x + y + z + w = 1$
$y + z = 1$
$z + w = -6$
$x + w = 0$

16. $4x - 3y + 2w = 9$
$2x + 6z = 28$
$-2y + 4w = 14$
$3x + 4w = 26$

17. $x + y + z + 4w = 0$
$2x + 4y + z + 8w = 4$
$4x + y + 2z + 13w = 5$
$2x + 4y + 2z + 11w = 3$

18. $x + y + z + w = a$
$x + y + z - w = b$
$x + y - z - w = c$
$x - y - z - w = d$

19. $x + 2y + 2z + 2w = 1$
$x + y + 2z + 2w = 2$
$x + y + z + 2w = 3$
$x + y + z + w = 4$

20. $x + y + z + w + t = 1$
$x + y + z + w + v = 1$
$x + y + z + t + v = 1$
$x + y + w + t + v = 1$
$x + z + w + t + v = 1$
$y + z + w + t + v = 1$

CHAPTER REVIEW

Evaluate the following determinants (Exercises 1–6).

1. $\begin{vmatrix} 0 & 3 & 0 \\ 5 & 9 & 8 \\ 2 & 7 & 3 \end{vmatrix}$

2. $\begin{vmatrix} 3 & -4 & 5 \\ 1 & 0 & -3 \\ -2 & 6 & 1 \end{vmatrix}$

3. $\begin{vmatrix} 1 & 2 & 3 \\ 5 & 2 & 0 \\ 3 & 2 & 7 \end{vmatrix}$

4. $\begin{vmatrix} 1 & 1 & 1 \\ 1 & 3 & 2 \\ 1 & 6 & 3 \end{vmatrix}$

5. $\begin{vmatrix} 1 & a^2 & a^3 \\ 1 & b^2 & b^3 \\ 1 & c^2 & c^3 \end{vmatrix}$

6. $\begin{vmatrix} 1 & 1 & 1 & 1 \\ 1 & 2 & 3 & 4 \\ 1 & 3 & 6 & 10 \\ 1 & 4 & 10 & 20 \end{vmatrix}$

Verify the following identities (Exercises 7–10).

7. $\begin{vmatrix} \cos x & \sin x \\ \sin y & \cos y \end{vmatrix} = \cos(x + y).$

8. $\begin{vmatrix} \sin x & \cos x \\ \cos x & \sin x \end{vmatrix} = -\cos 2x.$

9. $\begin{vmatrix} \tan x & \cot y \\ \tan y & \cot x \end{vmatrix} = 0.$

10. $\begin{vmatrix} 1 & 1 & 1 \\ \sin x & \cos x & \tan x \\ \sin^3 x & \cos^3 x & \tan^3 x \end{vmatrix}$

$$= (\sin x - \cos x)(\cos x - \tan x)(\tan x - \sin x)(\sin x + \cos x + \tan x).$$

Find the solution sets of the following systems of equations using determinants (Exercises 11–14).

11. $3x + 4y = 10$
    $4x + y = 9$

12. $x \cos A - y \sin A = 0$
    $x \sin A + y \cos A = 1$

13. $x + y + z = 0$
    $3x + 4y + 2z = 0$
    $2x - 6y - z = 0$

14. $x + 2z = 30$
    $2y + 2z = 18$
    $2x + 3y = 21$

15. Find the solution set of:

$$\begin{vmatrix} 0 & x & 4 \\ -2 & 5 & -1 \\ 4 & 1 & -3 \end{vmatrix} = -18.$$

16. Find the solution set of:

$$\begin{vmatrix} x^2 & x & 1 \\ 3 & -2 & 1 \\ -4 & 0 & -2 \end{vmatrix} = 0.$$

17. Write the following as a third order determinant:

$$3\begin{vmatrix} 1 & 0 \\ -1 & 0 \end{vmatrix} - 2\begin{vmatrix} 1 & -1 \\ -1 & 0 \end{vmatrix} + 6\begin{vmatrix} -1 & -1 \\ 1 & 0 \end{vmatrix}$$

18. Given:
$$\begin{vmatrix} 3 & 2 & 1 & 4 \\ -2 & -3 & -1 & -5 \\ 0 & \frac{1}{2} & 7 & 6 \\ 8 & 9 & -6 & -9 \end{vmatrix}$$

(a) Write the cofactor of element -3.

(b) Write the minor of element $\frac{1}{2}$.

19. Given:
$$\begin{vmatrix} a_{11} & a_{12} & a_{13} & a_{14} \\ a_{21} & a_{22} & a_{23} & a_{24} \\ a_{31} & a_{32} & a_{33} & a_{34} \\ a_{41} & a_{42} & a_{43} & a_{44} \end{vmatrix}$$

(a) Write the minors of the elements of the second row.

(b) Write the cofactors of the elements of the first row.

20. Show that

$$\begin{vmatrix} a_{11} & a_{12} & a_{13} & a_{14} \\ a_{21} & a_{22} & a_{23} & a_{24} \\ 0 & 0 & a_{33} & a_{34} \\ 0 & 0 & a_{43} & a_{44} \end{vmatrix} = \begin{vmatrix} a_{11} & a_{12} \\ a_{21} & a_{22} \end{vmatrix} \cdot \begin{vmatrix} a_{33} & a_{34} \\ a_{43} & a_{44} \end{vmatrix}$$

CHAPTER 12

Complex Numbers

12.1 ◆ Definitions

Although real numbers are adequate for many mathematical problems, they are not adequate for solving some equations. For example, in the field of real numbers the solution set of the equation $x^2 + 1 = 0$ is the empty set, ϕ. In certain applications, it is important that equations such as $x^2 + 1 = 0$ have solutions. It is possible to construct a system in which such equations do have solutions. This is called the system of **complex numbers.**

Let us consider the problem of inventing a new mathematical system which contains the real number system and which has certain additional properties.

Let us consider the set, C, of all ordered pairs, (a, b), of real numbers. We shall use z to denote such an ordered pair. We establish the properties of this set in the following definitions.

DEFINITION 12.1. The set C of **complex numbers** is the set of all ordered pairs (a, b) of real numbers.

Using set notation, we write

$$C = \{(a, b) \mid a \in R \text{ and } b \in R\}$$

We now define equality of the numbers in set C.

DEFINITION 12.2. If $z_1 = (a, b)$ and $z_2 = (c, d)$, and z_1 and z_2 are elements of C, then z_1 is equal to z_2, denoted by $z_1 = z_2$, if and only if $a = c$ and $b = d$.

Now that equality of the numbers of C has been established, we define the operations of addition and multiplication of these numbers.

DEFINITION 12.3. If $z_1 = (a, b)$ and $z_2 = (c, d)$ are complex numbers, then their sum, denoted by $z_1 + z_2$, and their product, denoted by $z_1 z_2$, are defined as follows:

$$\text{(a)} \quad z_1 + z_2 = (a, b) + (c, d) = (a + c, b + d)$$
$$\text{(b)} \quad z_1 z_2 = (a, b)(c, d) = (ac - bd, ad + bc).$$

EXAMPLE 1. Find the sum $(2, 4) + (-3, 5)$.

SOLUTION. By Definition 12.3a we have

$$(2, 4) + (-3, 5) = (2 + (-3), 4 + 5)$$
$$= (-1, 9)$$

EXAMPLE 2. Find the product $(2, -1)(3, 2)$.

SOLUTION. By Definition 12.3b, we have

$$(2, -1)(3, 2) = (2 \cdot 3 - (-1) \cdot 2, 2 \cdot 2 + (-1) \cdot 3)$$
$$= (6 + 2, 4 - 3)$$
$$= (8, 1)$$

We now prove that the set C of complex numbers with the operations of addition and multiplication defined in Definition 12.3 is a field called the **complex number field.**

THEOREM 12.1. The set C of complex numbers with the operations of addition and multiplication is a field.

To prove Theorem 12.1 we must establish the following field properties.

C-1. (*Closure Property of Addition*)   If $z_1 \in C$ and $z_2 \in C$, then their sum $z_1 + z_2 \in C$.

C-2. (*Commutative Property of Addition*)   If $z_1 \in C$ and $z_2 \in C$, then $z_1 + z_2 = z_2 + z_1$.

C-3. (*Associative Property of Addition*)   If $z_1 \in C$, $z_2 \in C$ and $z_3 \in C$, then $(z_1 + z_2) + z_3 = z_1 + (z_2 + z_3)$.

C-4. (*Identity Element for Addition*)   There exists an element $z_0 \in C$ called the **identity element for addition** or the **additive identity** such that for all $z \in C$, $z + z_0 = z_0 + z = z$.

C-5. (*Additive Inverse Property*)   For each $z \in C$, there exists a unique element $-z \in C$, called the **additive inverse** of z such that $z + (-z) = (-z) + z = z_0$.

C-6. (*Closure Property of Multiplication*)   If $z_1 \in C$ and $z_2 \in C$, then their product $z_1 z_2 \in C$.

C-7. (*Commutative Property of Multiplication*)   If $z_1 \in C$ and $z_2 \in C$, then $z_1 z_2 = z_2 z_1$.

C-8. (*Associative Property of Multiplication*)   If $z_1 \in C$, $z_2 \in C$ and $z_3 \in C$, then $(z_1 z_2)z_3 = z_1(z_2 z_3)$.

C-9. (*Identity Element for Multiplication*)   There exists an element $z_u \in C$ called the **identity element for multiplication** or the **multiplicative identity** such that for all $z \in C$, $zz_u = z_u z = z$.

C-10. (*Multiplicative Inverse Property*)   For each $z \in C$, $z \neq z_0$, there exists a unique element $z^{-1} \in C$ called the **multiplicative inverse** of z such that $zz^{-1} = z^{-1}z = z_u$.

C-11. (*Distributive Property*)   For all $z_1 \in C$, $z_2 \in C$ and $z_3 \in C$, $z_1(z_2 + z_3) = z_1 z_2 + z_1 z_3$.

We shall prove some of the field properties for the complex numbers. The proofs of the other properties are left to the reader.

Proof of C-1. Let $z_1 = (a_1, b_1)$ and $z_2 = (a_2, b_2)$. By Definition 12.3a,

$$z_1 + z_2 = (a_1, b_1) + (a_2, b_2)$$
$$= (a_1 + a_2, b_1 + b_2)$$

Since the set of real numbers is closed under addition and since a_1, a_2, b_1, and b_2 are real numbers by Definition 12.1, the sums $a_1 + a_2$ and $b_1 + b_2$ are real numbers. Hence, $z_1 + z_2$ is a complex number by Definition 12.1.

Proof of C-4. Let us consider the complex number $z_0 = (0, 0)$. If $z = (a, b)$ is any complex number, we see that

$$z + z_0 = (a, b) + (0, 0) = (a + 0, b + 0) \qquad \text{Definition 12.3a}$$
$$= (a, b) \qquad\qquad\qquad\qquad\qquad \text{additive identity}$$
$$\text{axiom for real num-}$$
$$\text{bers}$$

$$= z$$

In a similar manner we can prove that $z_0 + z = z$. We called $z_0 = (0, 0)$ the **additive identity** or the **identity element of addition** for the set C of complex numbers.

Proof of C-5. Consider the complex number $(-a, -b)$ which we shall denote by $-z$. Then

$$z + (-z) = (a, b) + (-a, -b)$$
$$= [(a + (-a), b + (-b))] \quad \text{Definition 12.3a}$$
$$= (0, 0) \quad \text{additive inverse axiom}$$
$$\text{for real numbers}$$
$$= z_0$$

We called $-z = (-a, -b)$ the **additive inverse** of the complex number $z = (a, b)$.

The reader should prove that $(-z) + z = z_0$.

Proof of C-9. Let $z = (a, b)$ and consider the complex number $z_u = (1, 0)$. Then

$$zz_u = (a, b)(1, 0)$$
$$= (a \cdot 1 - b \cdot 0, a \cdot 0 + b \cdot 1) \quad \text{Definition 12.3b}$$
$$= (a - b \cdot 0, a \cdot 0 + b) \quad \text{multiplicative identity axiom}$$
$$\text{for real numbers}$$
$$= (a - 0, 0 + b) \quad \text{Theorem 1.4}$$
$$= (a + 0, 0 + b) \quad \text{definition of subtraction of}$$
$$\text{real numbers and 0 is its own}$$
$$\text{additive inverse}$$
$$= (a, b) \quad \text{additive identity axiom for}$$
$$\text{real numbers}$$
$$= z$$

The reader should prove that $z_u z = z$.

Proof of C-10. Let $z = (a, b)$, a and b not both zero. Since $z \neq z_0$ and a and b are real numbers not both equal to zero, we have $a^2 + b^2 \neq 0$. Let us consider the complex number

$$z^{-1} = \left(\frac{a}{a^2 + b^2}, \frac{-b}{a^2 + b^2} \right)$$

By Definition 12.3b,

$$zz^{-1} = (a, b) \left(\frac{a}{a^2 + b^2}, \frac{-b}{a^2 + b^2} \right)$$
$$= \left(\frac{a^2 + b^2}{a^2 + b^2}, \frac{-ab + ab}{a^2 + b^2} \right)$$
$$= (1, 0) = z_u$$

In a similar manner we can show that $z^{-1} z = z_u$. The complex number z^{-1} is called the **multiplicative inverse** of $z \neq z_0$.

As in the case of real numbers, subtraction in the complex number field is defined as adding the additive inverse and division is defined as multiplying by the multiplicative inverse.

DEFINITION 12.5. If $z_1 \in C$ and $z_2 \in C$, then their **difference,** denoted by $z_1 - z_2$, is $z_1 + (-z_2)$.

The operation of finding the difference is called **subtraction.**

DEFINITION 12.6. If $z_1 \in C$ and $z_2 \in C$, $z_2 \neq (0, 0)$, then their **quotient** denoted by $\dfrac{z_1}{z_2}$ is $z_1 \cdot z_2^{-1}$.

The operation of finding the quotient is called **division.**

EXAMPLE 3. Find the difference $(3, 4) - (5, 2)$.

SOLUTION. By Definition 12.4, we obtain

$$\begin{aligned} (3, 4) - (5, 2) &= (3, 4) + (-5, -2) \\ &= (3 + (-5), 4 + (-2)) \\ &= (-2, 2) \end{aligned}$$

EXAMPLE 4. Find the quotient $\dfrac{(4, 2)}{(1, 3)}$.

SOLUTION. By Definition 12.5,

$$\begin{aligned} \frac{(4, 2)}{(1, 3)} &= (4, 2)\left(\frac{1}{1^2 + 3^2}, \frac{-3}{1^2 + 3^2} \right) \\ &= (4, 2)(\tfrac{1}{10}, -\tfrac{3}{10}) \\ &= (4 \cdot \tfrac{1}{10} - 2(-\tfrac{3}{10}), 4(-\tfrac{3}{10}) + 2(\tfrac{1}{10})) \\ &= (\tfrac{4}{10} + \tfrac{6}{10}, -\tfrac{12}{12} + \tfrac{2}{10}) \\ &= (1, -1) \end{aligned}$$

EXERCISES 12.1

Perform the following operations (Exercises 1–21).

1. $(3, 4) + (4, 5)$
2. $(2, -1) + (-3, -4)$
3. $(7, -3) + (3, -5)$
4. $(5, 0) + (-5, 0)$
5. $(3, -2) - (1, -3)$
6. $(7, -2) - (2, -1)$
7. $(3, -5) - (-2, -4)$
8. $(7, -7) - (-4, 2)$

9. $(3, 4) \cdot (5, 3)$
10. $(2, -1) \cdot (-1, -2)$
11. $(-2, -1) \cdot (-4, 2)$
12. $(7, -3) \cdot (-3, -4)$
13. $\dfrac{(7, 6)}{(8, 3)}$
14. $\dfrac{(4, 5)}{(-1, -3)}$

15. $\dfrac{(-1, -6)}{(3, 0)}$

16. $\dfrac{(-2, -4)}{(0, 5)}$

17. $\dfrac{(-7, -1)}{(-2, -1)}$

18. $\dfrac{(0, 4)}{(-3, 2)}$

19. $\dfrac{(0, 2)}{(-1, 1)}$

20. $\dfrac{(-4, 0)}{(3, -1)}$

21. $\dfrac{(-5, -1)}{(0, -1)}$

Name the additive inverse of each of the following complex numbers (Exercises 22–27).

22. $(3, 4)$
23. $(5, -1)$
24. $(7, -3)$
25. $(-3, 5)$
26. $(-1, -5)$
27. $(-2, -7)$

Name the multiplicative inverse of each of the following complex numbers (Exercises 28–33).

28. $(-1, 1)$
29. $(0, -3)$
30. $(-6, -2)$
31. $(-5, 0)$
32. $(7, -2)$
33. $(-8, -5)$

34. Prove: If z_1 and z_2 are elements of C, then $z_1 + z_2 = z_2 + z_1$.
35. Prove: If z_1, z_2, and z_3 are elements of C, then $(z_1 + z_2) + z_3 = z_1 + (z_2 + z_3)$.
36. Prove: If z_1 and z_2 are elements of C, then $z_1 z_2 = z_2 z_1$.
37. Prove: If z_1 and z_2 are elements of C, then $z_1 z_2$ is an element of C.
38. Prove: If z_1, z_2, and z_3 are elements of C, then $(z_1 z_2) z_3 = z_1 (z_2 z_3)$.
39. Prove: If z_1, z_2, and z_3 are elements of C, then $z_1(z_2 + z_3) = z_1 z_2 + z_1 z_3$.

12.2 ◆ Standard Form of Complex Numbers

Let us consider the subset R' of C consisting of all ordered pairs of the form $(a, 0)$. Using Definitions 12.3, 12.4, and 12.5, it follows that

$$(a, 0) + (b, 0) = (a + b, 0)$$
$$(a, 0) - (b, 0) = (a - b, 0)$$
$$(a, 0) \cdot (b, 0) = (ab, 0)$$
$$\frac{(a, 0)}{(b, 0)} = \left(\frac{a}{b}, 0\right) \qquad b \neq 0$$

We can verify that the subset R' of C is a field with respect to addition and multiplication of complex numbers (Definition 12.3). We see that if complex numbers of the form $(a, 0)$, that is, complex numbers with second component 0, are subjected to the arithmetic operations of addition, subtraction, multiplication, and division, each repeated any number of times, the resulting complex number will again be of the form $(a, 0)$, whose first component results from performing the prescribed operation on the first components of the complex numbers involved. This means that complex numbers of the form $(a, 0)$ behave with respect to the four arithmetic operations exactly as their first components, which are real numbers. Accordingly, we shall agree from now on to denote a complex number of the type $(a, 0)$ simply by a.

DEFINITION 12.6. The complex number $(a, 0)$ is called a **purely real number** and is denoted by a.

We see, then, that the symbol a has two different meanings: one as a symbol of a real number and another as a symbol of a complex number $(a, 0)$. With these two meanings in mind, we see that the set of real numbers may be thought of as a subset of the set of complex numbers.

Now let us investigate the set of complex numbers of the form $(0, b)$, $b \neq 0$. These numbers are sometimes called the **purely imaginary numbers.** Observe

$$(0, b) \cdot (0, d) = (-bd, 0) = -bd$$

We see that the product of two purely imaginary numbers is a purely real number. As a special case, taking $b = d = 1$, we have

$$(0, 1) \cdot (0, 1) = (-1, 0) = -1$$

We denote the complex number $(0, 1)$ by the letter i. Then, from the above, we have

$$i^2 = -1$$

It is now easy to see that equations of the form $x^2 = -k$, where k is a positive real number, have solutions in C. In fact, one solution is $\sqrt{k} \cdot i$ since

$$(\sqrt{k} \cdot i)^2 = (\sqrt{k})^2 \cdot i^2 = ki^2 = (k)(-1) = -k$$

Similarly, $-\sqrt{k} \cdot i$ is a solution since

$$(-\sqrt{k} \cdot i)^2 = (-\sqrt{k})^2 \cdot i^2 = ki^2 = (k)(-1) = -k$$

We see from the above that negative real numbers have two square roots in the field of complex numbers. Thus, since

$$(-2i)^2 = -4 \qquad \text{and} \qquad (2i)^2 = -4$$

we see that $2i$ and $-2i$ are both square roots of -4.

If $b \in R$, then by Definitions 12.3, 12.6 and the fact that $(0, 1) = i$ we have

$$bi = (b, 0) \cdot (0, 1) = (0, b)$$

This shows that every purely imaginary number $(0, b)$ can be written as the product of the real number b and the complex number i.

Since every complex number (a, b) can be written as the sum

$$(a, b) = (a, 0) + (0, b)$$

and since $(a, 0)$ may be denoted by a and $(0, b)$ may be denoted by bi, we have

$$(a, b) = (a, 0) + (0, b)$$
$$= a + bi$$

In the set C of complex numbers we call the representation $a + bi$ the **standard form** of the complex number (a, b). When written in standard form a is called the **real part** and b is called the **imaginary part**. The complex number i is called the **imaginary unit**.

Definitions 12.2, 12.3, 12.4, and 12.5 may now be expressed using the standard form of the complex numbers as follows.

THEOREM 12.2. If $a \in R$, $b \in R$, and $i^2 = -1$ then:

1. $a + bi = c + di$ if and only if $a = c$ and $b = d$.
2. $(a + bi) + (c + di) = (a + c) + (b + d)i$.
3. $(a + bi)(c + di) = (ac - bd) + (ad + bc)i$.
4. $(a + bi) - (c + di) = (a - c) + (b - d)i$.
5. $\dfrac{a + bi}{c + di} = \dfrac{ac + bd}{c^2 + d^2} + \dfrac{bc - ad}{c^2 + d^2}\,i$ (c and d not both 0).

The main advantage of standard form for complex numbers is that it enables us to work with complex numbers just as though the symbols represented real numbers. The only change is that when the symbol i^2 appears, it should be replaced by -1.

EXAMPLE 1. Find the product $(3 + 2i)^2$. Give the result in the form $a + bi$.

SOLUTION

$$(3 + 2i)^2 = 3^2 + (2)(3)(2i) + (2i)^2$$
$$= 9 + 12i + 4i^2$$

$$= 9 + 12i - 4$$
$$= 5 + 12i$$

EXAMPLE 2. Find the product $(3 + \sqrt{2}\,i)(3 - \sqrt{2}\,i)$. Give the result in the form $a + bi$.

SOLUTION

$$(3 + \sqrt{2}\,i)(3 - \sqrt{2}\,i) = 3^2 - (\sqrt{2}\,i)^2$$
$$= 9 - 2i^2$$
$$= 9 + 2 = 11$$

EXAMPLE 3. Find the product $(3 + 4i)(2 - 3i)$. Give the result in the form $a + bi$.

SOLUTION

$$(3 + 4i)(2 - 3i) = 6 - i - 12i^2$$
$$= 6 - i + 12$$
$$= 18 - i$$

Using the above ideas simplifies finding the quotient of two complex numbers.

$$\frac{a + bi}{c + di} = \frac{a + bi}{c + di} \cdot \frac{c - di}{c - di} \qquad c + di \neq 0$$
$$= \frac{(a + bi)(c - di)}{(c + di)(c - di)}$$
$$= \frac{(ac + bd) + (bc - ad)i}{c^2 - (di)^2}$$
$$= \frac{ac + bd}{c^2 + d^2} + \frac{(bc - ad)}{c^2 + d^2}i$$

EXAMPLE 4. Find the quotient $\dfrac{3 + 4i}{5 - 2i}$. Write the quotient in the form $a + bi$.

SOLUTION

$$\frac{3 + 4i}{5 - 2i} = \frac{3 + 4i}{5 - 2i} \cdot \frac{5 + 2i}{5 + 2i}$$
$$= \frac{15 + 26i + 8i^2}{25 - 4i^2}$$
$$= \frac{15 + 26i - 8}{25 + 4}$$
$$= \frac{7}{29} + \frac{26}{29}i$$

The complex number $c - di$ is called the **conjugate** of the complex number $c + di$. We designate the conjugate of the complex number z by $\bar{z}$. Thus, if $z = c + di$, then $\bar{z} = c - di$.

We now prove some theorems about complex numbers and their conjugates.

DEFINITION 12.7. The **conjugate** of the complex number $z = a + bi$ is defined to be $a - bi$ and is denoted by z.

THEOREM 12.3. If z_1 and z_2 are complex numbers, then:

$$\text{(a) } \overline{z_1 + z_2} = \bar{z}_1 + \bar{z}_2$$
$$\text{(b) } \overline{z_1 \cdot z_2} = \bar{z}_1 \bar{z}_2$$

PROOF. (a) Let $z_1 = a + bi$ and $z_2 = c + di$, where a, b, c, $d \in R$. Then, by Theorem 12.2,

$$z_1 + z_2 = (a + bi) + (c + di)$$
$$= (a + c) + (b + d)i$$

Then by Definition 12.7

$$\overline{z_1 + z_2} = (a + c) - (b + d)i$$
$$= (a - bi) + (c - di)$$
$$= \bar{z}_1 + \bar{z}_2$$

(b) Let $z_1 = a + bi$ and $z_2 = c + di$ where a, b, c, and $d \in R$. Then, by Theorem 12.2,

$$z_1 \cdot z_2 = (a + bi)(c + di)$$
$$= (ac - bd) + (ad + bc)i$$

By Definition 12.7

$$\overline{z_1 \cdot z_2} = (ac - bd) - (ad + bc)i$$

But

$$\bar{z}_1 \cdot \bar{z}_2 = (a - bi)(c - di)$$
$$= (ac - bd) + (-ad - bc)i$$
$$= (ac - bd) - (ad + bc)i$$
$$= \overline{z_1 \cdot z_2}$$

THEOREM 12.4. The product of a complex number and its conjugate is a real number.

PROOF. Let $z = a + bi$; then by Definition 12.7 the conjugate of z is $\bar{z} = a - bi$.

Then, by Theorem 12.2,

$$z \cdot \bar{z} = (a + bi)(a - bi)$$
$$= a^2 + b^2$$

which is a purely real number.

EXAMPLE 5. Find the solution set over the set C of complex numbers of $z^2 = -9$.

SOLUTION. We want $z = a + bi$ to be a solution of this equation where a and b are real numbers. This means

$$(a + bi)^2 = -9$$

which is equivalent to

$$a^2 + 2abi + (bi)^2 = -9$$
$$a^2 - b^2 + 2abi = -9$$

By Definition 12.2,

$$a^2 - b^2 = -9 \qquad \text{and} \qquad 2ab = 0.$$

Since a and b are real numbers and $2ab = 0$, either $a = 0$ or $b = 0$ by Theorem 1.7. If $b = 0$, then $a^2 = -9$ which is impossible since a is a real number. Hence, $a = 0$ and we have

$$-b^2 = -9$$
$$b^2 = 9$$
$$b = 3 \qquad \text{or} \qquad b = -3$$

Thus, $0 - 3i = -3i$ and $0 + 3i = 3i$ are the solutions of the given equation and the solution set is $\{-3i, 3i\}$.

EXERCISES 12.2

Name the conjugate of each of the following complex numbers (Exercises 1–10).

1. $3 + 4i$
2. $2 + 3i$
3. $5 - 7i$
4. $4 - 5i$
5. $2\sqrt{2} + 3\sqrt{3}\,i$
6. $7 + \sqrt{3}\,i$
7. $-4 - \sqrt{5}\,i$
8. $8 + 3i$
9. $4i$
10. $-2i$

Simplify. Give each result in the form $a + bi$ (Exercises 11–30).

11. $(3 + 2i) + (5 + 6i)$
12. $(4 - 2i) + (-3 + 4i)$
13. $(-2 - i) + (-7 + 2i)$
14. $(2 - 5i) + (3 - 2i)$
15. $(2 + 5i) - (3 - 7i)$
16. $(-4 - i) - (8 - i)$

17. $(-2 + 5i) - (7 - 3i)$

18. $(3 + \sqrt{2}\,i) - (5 - \sqrt{2}\,i)$

19. $(3 + 2i)(3 - 2i)$

20. $(4 + 5i)^2$

21. $(4 - 2i)(3 - 5i)$

22. $(3 + 4i)(2 - 3i)$

23. $(\sqrt{2} + \sqrt{3}\,i)(\sqrt{2} - \sqrt{3}\,i)$

24. $(2\sqrt{3} + i)(4\sqrt{3} - 2i)$

25. $\dfrac{2 + i}{2 - i}$

26. $\dfrac{3 + 4i}{3 + 2i}$

27. $\dfrac{2 - \sqrt{2}\,i}{3 - \sqrt{2}\,i}$

28. $\dfrac{5 - 3i}{3 + 2i}$

29. $\dfrac{4 - 5i}{3 + 4i}$

30. $\dfrac{7 - i}{3i}$

Find the solution sets over the set C of complex numbers of the following equations (Exercises 31–38).

31. $z^2 = -4$

32. $z^2 = -25$

33. $z^2 = -20$

34. $z^2 = -32$

35. $z^2 = -56$

36. $z^2 = -54$

37. $z^2 = -48$

38. $z^2 = -72$

12.3 ◦ Square Roots which are Complex Numbers

We observe that

(1) $(4i)(4i) = 16i^2 = -16$

(2) $(-4i)(-4i) = 16i^2 = -16$

(3) $(\sqrt{2}\,i)(\sqrt{2}\,i) = 2i^2 = -2$

(4) $(-\sqrt{2}\,i)(-\sqrt{2}\,i) = 2i^2 = -2$

From Equation 1 above it is evident that $4i$ is a square root of -16 since $(4i)(4i) = -16$. From Equation 2 it is evident that $-4i$ is also a square root of -16. From Equations 3 and 4 it is evident that $\sqrt{2}\,i$ and $-\sqrt{2}\,i$ are square roots of -2.

It appears from the above that for each real number $a < 0$, $\sqrt{-a}\,i$ is a square root of a and $-\sqrt{-a}\,i$ is a square root of a. Observe that in this statement, $a < 0$ and, therefore, $-a > 0$, so that $\sqrt{-a} > 0$. It also follows that $(\sqrt{-a})^2\,i$ and $(-\sqrt{-a})^2\,i = a$. For $a < 0$ we do not call either $\sqrt{-a}\,i$ or $-\sqrt{-a}\,i$ the principal square root of a although we will use the square root symbol in the following definition.

DEFINITION 12.8. Each real number $x > 0$, has two **square roots:**

$$\sqrt{-x} = \sqrt{x}\,i \qquad \text{and} \qquad -\sqrt{-x} = -\sqrt{x}\,i.$$

Using Definition 12.8 we see that

$$(\sqrt{-4})(\sqrt{-9}) = (2i)(3i) = 6i^2 = -6.$$

Notice that $\sqrt{-4}\sqrt{-9} \neq \sqrt{(-4)(-9)}$ since $\sqrt{(-4)(-9)} = \sqrt{36} = 6$.

EXAMPLE 1. Write $\sqrt{-25}$ in the form $a + bi$, where a and b are real numbers.

SOLUTION

$$\sqrt{-25} = i\sqrt{25} = i \cdot 5 = 5i = 0 + 5i$$

EXAMPLE 2. Write $-\sqrt{-4}$ in the form $a + bi$, where a and b are real numbers.

SOLUTION

$$-\sqrt{-4} = -i\sqrt{4} = -i \cdot 2 = -2i = 0 - 2i$$

EXAMPLE 3. Write $\sqrt{-27}$ in the form $a + bi$, where a and b are real numbers.

SOLUTION

$$\sqrt{-27} = i\sqrt{27} = i\sqrt{9}\sqrt{3} = 3\sqrt{3}\,i = 0 + 3\sqrt{3}\,i$$

EXAMPLE 4. Write $(2 + \sqrt{-3})(2 - \sqrt{-3})$ in the form $a + bi$, where a and b are real numbers.

SOLUTION

$$\begin{aligned}
(2 + \sqrt{-3})(2 - \sqrt{-3}) &= (2 + i\sqrt{3})(2 - i\sqrt{3}) \\
&= 2^2 - i^2 \cdot 3 \\
&= 4 + 3 = 7 \\
&= 7 + 0 \cdot i
\end{aligned}$$

EXAMPLE 5. Write $\dfrac{1}{1 + \sqrt{-9}}$ in the form $a + bi$, where a and b are real numbers.

SOLUTION

$$\begin{aligned}
\frac{1}{1 + \sqrt{-9}} &= \frac{1}{1 + 3i} \\
&= \frac{1}{1 + 3i} \cdot \frac{1 - 3i}{1 - 3i} \\
&= \frac{1 - 3i}{1^2 - i^2 \cdot 3^2}
\end{aligned}$$

$$= \frac{1 - 3i}{10}$$

$$= \frac{1}{10} - \frac{3}{10}i$$

EXERCISES 12.3

Write each of the following in the form $a + bi$, where a and b are real numbers.

1. $\sqrt{-9}$
2. $-\sqrt{-25}$
3. $-\sqrt{-16}$
4. $\sqrt{-27}$
5. $-\sqrt{-27}$
6. $2 + \sqrt{-5}$
7. $\sqrt{-4} \cdot \sqrt{-9}$
8. $\sqrt{-27} \cdot \sqrt{4}$
9. $(2 + \sqrt{-3})(2 - \sqrt{-3})$
10. $(3 - \sqrt{-5})^2$
11. $\dfrac{1}{2 + \sqrt{-2}}$
12. $\dfrac{1}{3 - \sqrt{-4}}$
13. $\dfrac{\sqrt{-5}}{1 - \sqrt{-5}}$
14. $\dfrac{2 + \sqrt{-1}}{2 + \sqrt{-4}}$
15. $\dfrac{3 + \sqrt{-3}}{5 + \sqrt{-3}}$
16. $(2 - \sqrt{-3})^2$
17. $(2\sqrt{2} + 3\sqrt{-2})^2$

18. $(2 + \sqrt{-3})(3 - \sqrt{-3})$
19. $(7 - 2\sqrt{-5})(3 + 2\sqrt{-5})$
20. $(2\sqrt{2} + 3\sqrt{-3})(5\sqrt{2} - 3\sqrt{-3})$
21. $(7\sqrt{2} + 3\sqrt{-5})(5\sqrt{2} + 2\sqrt{-5})$
22. $(3\sqrt{5} - 5\sqrt{-3})(2\sqrt{5} - 2\sqrt{-3})$
23. $\dfrac{2 + \sqrt{-3}}{3 + \sqrt{-2}}$
24. $\dfrac{5 - \sqrt{-5}}{5 + \sqrt{-5}}$
25. $\dfrac{7 + \sqrt{-3}}{2 - \sqrt{-3}}$
26. $\dfrac{6 + \sqrt{-6}}{4 + 3\sqrt{-6}}$
27. $\dfrac{3\sqrt{2} + \sqrt{-5}}{\sqrt{2} + \sqrt{-5}}$
28. $\dfrac{6\sqrt{5} - 5\sqrt{-6}}{7\sqrt{5} - 6\sqrt{-6}}$
29. $\dfrac{\sqrt{7}}{1 - 3\sqrt{-5}}$
30. $\dfrac{\sqrt{-7}}{\sqrt{2} + 3\sqrt{-7}}$

12.4 ◆ Complex Numbers and Quadratic Equations

In the real number system a quadratic equation of the form

$$ax^2 + bx + c = 0 \qquad (a \neq 0)$$

has two solutions (in the case $b^2 - 4ac > 0$), one solution (in the case $b^2 - 4ac = 0$), or no solutions (in the case $b^2 - 4ac < 0$). In the case where $b^2 - 4ac < 0$ there are no real number solutions, since $\sqrt{b^2 - 4ac}$ is not a real number. But if $b^2 - 4ac < 0$, then $\sqrt{b^2 - 4ac}$ is a complex number and the solution set over the set C of complex numbers will have two complex numbers in it, namely,

$$\frac{-b + \sqrt{b^2 - 4ac}}{2a} \quad \text{and} \quad \frac{-b - \sqrt{b^2 - 4ac}}{2a}$$

EXAMPLE 1. Find the solution set over the set C of complex number of

$$x^2 + 2x + 5 = 0$$

SOLUTION. Here a is 1, b is 2, and c is 5. Since $b^2 - 4ac = (2)^2 - 4(1)(5) = -16$, this equation has two complex number solutions:

$$x = \frac{-2 \pm \sqrt{2^2 - 4(1)(5)}}{2(1)}$$
$$= \frac{-2 \pm \sqrt{-16}}{2}$$
$$= \frac{-2 \pm 4i}{2}$$
$$= -1 \pm 2i$$

The solution set is $\{-1 + 2i, -1 - 2i\}$.

EXAMPLE 2. Find the solution set over the set C of complex numbers of

$$x^2 + 2x + 4 = 0$$

SOLUTION. Here a is 1, b is 2, and c is 4. Since $b^2 - 4ac = 2^2 - 4(1)(4) = -12$, this equation has two complex number solutions:

$$x = \frac{-2 \pm \sqrt{2^2 - 4(1)(4)}}{2(1)}$$
$$x = \frac{-2 \pm \sqrt{-12}}{2}$$
$$x = \frac{-2 \pm 2\sqrt{3}\,i}{2}$$
$$x = -1 \pm \sqrt{3}\,i$$

The solution set is $\{-1 + \sqrt{3}\,i, -1 - \sqrt{3}\,i\}$.

EXERCISES 12.4

Find the solution sets over the set C of complex numbers of the following quadratic equations.

1. $x^2 - 2x + 40 = 0$

2. $x^2 - 4x + 20 = 0$

3. $x^2 + 6x + 10 = 0$

4. $x^2 - 10x + 50 = 0$

5. $x^2 + 8x + 25 = 0$

6. $2x^2 - 2x + 25 = 0$

7. $9x^2 - 6x + 5 = 0$

8. $4x^2 + 4x + 10 = 0$

9. $4x^2 + 3x + 10 = 0$

10. $2x^2 - 5x + 17 = 0$

11. $x^2 - x + 1 = 0$

12. $x^2 + x + 1 = 0$

13. $2x^2 - 3x + 6 = 0$

14. $x^2 - 4x + 9 = 0$

15. $3x^2 + 5x + 7 = 0$

16. $5x^2 + x + 6 = 0$

17. $4x^2 + x + 1 = 0$

18. $3x^2 - 2x + 1 = 0$

19. $2x^2 + 5x + 12 = 0$

20. $x^2 + 8x = -21$

21. $4x^2 + 7 - 8x = 0$

22. $2x^2 + 2x + 5 = 0$

23. $9x^2 - 12x + 5 = 0$

24. $36x - 9x^2 - 38 = 0$

25. $x^2 + 2x + 9 = 0$

26. $x^2 + 2\sqrt{3}\,x + 8 = 0$

27. $5x^2 - 3x + 7 = 0$

28. $x^2 + \frac{2}{3}x + \frac{2}{9} = 0$

29. $x^2 - 2x + 8 = 0$

30. $x^2 + \frac{1}{3}x + \frac{4}{3} = 0$

31. $x^2 - 2x + 10 = 0$

32. $9x^2 + 9x + \frac{13}{4} = 0$

33. $x^2 + 2\sqrt{2}\,x + 9 = 0$

34. $5x^2 - \sqrt{2}\,x + 2 = 0$

35. $6x^2 - 17x + 45 = 0$

36. $15x^2 + x + 1 = 0$

37. $x^2 + 3x + 9 = 0$

38. $2x^2 + \sqrt{2}\,x + 3 = 0$

39. $x^2 + \dfrac{\sqrt{3}}{2}x + \dfrac{2}{3} = 0$

40. $x^2 + \sqrt{3}\,x + 4 = 0$

12.5 ◆ Geometric Representation of Complex Numbers

A complex number is an ordered pair, (a, b), of real numbers and we know that every ordered pair of real numbers may be represented by a point in a plane with a Cartesian coordinate system. Since the standard form of a complex number, $a + bi$ (where a and b are real numbers), is merely another way of writing the ordered pair (a, b), we may consider $a + bi$ as representing the point (a, b) in the plane.

Since the real part, a, of the complex number $a + bi$, is taken as the abscissa or x-coordinate of the point (a, b) in this context, we called the x-axis the **real axis**. Similarly, since the imaginary part, b, of $a + bi$, is taken as the ordinate or y-coordinate of the point (a, b), we call the y-axis the **imaginary axis**. The plane is called the **complex plane.**

To emphasize that we are assigning complex numbers to points in the plane, we will label a point by using the complex number $a + bi$ rather than the ordered pair (a, b). In Figure 12.1 we have indicated the representation in the complex plane of several complex numbers.

Let us now consider the complex number $z = a + bi$ as the label of the point P in Figure 12.2. The distance from the origin $(0, 0) = 0 + 0i$ to the

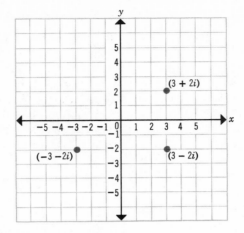

Figure 12.1

point P is called the **absolute value** or **modulus** of the complex number z. We see from Figure 12.2 that this distance, denoted by r, is $\sqrt{a^2 + b^2}$.

DEFINITION 12.9. The **absolute value** or **modulus** of the complex number $z = a + bi$, denoted by r, $|z|$, or $|a + bi|$ is $\sqrt{a^2 + b^2}$.

An **argument** of z is an angle measured from the positive real axis to the ray $\overrightarrow{OP}$.

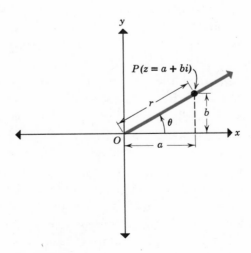

Figure 12.2

DEFINITION 12.10. An **argument** of the complex number $z = a + bi$, denoted by arg $(a + bi)$ is an angle θ with initial side the positive x-axis and terminal side the ray with endpoint the origin and containing the point representing $a + bi$.

We see from Definition 12.10 that if θ is an argument of $a + bi$, so is $\theta + 2\pi k$ or $\theta + 360°k$, k an integer. It also follows from Definition 12.10 that

$$\theta = \arg(a + bi) = \arctan\frac{b}{a} = \arcsin\frac{b}{r}$$

EXAMPLE 1. Find the absolute value of $z = 1 - \sqrt{3}\,i$.

SOLUTION. By Definition 12.9,

$$z = \sqrt{a^2 + b^2}$$
$$= \sqrt{1^2 + (\sqrt{3})^2}$$
$$= \sqrt{1 + 3} = 2$$

EXAMPLE 2. Find an argument of the complex number $1 - \sqrt{3}\,i$.

SOLUTION. By Definition 12.10,

$$\theta = \arctan\frac{b}{a} = \arcsin\frac{b}{r}$$

But $b = -\sqrt{3}$, $a = 1$, and from Example 1, $r = |z| = 2$. Then,

$$\theta = \arctan(-\sqrt{3}) = \arcsin\left(-\frac{\sqrt{3}}{2}\right)$$

Since $\tan\theta < 0$ and $\sin\theta < 0$, the argument θ must be in the fourth quadrant and one value is $300°$. Another possible value is $-60°$. Then,

$$\theta = 300° + 360k, \ k \text{ an integer}$$

EXAMPLE 3. Find an argument of the complex number $z = 3 + 4i$.

SOLUTION. By Definition 12.10,

$$\theta = \arctan\frac{b}{a} = \arcsin\frac{b}{r}$$

But $a = 3$, $b = 4$, and $r = \sqrt{a^2 + b^2} = \sqrt{3^2 + 4^2} = \sqrt{25} = 5$.

Then,

$$\theta = \text{arc tan } \tfrac{4}{3} = \text{arc sin } \tfrac{4}{5}$$
$$= \text{arc tan } 1.333 = \text{arc sin } 0.8000$$

Since $\tan \theta > 0$ and $\sin \theta > 0$, the argument θ must be in the first quadrant and one value is $53° \, 10'$. Another possible value is $413° \, 10'$.

EXERCISES 12.5

Find the absolute value of each of the following complex numbers (Exercises 1–10).

1. $2 + 3i$ 6. $\sqrt{5} - 2\sqrt{3}\,i$
2. $4 - 3i$ 7. $\sqrt{2} + 2\sqrt{5}\,i$
3. $\sqrt{2} - \sqrt{3}\,i$ 8. $\sqrt{2} - 3i$
4. $7 + 2\sqrt{5}\,i$ 9. $-4 - \sqrt{3}\,i$
5. $-3 - 4i$ 10. $-5 - 5\sqrt{2}\,i$

Find an argument between $0°$ and $360°$ of each of the following complex numbers (Exercises 11–20).

11. $-1 - i$ 16. $4i$
12. $2 - 2i$ 17. $1 - \sqrt{3}\,i$
13. 4 18. $-3 - 3i$
14. $1 + i$ 19. $-5i$
15. $-1 - \sqrt{3}\,i$ 20. $-\sqrt{3} - \sqrt{3}\,i$

Plot the points that represent the complex numbers given below (Exercises 21–30).

21. $2 + 3i$ 26. $-8 + 4i$
22. $4 - 3i$ 27. $-10 - 5i$
23. $\sqrt{3} - i$ 28. $1 - \sqrt{3}\,i$
24. 6 29. $\tfrac{1}{2}(3 - 4i)$
25. $-7i$ 30. $\sqrt{5} - 3i$
31. If $z_1 = 3 + 4i$ and $z_2 = 4 - 3i$, is $|z_1 + z_2|$ less than, equal to, or greater than $|z_1| + |z_2|$?
32. If $z_1 = 3 - 4i$ and $z_2 = 2 + i$, is $|z_1 - z_2|$ less than, equal to, or greater than $|z_1| - |z_2|$?
33. If $z = a + bi$, prove that $|z| = |-z|$.
34. If $z_1 = a + bi$ and $z_2 = c + di$, prove that $|z_1 z_2| = |z_1| \cdot |z_2|$.
35. If $z = a + bi$, show that $|z|^2 = |z \cdot \bar{z}|$.
36. If $z = a + bi$, show that $|z| = |\bar{z}|$.

12.6 ✦ Trigonometric Form of a Complex Number

Every ordered pair (a, b) of real numbers may be written $(r \cos \theta, r \sin \theta)$ where r is the modulus $\sqrt{a^2 + b^2}$ of $a + bi$ and θ is an argument of $a + bi$.

Hence, every complex number $a + bi$ may be written in the form
$$r \cos \theta + ir \sin \theta$$
or
$$r(\cos \theta + i \sin \theta)$$

This form is called the **trigonometric form** or **polar form** of the complex number.

Since $\cos \theta = \cos (\theta + 360°k)$ and $\sin \theta = \sin (\theta + 360°k)$, where k is an integer, we have, in general,

$$a + bi = r \left[\cos (\theta + 360°k) + i \sin (\theta + 360°k)\right] \qquad (k \text{ an integer})$$

We customarily use the angle of smallest nonnegative measure for θ.

EXAMPLE 1. Express $1 - i$ in trigonometric form.

SOLUTION. Since $a = 1$ and $b = -1$, we have

$$r = \sqrt{a^2 + b^2} = \sqrt{1^2 + (-1)^2} = \sqrt{2}$$

Since $1 - i$ is in the fourth quadrant and since $\tan \theta = \dfrac{-1}{1} = -1$,

we have $\theta = 315°$. Therefore,

$$1 - i = \sqrt{2} \, (\cos 315° + i \sin 315°).$$

EXAMPLE 2. Write $3 \, (\cos 120° + i \sin 120°)$ in standard form.

SOLUTION

$$a = 3 \cos 120° = 3 \left(-\frac{1}{2}\right) = -\frac{3}{2}$$

$$b = 3 \sin 120° = 3 \left(\frac{\sqrt{3}}{2}\right) = \frac{3\sqrt{3}}{2}$$

and

$$a + bi = -\frac{3}{2} + \frac{3\sqrt{3}}{2} i$$

EXERCISES 12.6

Express the following complex numbers in trigonometric form.

1. $1 + i$

2. $-1 - i$

3. 3

4. $1 - \sqrt{3}\,i$

5. $1 - i$

6. $-1 + i$

7. $\sqrt{3} + i$

8. $-\sqrt{3} - \sqrt{3}\,i$

9. $7 - 7i$

10. -4

11. $-2i$

12. $3i$

13. -5

14. $-3 + 3i$

15. $1 + \sqrt{3}\,i$

16. $\dfrac{1}{2} + \dfrac{\sqrt{3}}{2}\,i$

17. 1

18. $5 + 5i$

19. $\dfrac{1}{2} - \dfrac{\sqrt{3}}{2}\,i$

20. $2\sqrt{3} - 2i$

12.7 ◆ Multiplication and Division of Complex Numbers

Products and quotients of complex numbers can be found easily when the numbers are expressed in trigonometric form. Let

$$z_1 = r_1 \left(\cos \theta_1 + i \sin \theta_1 \right) \qquad \text{and} \qquad z_2 = r_2 \left(\cos \theta_2 + i \sin \theta_2 \right)$$

Then

$$
\begin{aligned}
z_1 z_2 &= [r_1 \left(\cos \theta_1 + i \sin \theta_1 \right)][r_2 \left(\cos \theta_2 + i \sin \theta_2 \right)] \\
&= r_1 r_2 \left[\cos \theta_1 \cos \theta_2 + i \cos \theta_1 \sin \theta_2 + i \sin \theta_1 \cos \theta_2 + i^2 \sin \theta_1 \sin \theta_2 \right] \\
&= r_1 r_2 \left[(\cos \theta_1 \cos \theta_2 - \sin \theta_1 \sin \theta_2) + (\cos \theta_1 \sin \theta_2 + \sin \theta_1 \cos \theta_2)i \right]
\end{aligned}
$$

But

$$\cos \theta_1 \cos \theta_2 - \sin \theta_1 \sin \theta_2 = \cos \left(\theta_1 + \theta_2 \right)$$

and

$$\cos \theta_1 \sin \theta_2 + \sin \theta_1 \cos \theta_2 = \sin \left(\theta_1 + \theta_2 \right)$$

Hence

$$z_1 z_2 = r_1 r_2 \left[\cos \left(\theta_1 + \theta_2 \right) + i \sin \left(\theta_1 + \theta_2 \right) \right]$$

We have proved the following theorem.

THEOREM 12.4. If $z_1 = r_1 \left(\cos \theta_1 + i \sin \theta_1 \right)$ and $z_2 = r_2 \left(\cos \theta_2 + i \sin \theta_2 \right)$ then

$$z_1 z_2 = r_1 r_2 \left[\cos \left(\theta_1 + \theta_2 \right) + i \sin \left(\theta_1 + \theta_2 \right) \right]$$

EXAMPLE 1. Find the product of $z_1 = 2 \left(\cos 80° + i \sin 80° \right)$ and $z_2 = 3 \left(\cos 40° + i \sin 40° \right)$.

SOLUTION. Using Theorem 12.4, we obtain

$$
\begin{aligned}
z_1 z_2 &= (2)(3)[\cos \left(80° + 40° \right) + i \sin \left(80° + 40° \right)] \\
&= 6 \left(\cos 120° + i \sin 120° \right) \\
&= 6 \left(-\frac{1}{2} + i \frac{\sqrt{3}}{2} \right) \\
&= -3 + 3\sqrt{3}\,i
\end{aligned}
$$

Now let us find the quotient $\dfrac{z_1}{z_2}$, $z_2 \neq 0$

$$
\begin{aligned}
\frac{z_1}{z_2} &= \frac{r_1\left(\cos\theta_1 + i\sin\theta_1\right)}{r_2\left(\cos\theta_2 + i\sin\theta_2\right)} \\[6pt]
&= \frac{r_1\left(\cos\theta_1 + i\sin\theta_1\right)}{r_2\left(\cos\theta_2 + i\sin\theta_2\right)} \cdot \frac{\cos\theta_2 - i\sin\theta_2}{\cos\theta_2 - i\sin\theta_2} \\[6pt]
&= \frac{r_1}{r_2} \cdot \frac{\cos\theta_1\cos\theta_2 - i\cos\theta_1\sin\theta_2 + i\sin\theta_1\cos\theta_2 - i^2\sin\theta_1\sin\theta_2}{\cos^2\theta_2 + \sin^2\theta_2} \\[6pt]
&= \frac{r_1}{r_2} \cdot \frac{\left(\cos\theta_1\cos\theta_2 + \sin\theta_1\sin\theta_2\right) + i\left(\sin\theta_1\cos\theta_2 - \cos\theta_1\sin\theta_2\right)}{\cos^2\theta_2 + \sin^2\theta_2} \\[6pt]
&= \frac{r_1}{r_2} \cdot \frac{\cos\left(\theta_1 - \theta_2\right) + i\sin\left(\theta_1 - \theta_2\right)}{1}
\end{aligned}
$$

We have proved the following theorem.

THEOREM 12.5. If $z_1 = r_1\left(\cos\theta_1 + i\sin\theta_1\right)$ and $z_2 = r_2\left(\cos\theta_2 + i\sin\theta_2\right)$, $z_2 \neq 0$, then

$$
\frac{z_1}{z_2} = \frac{r_1}{r_2}\left[\cos\left(\theta_1 - \theta_2\right) + i\sin\left(\theta_1 - \theta_2\right)\right]
$$

EXAMPLE 2. Find the quotient $\dfrac{6\left(\cos 240° + i\sin 240°\right)}{3\left(\cos 150° + i\sin 150°\right)}$

SOLUTION. By Theorem 12.5,

$$
\begin{aligned}
\frac{6\left(\cos 240° + i\sin 240°\right)}{3\left(\cos 150° + i\sin 150°\right)} &= \frac{6}{3}\left[\cos\left(240 - 150\right)°\right. \\
&\qquad\qquad\left. + i\sin\left(240 - 150\right)\right]° \\
&= 2\left(\cos 90° + i\sin 90°\right) \\
&= 2\left(0 + i\cdot 1\right) \\
&= 2i
\end{aligned}
$$

12.8 ◆ DeMoivre's Theorem

We now state a simple formula by means of which the nth power of any complex number z can be found at once when the complex number is expressed in trigonometric form.

Let $z = r\left(\cos\theta + i\sin\theta\right)$. Then, using Theorem 12.4, we obtain

$$
\begin{aligned}
z^2 &= r^2\left(\cos 2\theta + i\sin 2\theta\right) \\
z^3 &= r^3\left(\cos 3\theta + i\sin 3\theta\right) \\
z^4 &= r^4\left(\cos 4\theta + i\sin 4\theta\right)
\end{aligned}
$$

and, in general,

$$z^n = r^n \left(\cos n\theta + i \sin n\theta \right)$$

where n is any positive integer.

This formula for z^n can be established by applying deMoivre's theorem, which we shall accept without proof.

THEOREM 12.6. DeMoivre's Theorem. If $z = r \left(\cos \theta + i \sin \theta \right)$ and n is a positive integer then $z^n = r^n \left(\cos n\theta + i \sin n\theta \right)$.

We now define z^{-p}:

DEFINITION 12.11. If z is a complex number, $z \neq 0 + 0 \cdot i$, then for p a positive integer

$$z^{-p} = \frac{1}{z^p}$$

Observing that

$$
\begin{aligned}
\left[r \left(\cos \theta + i \sin \theta \right) \right]^{-p} &= \frac{1}{\left[r \left(\cos \theta + i \sin \theta \right) \right]^p} \\
&= \frac{1}{r^p \left(\cos \theta + i \sin \theta \right)^p} \\
&= \frac{1}{r^p \left(\cos p\theta + i \sin p\theta \right)} \\
&= \frac{r^{-p} \left[\cos (p\theta) - i \sin (p\theta) \right]}{\cos^2 p\theta + \sin^2 p\theta} \\
&= r^{-p} \left(\cos p\theta - i \sin p\theta \right) \\
&= r^{-p} \left[\cos (-p\theta) + i \sin (-p\theta) \right]
\end{aligned}
$$

we see that we have established deMoivre's theorem for negative integers.

We now define the complex number z to the zero power.

DEFINITION 12.12. If z is a complex number, $z \neq 0 + 0 \cdot i$, then

$$z^0 = 1 + 0 \cdot i = 1$$

By defining z^0 as $1 + 0i$ when $z \neq 0 + 0i$, we can extend deMoivre's theorem to include all integral exponents, since

$$
\begin{aligned}
z^0 &= r^0 \left(\cos 0 \cdot \theta + i \sin 0 \cdot \theta \right) \\
&= 1 \left(\cos 0 + i \sin 0 \right) \\
&= 1
\end{aligned}
$$

EXAMPLE 1. Evaluate $(-1 - i)^6$.

SOLUTION. Since $a = -1$ and $b = -1$, $r = \sqrt{(1)^2 + (-1)^2} = \sqrt{2}$, $\tan \theta = \dfrac{-1}{-1}$ and $\theta = 225°$. Using deMoivre's theorem, we have

$$
\begin{aligned}
(-1 - i)^6 &= [2^{1/2} (\cos 225° + i \sin 225°)]^6 \\
&= 2^3 (\cos 6 \cdot 225 + i \sin 6 \cdot 225) \\
&= 8 (\cos 1350° + i \sin 1350°) \\
&= 8 (\cos 270° + i \sin 270°) \\
&= 8 [0 + i(-1)] \\
&= -8i
\end{aligned}
$$

EXAMPLE 2. Evaluate $(1 - \sqrt{3}\,i)^{-5}$.

SOLUTION. Since $a = 1$ and $b = -\sqrt{3}$, $r = \sqrt{1^2 + (-\sqrt{3})^2} = \sqrt{1 + 3} = 2$. Thus $\theta = \arctan(-\sqrt{3}) = \arcsin(-\frac{1}{2}\sqrt{3})$. Since $\sin \theta$ and $\tan \theta$ are both negative, θ is in the fourth quadrant. Thus $\theta = 300°$. By deMoivre's theorem,

$$
\begin{aligned}
(1 - \sqrt{3}\,i)^{-5} &= [2 (\cos 300° + i \sin 300°)]^{-5} \\
&= 2^{-5} [\cos (-5)(300°) + i \sin (-5)(300°)] \\
&= \tfrac{1}{32} [\cos (-1500°) + i \sin (-1500°)] \\
&= \tfrac{1}{32} [\cos (-60°) + i \sin (-60°)] \\
&= \tfrac{1}{32} (\tfrac{1}{2} - \tfrac{1}{2} \sqrt{3}\,i) \\
&= \frac{1}{64} - \frac{\sqrt{3}}{64}\,i
\end{aligned}
$$

EXERCISES 12.7

Perform the indicated operations. Give the results in trigonometric form.

1. $[3 (\cos 30° + i \sin 30°)]^3$
2. $[5 (\cos 60° + i \sin 60°)]^4$
3. $[2 (\cos 135° + i \sin 135°)]^{10}$
4. $(\cos 215° + i \sin 215°)^{12}$
5. $(\cos 320° + i \sin 320°)^{-1/2}$
6. $4 (\cos 136° + i \sin 136°)^{-1/4}$
7. $(\cos (-15°) + i \sin (-15°))^4$
8. $(\cos 30° - i \sin 30°)^5$
9. $[2 (\cos 135° + i \sin 135°)]^6$
10. $[3 (\cos 35° - i \sin 35°)]^4$
11. $(\sqrt{2} - \sqrt{2}\,i)^5$
12. $(1 - i)^7$
13. $(1 - \sqrt{3}\,i)^4$
14. $(\sqrt{3} - i)^8$
15. $(-i)^{12}$
16. $(1 - \sqrt{3}\,i)^{10}$
17. $(-\sqrt{2} + \sqrt{2}\,i)^{-4}$
18. $(-1 + \sqrt{3}\,i)^{-2}$
19. $(-1 - i)^5$
20. $(5 + 12i)^{-2}$

12.9 ◆ Roots of Complex Numbers

A complex number w is called an **nth root** of $z = r(\cos \theta + i \sin \theta)$ provided $w^n = z$, n a positive integer. To find all the nth roots of a complex number z, we must solve the equation

$$w^n = z$$

where z is a fixed complex number, $z \neq 0 + 0i$, and n is a positive integer.

We can use deMoivre's theorem to show that there are exactly n distinct nth roots of any nonzero complex number z.

If θ is an argument of z and r is the modulus of z, then z can be written in the form

$$z = r(\cos \theta + i \sin \theta)$$

But if k is any integer, we may also write

$$z = r[\cos \theta + 2\pi k) + i \sin (\theta + 2\pi k)]$$

since we know that the trigonometric functions cosine and sine have period 2π. Applying deMoivre's theorem, we have

$$\left[r^{1/n} \left(\cos \frac{\theta + 2\pi k}{n} + i \sin \frac{\theta + 2\pi k}{n} \right) \right]^n = r[\cos (\theta + 2\pi k) + i \sin (\theta + 2\pi k)]$$

$$= z$$

Therefore, every complex number of the form

$$r^{1/n} \left[\cos \frac{\theta + 2\pi k}{n} + i \sin \frac{\theta + 2\pi k}{n} \right]$$

where $k = 0, 1, 2, \ldots, n - 1$, is an nth root of z. We thus obtain the n distinct nth roots of z. Since any other integer value of k will differ from one of these by an integer multiple of n, the corresponding argument will differ by a multiple of 2π and the same nth root will result. We have thus proved:

THEOREM 12.7. The n distinct nth roots of the complex number

$$z = r(\cos \theta + i \sin \theta) \qquad (r \neq 0)$$

are the complex numbers

$$r^{1/n} \left[\cos \frac{\theta + 2\pi k}{n} + i \sin \frac{\theta + 2\pi k}{n} \right], k = 0, 1, 2, \ldots, n - 1$$

EXAMPLE 1. Find the five fifth roots of $1 - i$.

SOLUTION. For $z = a + bi$ we have $a = 1$ and $b = -1$. Thus, $r = \sqrt{1^2 + (-1)^2} = \sqrt{2}$. We have $\theta = \text{arc tan} - 1 = \text{arc sin}$ $(-\frac{1}{2}\sqrt{2})$; θ is then in the fourth quadrant so $\theta = 315°$. Therefore,

$$1 - i = \sqrt{2} \, (\cos 315° + i \sin 315°)$$

By Theorem 12.7 we obtain as the five fifth roots of $1 - i$:

$$(\sqrt{2})^{1/5} \left[\cos \left(\frac{315 + 360k}{5} \right)^\circ + i \sin \left(\frac{315 + 360k}{5} \right)^\circ \right],$$
$$k = 0, 1, 2, 3, 4$$

Substituting the indicated values of k gives as the five fifth roots of $1 - i$

$$2^{1/10} \, (\cos 63° + i \sin 63°)$$
$$2^{1/10} \, (\cos 135° + i \sin 135°)$$
$$2^{1/10} \, (\cos 207° + i \sin 207°)$$
$$2^{1/10} \, (\cos 279° + i \sin 279°)$$
$$2^{1/10} \, (\cos 351° + i \sin 351°)$$

EXAMPLE 2. Find the four fourth roots of 16,

SOLUTION. For $16 = a + bi$ we have $a = 16$ and $b = 0$. Thus, $r = 16$ and $\theta = \text{arc tan } 0 = \text{arc sin } 0 = 0$. We can write

$$16 = 16 \, (\cos 0° + i \sin 0°)$$

Thus the four fourth roots of 16 are

$$(16)^{1/4} \left[\cos \left(\frac{0 + 360k}{4} \right)^\circ + i \sin \left(\frac{0 + 360k}{4} \right)^\circ \right], k = 0, 1, 2, 3$$

That is, the four fourth roots of 16 are

$$2 \, (\cos 0° + i \sin 0°) = 2 \, (1 + 0i) = 2$$
$$2 \, (\cos 90° + i \sin 90°) = 2 \, (0 + i) = 2i$$
$$2 \, (\cos 180° + i \sin 180°) = 2 \, (-1 + 0i) = -2$$
and $\quad 2 \, (\cos 270° + i \sin 270°) = 2 \, (0 - i) = -2i$

EXERCISES 12.8

Find all the indicated roots of the following. Express in trigonometric form.

1. The cube roots of 8.
2. The fourth roots of 1.
3. The fifth roots of $-\frac{1}{2} - \frac{1}{2}\sqrt{3}\,i$.

4. The sixth roots of i.
5. The cube roots of -1.
6. The fifth roots of $-i$.
7. The fourth roots of -16.
8. The square roots of $1 + i$.
9. The cube roots of -27.
10. The cube roots of 27.
11. The square roots of $3 - \sqrt{3}\,i$.
12. The fifth roots of $24 - 7i$.
13. The fifth roots of $-1 - \sqrt{3}\,i$.
14. The cube roots of $\sqrt{3} + i$.
15. The cube roots of $4\sqrt{2} + 4\sqrt{2}\,i$.

CHAPTER REVIEW

1. Perform the following operations.
   (a) $(3, -2) + (-4, -5)$
   (b) $(-2, -5) - (3, -2)$
   (c) $(8, -6) \cdot (-2, 6)$
   (d) $\dfrac{(-2, 1)}{(3, 4)}$

2. Simplify. Give each result in the form $a + bi$.
   (a) $(3 + 4i) + (-3 - 5i)$
   (b) $(-2 + i) - (3 - 4i)$
   (c) $(6 + 3i) \cdot (-2 - i)$
   (d) $\dfrac{4 + 3i}{4 - 3i}$

3. Write each of the following in the form $a + bi$.
   (a) $\sqrt{-9}$ (d) $\sqrt{-27}$
   (b) $\sqrt{-12}$ (e) $-\sqrt{-81}$
   (c) $-\sqrt{-25}$ (f) $-\sqrt{-64}$

4. Find the solution sets of the following quadratic equations over the set C of complex numbers.
   (a) $x^2 - x + 4 = 0$ (c) $4x^2 - x + 1 = 0$
   (b) $2x^2 + 3x + 5 = 0$ (d) $3x^2 - 2x + 7 = 0$

5. Plot the points that represent the complex numbers given below. Give the modulus and an argument of each.
   (a) $-1 + i$ (c) $2i$
   (b) $-3 - 3i$ (d) $2 + 2i$

6. Express the following numbers in trigonometric form.

(a) $\dfrac{1}{2} - \dfrac{\sqrt{3}}{2}i$ (d) $-3i$

(b) $4i$ (e) 6

(c) -2 (f) $\sqrt{3} - i$

7. Perform the indicated operations. Give the result in trigonometric form.
   (a) $[3(\cos 30° + i \sin 30°)][2(\cos 40° + i \sin 40°)]$
   (b) $[\tfrac{1}{2}(\cos 45° + i \sin 45°)][\tfrac{1}{4}(\cos 60° + i \sin 60°)]$
   (c) $\dfrac{8(\cos 150° + i \sin 150°)}{2(\cos 30° + i \sin 30°)}$
   (d) $\dfrac{125(\cos 40° + i \sin 40°)}{25(\cos 120° + i \sin 120°)}$

8. State deMoivre's theorem.

9. Evaluate:
   (a) $[2(\cos 30° + i \sin 30°)]^3$
   (b) $[8(\cos 75° + i \sin 75°)]^{-1/3}$

10. Evaluate:
    (a) $(1 + 5i)^3$ (b) $(-1 - \sqrt{3}\,i)^{-4}$

11. Find the three cube roots of -8.

12. Find the five fifth roots of i.

13. Find the square roots of $1 - i$.

14. Find the square roots of $-i$.

15. Given four complex numbers $z_1 = 2 + 3i$, $z_2 = 1 + i$, $z_3 = 5$, and $z_4 = 2i$, determine the following.
    (a) $z_1 + z_2$
    (b) $z_1 \cdot z_2$
    (c) $z_1 - z_2$
    (d) $\dfrac{z_1}{z_3}$
    (e) the additive inverse of z_1
    (f) the multiplicative inverse of z_2
    (g) the trigonometric form of z_2
    (h) the conjugate of z_2
    (i) the fourth power of z_2
    (j) the square roots of z_4
    (k) the cube roots of z_3

16. Compute and express your answer in simplest form.
    (a) i^2 (f) i^7
    (b) i^3 (g) i^8
    (c) i^4 (h) i^9
    (d) i^5 (i) i^{10}
    (e) i^6 (j) i^{11}

17. Is the set $\{1, -1, i, -i\}$ closed under the operation of (a) addition; (b) multiplication?

18. Which of the following statements are true?
    (a) There is a one-to-one correspondence between the complex numbers and the points in a plane.
    (b) For each real number x, the graph of $x + 0 \cdot i$ is the x-axis.

(c) For each real number y, the graph of $0 + yi$ is the x-axis.

19. Which of the following statements are true? For all complex numbers z_1, z_2, and z_3;

(a) $z_1 \cdot z_2 = (0, 1)$

(b) $z_1 \cdot z_2 = z_2 \cdot z_1$

(c) $z_1 (z_2 + z_3) = z_1 z_2 + z_1 z_3$

(d) $z_1 \cdot \dfrac{1}{z_1} = (1, 0)$, $z_1 \neq (0, 0)$

CHAPTER 13

Theory of Equations

13.1 ◆ Introduction

A **polynomial** of degree n over a field F is an algebraic expression of the form

$$a_0x^n + a_1x^{n-1} + \cdots + a_{n-1}x + a_n$$

where $a_0 \neq 0$ and $a_0, a_1, \ldots, a_n$ are elements of F. It is customary to denote a polynomial by the symbol $P(x)$ read: P at x. When the variable x is replaced by a real number, a polynomial denotes a real number. When a particular real number c is substituted for x in $P(x)$, the real number named by $P(x)$ is called the **value** of $P(x)$ and denoted by $P(c)$. Thus, if

$$P(x) = x^2 + 2x + 1$$

then

$$P(2) = 2^2 + 2 \cdot 2 + 1 = 9$$
and
$$P(3) = 3^2 + 2 \cdot 3 + 1 = 16$$

The equation

$$a_0x^n + a_1x^{n-1} + \cdots + a_{n-1}x + a_n = 0$$

that results from equating a polynomial $P(x)$ to zero is called a **polynomial equation** of degree n. If $P(x)$ denotes a polynomial and if $P(r) = 0$, then r is called a **zero** of the polynomial $P(x)$ and a **root** of the equation $P(x) = 0$.

13.2 ◆ Synthetic Division

If $P(x)$ and $D(x)$ are polynomials of degrees n and m, respectively, $n > m$, over a field of F, there exist polynomials $Q(x)$ and $R(x)$ over the field F such that

$$P(x) = D(x)Q(x) + R(x)$$

where $R(x)$ is of degree less than $D(x)$.

If we are given $P(x)$ and $D(x)$ we find $Q(x)$ and $R(x)$ by dividing $P(x)$ by $D(x)$. In performing the division of one polynomial by another to find $Q(x)$ and $R(x)$, we usually arrange both the **dividend,** $P(x)$, and the **divisor,** $D(x)$, in descending powers of the variable and supply any missing terms, that is, terms with coefficients zero. The division of $P(x) = x^4 - 7x^3 + 5x - 7$ by $D(x) = x - 3$ is shown below.

$$
\begin{array}{r}
x^3 - 4x^2 - 12x\ - 31 \\
\hline
x - 3)\overline{x^4 - 7x^3 +\ 0x^2 +\ 5x -\ 7} \\
x^4 - 3x^3 \\
\hline
-\ 4x^3 +\ 0x^2 \\
-\ 4x^3 + 12x^2 \\
\hline
-\ 12x^2 +\ 5x \\
-\ 12x^2 + 36x \\
\hline
-\ 31x -\ 7 \\
-\ 31x + 93 \\
\hline
-100 \qquad \text{(remainder)}
\end{array}
$$

We see from the above division that the **quotient** is $Q(x) = x^3 - 4x^2 - 12x - 31$ and the **remainder** is $R(x) = -100$, hence

$$x^4 - 7x^3 + 5x - 7 = (x - 3)(x^3 - 4x^2 - 12x - 31) + (-100)$$

The quotient in the division of a polynomial by $(x - r)$, a polynomial of degree one, can be found much easier than the division shown above by a process called **synthetic division.**

In the above example, omitting variables and writing the coefficients of the terms, using zero coefficients for any missing terms, we have

$$
\begin{array}{r}
1 - 4 -\ 12 - 31 \\
\hline
1 - 3)\overline{1 - 7 +\ 0 +\ 5 -\ 7} \\
1 - 3 \\
\hline
-\ 4 +\ 0 \\
-\ 4 + 12 \\
\hline
-\ 12 +\ 5 \\
-\ 12 + 36 \\
\hline
-\ 31 -\ 7 \\
-\ 31 + 93 \\
\hline
-\ 100
\end{array}
$$

We see that it is repetitious to write the first term of every second line and the second term in the third, fifth, and seventh lines, since they are repetitions of the numerals written directly above them and of the coefficients of the associated variable in the quotient and of the coefficients in the dividend, respectively. Hence we can compress the process into the compact form below:

(1) $\qquad$ $-3\rfloor\ 1\quad -7\quad +\ 0\quad +\ 5\quad -\quad 7$

(2) $\qquad\qquad\qquad -3\quad +12\quad +36\quad +\ 93$

(3) $\qquad\qquad\ \ 1\quad -4\quad -12\quad -31\quad -100$ $\quad$ (remainder: -100)

In this form the repetitions are omitted and 1, the coefficient of x, in the divisor is also omitted.

The numbers named in line 3 are the coefficients of descending powers of the variables in the quotient and the remainder. They have been obtained by subtracting the detached coefficients in line 2 from the detached coefficients of the terms of the same degree in line 1. We could have obtained the same result by replacing -3 by 3 in the divisor and adding, instead of subtracting, at each step. This is precisely what is done in the synthetic division process. The final form is then

$\qquad$ $3\rfloor\ 1\quad -7\quad +\ 0\quad +\ 5\quad -\quad 7$

$\qquad\qquad\qquad +3\quad -12\quad -36\quad -\ 93$

$\qquad\qquad\ \ 1\quad -4\quad -12\quad -31\quad -100$ $\quad$ (remainder: -100)

The steps in the synthetic division of $P(x)$, a polynomial of degree n by $x - r$ are as follows.

1. Arrange $P(x)$ in descending powers of x, inserting any missing term with a coefficient of zero.

2. Write in order in a line the coefficients $a_0, a_1, a_2, \ldots, a_n$ of $P(x)$, retaining each sign. Write r to the left, separated by a vertical line.

3. Leave the next line blank and write a_0 in line 3 directly below a_0 in line 1.

4. Multiply a_0 by r and insert the product in line 2 under a_1. Add the product $a_0 r$ to a_1 and insert the result in line 3. Multiply this result by r and insert the product in line 2 under a_2. Add the product to a_2 and insert the result in line 3. Continue this process until a product is added to a_n and the result inserted in line 3.

5. The first n numbers named in line 3 are the coefficients of descending powers of the variables in the quotient, starting with x^{n-1}, and the $(n + 1)$th number is the remainder.

EXAMPLE 1. Divide $3x^4 - 5x^3 - 3x^2 + 11x - 10$ by $x - 5$.

SOLUTION. Using synthetic division, we write

$$
\begin{array}{r|rrrrr}
5 & 3 & -5 & -3 & +11 & -10 \\
  &   & +15 & +50 & +235 & +1230 \\
\hline
  & 3 & +10 & +47 & +246 & +1220
\end{array}
$$

The quotient is $3x^3 + 10x^2 + 47x + 246$. The remainder is 1220.

EXAMPLE 2. Divide $x^4 + 2x^3 + 10x - 12$ by $x + 3$.

SOLUTION. Since $x + 3 = x - (-3)$, the value of r is -3. Using synthetic division, we write

$$
\begin{array}{r|rrrrr}
-3 & 1 & +2 & +0 & +10 & -12 \\
   &   & -3 & +3 & -9 & -3 \\
\hline
   & 1 & -1 & +3 & +1 & -15
\end{array}
$$

The quotient is $x^3 - x^2 + 3x + 1$. The remainder is -15.

EXERCISES 13.1

Use synthetic division to determine the quotient and remainder in each division.

1. $(4x^3 + 5x^2 - 23x + 6) \div (x - 2)$
2. $(2x^4 - 8x^2 + 4x - 7) \div (x - 1)$
3. $(x^4 + 5x^3 + 7x^2 - 8x + 9) \div (x + 5)$
4. $(2x^5 + 7x^2 + 8x - 12) \div (x - 3)$
5. $(x + 3x^3 - 7x^4 + 2) \div (x - 1)$
6. $(x^2 - 12x + 7x^3 - 97) \div (x - 4)$
7. $(15x - 4x^5 + 26x^3 - 12) \div (x + 1)$
8. $(x^5 - 1) \div (x + 1)$
9. $(x^9 + 1) \div (x - 1)$
10. $(x^5 - 2x + 1) \div (x + 2)$
11. $(x^6 + x^5 - x + 2) \div (x - 4)$
12. $(x^5 - 3x^4 + 3x^3 - 2x + 1) \div (x + 1)$
13. $(11x^8 - 2x^5 + 3x^2 - 23) \div (x + 1)$
14. $(21x^7 - 12x^6 + 11x^5 - 3x^3 + 2x^2 - x) \div (x + 1)$
15. $(14x^7 - 3x^6 + 17x^5 - 11x^4 + 9x^3 + 1) \div (x - 2)$
16. $(5x^4 + 4x^3a - 6x^2a^2 - 7a^4) \div (x + 2a)$
17. $(6x^6 - 4x^5 - 9x^3 + 6x^2 - 6) \div (x - \frac{2}{3})$
18. $(4x^4 - 4x^3 + 3x^2 - 4x - 8) \div (x - \frac{1}{2})$

19. $(6x^4 - 3x^3 + 8x^2 + 8) \div (x - \frac{1}{2})$
20. $(6x^4 + 2x^3a - x^2a^2 + 7xa^3 + 28a^4) \div (x - \frac{2}{3}a)$

13.3 ◆ The Remainder Theorem

Let the polynomial $P(x)$ be divided by the polynomial $D(x)$. This division produces polynomials $Q(x)$, called the **quotient,** and $R(x)$, called the **remainder,** such that

$$P(x) = D(x) \cdot Q(x) + R(x)$$

and the degree of $R(x)$ is less than the degree of $D(x)$, the **divisor.**
Now let $P(x)$ be divided by $D(x) = (x - r)$. Then

(1) $$P(x) = (x - r) \cdot Q(x) + R(x)$$

The degree of $R(x)$ must be less than the degree of $x - r$, hence the degree of $R(x)$ must be zero and $R(x) = R$, a constant.

Equation 1 holds for all values of x. In particular, for $x = r$, we have

$$\begin{aligned} P(r) &= (r - r) \cdot Q(x) + R \\ &= 0 + R \\ &= R \end{aligned}$$

We see from the above that the value of the polynomial $P(x)$, when x has the value r, is equal to the remainder R, when $P(x)$ is divided by $(x - r)$. This is called the **Remainder Theorem** which we now state as Theorem 13.1.

THEOREM 13.1 (Remainder Theorem) If a polynomial $P(x)$ is divided by $(x - r)$, where r is any number, the resulting remainder is equal to $P(r)$.

EXAMPLE 1. Find the value of $x^4 + 8x + 2$ when $x = -1$.

SOLUTION. Using synthetic division to divide $x^4 + 8x + 2$ by $[x - (-1)]$, we have

$$\begin{array}{r|rrrrr} -1 & 1 & +0 & +0 & +8 & +2 \\ & & -1 & +1 & -1 & -7 \\ \hline & 1 & -1 & +1 & +7 & -5 \end{array}$$

By the remainder theorem, $P(-1) = -5$.

EXAMPLE 2. Find the remainder when $x^3 + 7x^2 + 3x + 3$ is divided by $x + 1$.

SOLUTION. Using synthetic division, we divide $x^3 + 7x^2 + 3x + 3$ by $x - (-1)$,

$$
\begin{array}{r|rrrr}
-1 & 1 & +7 & +3 & +3 \\
   &   & -1 & -6 & +3 \\
\hline
   & 1 & +6 & -3 & +6
\end{array}
$$

By the remainder theorem the remainder is 6.

As a corollary of the remainder theorem we have the **Factor Theorem.**

THEOREM 13.2 (Factor Theorem) If a polynomial $P(x)$ is divided by $x - r$ and if the remainder $R = P(r)$ is zero, then $x - r$ is a factor of $P(x)$.

PROOF. Since $P(r) = R = 0$, we have

$$P(x) = (x - r) \cdot Q(x) + 0$$

and clearly $(x - r)$ is a factor of $P(x)$.

We now prove the converse of the factor theorem.

THEOREM 13.2a. If $x - r$ is a factor of a polynomial $P(x)$, then $P(r) = 0$ and r is a zero of $P(x)$ and a root of $P(x) = 0$.

PROOF. It is given that $x - r$ is a factor of $P(x)$. Therefore,

$$
\begin{aligned}
P(x) &= (x - r) \cdot Q(x) \\
&= (x - r) \cdot Q(x) + 0
\end{aligned}
$$

Hence, $P(r) = 0$ by the remainder theorem.

EXAMPLE 3. Use the factor theorem to determine whether $x + 2$ is a factor of

$$x^4 - x^3 - 3x^2 + 4x + 4.$$

SOLUTION. Using synthetic division with $r = -2$, we have

$$
\begin{array}{r|rrrrr}
-2 & 1 & -1 & -3 & +4 & +4 \\
   &   & -2 & +6 & -6 & +4 \\
\hline
   & 1 & -3 & +3 & -2 & +8
\end{array}
$$

By the remainder theorem, $P(-2) = 8$ and $(x + 2)$ is not a factor of the given polynomial.

EXAMPLE 4. Use the factor theorem to determine whether $(x + 2a)$ is a factor of $3x^4 + 4ax^3 - 2a^3x - 20a^4$.

SOLUTION. Using synthetic division with $r = -2a$, we obtain

$$
\begin{array}{r|rrrr}
-2a & 3 & +4a & +0 & -2a^3 & -20a^4 \\
& & -6a & +4a^2 & -8a^3 & +20a^4 \\
\hline
& 3 & -2a & +4a^2 & -10a^3 & +0
\end{array}
$$

The remainder is 0, hence $(x + 2a)$ is a factor of the given polynomial.

EXERCISES 13.2

Use synthetic division to find the values of the polynomials for the specified values (Exercises 1–14).

1. $P(x) = x^4 - x^3 + x^2 - 4$; find $P(1)$, $P(2)$, and $P(3)$.
2. $P(x) = 4x^3 - 5x^2 - 23x + 6$; find $P(2)$, $P(-1)$, and $P(-3)$.
3. $P(x) = 2x^6 - 3x^5 + x^3 - 10$; find $P(-2)$, $P(-3)$, and $P(1)$.
4. $P(x) = 2x^4 + 7x^3 - 12x - 9$; find $P(1)$, $P(-2)$, and $P(-3)$.
5. $P(x) = 7x^3 - 14x^2 + 25x - 21$; find $P(-1)$, $P(-2)$, and $P(2)$.
6. $P(x) = 3x^4 - 2x^3 + 5x^2 - 4x + 1$; find $P(1)$, $P(-1)$, and $P(2)$.
7. $P(x) = 2x^5 - 3x^3 + x$; find $P(0)$, $P(-1)$, and $P(1)$.
8. $P(x) = 3x^4 - 3x^3 + x^2 - x + 1$; find $P(-1)$, $P(-2)$, and $P(-3)$.
9. $P(x) = 3x^5 - 2x^4 + 7x^3 + 5$; find $P(1)$, $P(2)$, and $P(3)$.
10. $P(x) = 12x^{10} - 3x^6 + x^2 - 11$; find $P(2)$, $P(1)$, and $P(-1)$.
11. $P(x) = 9x^6 + 7x^4 - 3x + 9$; find $P(-1)$, $P(2)$ and $P(-3)$.
12. $P(x) = 4x^3 + 6x^2 + 8x - 4$; find $P(\frac{1}{2})$, $P(-\frac{1}{2})$, and $P(\frac{3}{2})$.
13. $P(x) = 6x^5 + 15x^4 - 23x^3 + 110x^2 + 3x - 1$; find $P(-\frac{3}{2})$, $P(\frac{1}{2})$, and $P(-1)$.
14. $P(x) = 24x^8 - 27x^7 - 210x^6 + 98x^5 + 451x^4 - 57x^3 - 120x^2 - 54x$. Find $P(2)$, $P(1)$, and $P(-1)$.

Determine whether the linear expressions given are factors of the accompanying polynomials (Exercises 15–29).

15. $x^3 - 3x^2 + 3x - 1$; $x - 1$, $x + 1$, and $x - 2$.
16. $x^4 - 1$; $x - 1$, $x + 1$, $x - 2$, and $x + 2$.
17. $x^5 - 2x^4 + 3x^3 - 3x^2 + x$, $x - 1$, $x + 1$, and $x - 3$.
18. $x^4 + 3x^3 - 5x^2 + 7x - 6$; $x - 1$, $x + 1$, and $x + 4$.
19. $x^5 - 11x^3 + 12x - 3$; $x + 2$, $x - 3$, $x + 5$.
20. $3x^4 + 5x^3 - 17x^2 + x - 14$; $x + 3$, $x - 4$, $x + 5$.
21. $2x^5 - 6x^4 + 5x^3 - 1$; $x + 2$, $x - 3$, $x + 5$.
22. $3x^3 - 4x^2 + x - 20$; $x - 2$, $x + 3$, $x - 4$.
23. $5x^4 + 5x^2 - 10$; $x + 1$, $x - 1$, x.
24. $35x^5 - 7x^4 + 11x^3 + 37$; $x - 3$, $x + 4$, $x - 6$.
25. $3x^4 + 13x^3 + 7x^2 + 11x + 4$; $x + 2$, $x + 4$, $x - 1$.
26. $3x^5 - 4x^4 + 5x^3 - 7x + 2$; $x - 1$, $x - 2$, $x + \frac{2}{3}$.
27. $x^n + a^n$; $x + a$ (n is a positive even integer).

28. $x^n + a^n$; $x + a$ (n is a positive odd integer).

29. $x^{2n} - a^{2n}$; $x - a$; (n is a positive integer).

13.4 ◆ Graphs of Polynomial Functions

If the coefficients of a polynomial $P(x)$ are real numbers, then $y = P(x)$ specifies a function, called a **polynomial function,** with the set of real numbers as domain. The graph of this function is called the **graph** of the polynomial $P(x)$. The first step in the construction of the graph of a polynomial function is a table of ordered pairs that belong to the function. Since by using synthetic division the values of the function may be obtained for any value assigned to the variable, a table of values can be constructed. From this table the graph can be sketched.

EXAMPLE 1. Draw a graph of $y = x^3 + x^2 - 3x - 4$.

SOLUTION. We first construct a table of values. For any arbitrary x, the corresponding value of y is found by synthetic division.

x	-3	-2	-1	0	1	2	3	4
y	-13	-2	-1	-4	-5	2	23	64

The points whose coordinates have the values given in the table have been plotted and the graph drawn as shown in Figure 13.1. Since the values of y are frequently larger than the corresponding values of x, it will often, as in this case, be convenient in drawing a graph to use a smaller scale on the y-axis than on the x-axis.

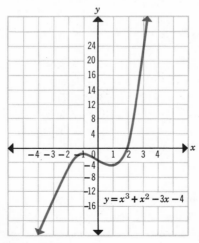

Figure 13.1

EXAMPLE 2. Draw the graph of $y = x^4 - 2x^3 - 7x^2 + 10x + 10$.

SOLUTION. We first construct a table of values.

x	-4	-3	-2	-1	0	1	2	$2\frac{1}{2}$	3	4
y	242	52	-6	-4	10	12	2	$-\frac{15}{16}$	4	66

The values from the table have been plotted as points and the graph drawn as shown in Figure 13.2. To get the shape of the curve it is sometimes necessary to assign fractional values to x, as in this case.

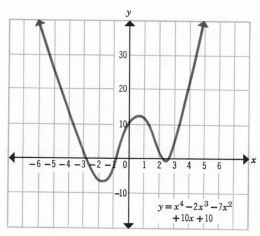

Figure 13.2

EXERCISES 13.3

Draw the graphs of the following equations.

1. $y = 2x^3 + 3x^2 - 12x + 5$.
2. $y = 2x^3 - 3x^2 - 6x + 7$.
3. $y = x^4 - 2x^2 + 1$.
4. $y = x^4 + 2x^2 - 3$.
5. $y = 3x^4 - 8x^3 - 6x^2 + 24x + 12$.
6. $y = 3x^4 - 4x^3 - 36x^2 + 25$.
7. $y = 3x^4 - 4x^3 - 60x^2 + 192x - 140$.
8. $y = 3x^4 - 8x^3 - 30x^2 + 72x - 25$.
9. $y = x^4 - 8x^3 + 22x^2 - 24x + 3$.
10. $y = x^4 - 38x^2 - 120x + 85$.
11. $y = x^3 - 9x^2 + 24x - 7$.
12. $y = x^7 - 2x^5 - 10x^2 + 1$.
13. $y = x^5 - 10x - 60$.
14. $y = x^4 - 2x^3 - 7x^2 + 10x + 10$
15. $y = x^5 + 2x^4 + x^3 - 4x^2 - 3x - 5$.

13.5 ✦ Complex Zeros of Polynomial Functions

We now state the fundamental theorem of algebra.

THEOREM 13.4 (The Fundamental Theorem of Algebra) Every poly-
nomial over the complex number field $P(x) = a_0x^n + a_1x^{n-1} + \cdots + a_{n-1}x + a_n$ with $n \geq 1$, $a_0 \neq 0$, has at least one (real or complex)
zero.

Since the proof of the Fundamental Theorem of Algebra is beyond the
scope of this book, we shall accept it without proof.

We recall that the zeros of a polynomial $P(x)$ are identical with the roots of
the polynomial equation $P(x) = 0$. Consequently, all remarks about the zeros
of a polynomial are also true about the roots of the associated polynomial
equation.

Let r_1 be a zero of $P(x)$. Then by the Factor Theorem, $(x - r_1)$ is a factor of
$P(x)$ and

$$P(x) = (x - r_1) \cdot P_1(x)$$

Now $P_1(x)$ has a zero r_2 by the Fundamental Theorem of Algebra, and hence
$x - r_2$ is a factor of $P_1(x)$ and

$$P(x) = (x - r_1)(x - r_2) \cdot P_2(x)$$

Again, $P_2(x)$ has a zero r_3, and $x - r_3$ is a factor of $P_2(x)$. We now have

$$P(x) = (x - r_1)(x - r_2)(x - r_3) \cdot P_3(x)$$

Each new quotient, $P_1(x)$, $P_2(x)$, $P_3(x)$, and so forth is of degree one less than
the degree of the previous quotient, so that by continuing in the same manner
we see that $P(x)$ has exactly n factors of the form $x - r$. It may happen that
the factors are not all different, in which case $P(x)$ will have factors of the
form $(x - r)^m$, $m > 1$. Thus, in general, $P(x)$ may be written

$$P(x) = a_0(x - r_1)^{m_1}(x - r_2)^{m_2} \ldots (x - r_k)^{m_k}$$

where the exponents m_1, $m_2, \ldots, m_k$ are positive integers such that
$m_1 + m_2 + \cdots + m_k = n$.

We say that r_1 is a zero of $P(x)$ of **multiplicity** m_1, r_2 is a zero of $P(x)$ of
multiplicity m_2, and so forth.

Similarly, we say that r_1 is a root of multiplicity m_1 of the polynomial
equation $P(x) = 0$, r_2 is a root of multiplicity m_2 of $P(x) = 0$, and so forth.
Taking into account the multiplicity of each root, we now say that the equation $P(x) = 0$, where $P(x) = 0$ is of degree n, has exactly n roots.

THEOREM 13.5. If $P(x)$ is a polynomial of degree n, it has exactly n (real or complex) zeros.

THEOREM 13.6. If $P(x) = 0$ is a polynomial equation of degree n, it has exactly n (real or complex) roots.

EXAMPLE 1. If $P(x) = 3(x - 3)^2(x + 1)(x + 7)^3$, name the zeros of $P(x)$.

SOLUTION. The zeros of $P(x)$ are 3, -1, and -7; 3 is a zero of multiplicity two or a double zero, -1 is a zero of multiplicity one or a simple zero, and -7 is a zero of multiplicity three.

In solving quadratic equations over the field C of complex numbers, we found that every quadratic equation has two roots. This is consistent with the discussion in Section 9.5. We see, then, that every quadratic function has two zeros.

EXAMPLE 2. Find all the zeros of $P(x) = x^2 - 4x + 5$.

SOLUTION. Setting $P(x) = x^2 - 4x + 5$ equal to zero and using the quadratic formula, we find the roots of $x^2 - 4x + 5 = 0$. We have

$$x = \frac{4 + \sqrt{16 - 20}}{2}$$

$$= \frac{4 + 2i}{2}$$

$$= 2 \pm i$$

The zeros of $P(x) = x^2 - 4x + 5$ are $2 + i$ and $2 - i$.

EXAMPLE 3. Find the zeros of $P(x) = x^2 + x + 1$.

SOLUTION. Setting $P(x) = x^2 + x + 1$ equal to zero and using the quadratic formula, we obtain

$$x = \frac{-1 + \sqrt{1 - 4}}{2}$$

$$= -\frac{1}{2} \pm \frac{\sqrt{3}}{2}i$$

The zeros of $P(x) = x^2 + x + 1$ are $\frac{1}{2} + \frac{\sqrt{3}}{2}i$ and $-\frac{1}{2} - \frac{\sqrt{3}}{2}i$.

Examples 2 and 3 reveal a very interesting fact about complex zeros of polynomials. Notice that the two complex zeros in each case are conjugates of one another. This is always true, not only for quadratic polynomials, but for polynomials of any degree.

THEOREM 13.7. If $P(x)$ is a polynomial over the field of real numbers and if $P(x)$ has a zero $z = a + bi$, with a and b real numbers and $b \neq 0$, then $\bar{z} = a - bi$ is also a zero of $P(x)$.

PROOF. Let

$$P(x) = a_0 x^n + a_1 x^{n-1} + a_2 x^{n-2} + \cdots + a_{n-1} x + a_n$$

where $a_0, a_1, a_2, \ldots, a_n$ are real numbers and $a_0 \neq 0$. Since $P(z) = 0$, we have

$$(1) \qquad a_0 z^n + a_1 z^{n-1} + \cdots + a_{n-1} z + a_n = 0$$

The conjugate of the left member of Equation 1 is equal to the conjugate of the right member and we have

$$(2) \qquad \overline{a_0 z^n + a_1 z_{n-1} + \cdots + a_{n-1} z + a_n} = \overline{0}$$

But $\overline{0} = 0$ since the conjugate of a real number is the real number. But Theorem 12.3a assures us that the conjugate of the sum of two complex numbers is the sum of the conjugates. Hence, Equation 2 may be written

$$(3) \qquad \overline{a_0 z^n} + \overline{a_1 z^{n-1}} + \cdots + \overline{a_{n-1} z} + \bar{a}_n = 0$$

We know from Theorem 12.3b that the conjugate of the product of two complex numbers is the product of the conjugates, hence Equation 3 may be written

$$(4) \qquad \bar{a}_0 \overline{z^n} + \bar{a}_1 \overline{z^{n-1}} + \cdots + \bar{a}_{n-1} \bar{z} + \bar{a}_n = 0$$

Since the conjugate of a real number is that real number and since $a_0, a_1, \ldots a_n$ are real numbers, we have

$$a_0 \overline{z^n} + a_1 \overline{z^{n-1}} + \cdots + a_{n-1} \bar{z} + a_n = 0$$

which tells us that

$$P(\bar{z}) = 0$$

and $\bar{z}$ is a zero of $P(x)$. This proves the theorem.

EXAMPLE 4. Verify that $2 - 2i$ is a root of $2x^3 - 11x^2 + 28x - 24 = 0$. Find the solution set of this equation.

SOLUTION. By direct substitution we can verify that $(2 - 2i)$ is a root of the given equation. Since $(2 - 2i)$ is a root, $(2 + 2i)$ is also a root by Theorem 13.5, and hence

$$[x - (2 - 2i)][x - (2 + 2i)] = (x - 2)^2 + 4 = x^2 - 4x + 8$$

is a factor of the polynomial $2x^3 - 11x^2 + 28x - 24$. But

$$2x^3 - 11x^2 + 28x - 24 = [(x - 2)^2 + 4](2x - 3)$$

and $\frac{3}{2}$ is also a zero of the polynomial and hence a root of the equation. The solution set of the given equation is $\{2 - 2i, 2 + 2i, \frac{3}{2}\}$.

EXERCISES 13.4

One or more roots are given for each of the polynomial equations below. Find the other roots. Verify by synthetic division. (Exercises 1–10)

1. $x^2 + 9 = 0$; $3i$ is a root.
2. $2x^2 + 8 = 0$; $-2i$ is a root.
3. $x^4 - 1 = 0$; i is a root.
4. $x^4 - 2x^3 + 4x^2 + 4x - 12 = 0$; $1 - \sqrt{5}\,i$ is a root.
5. $x^3 - 3x^2 - 6x - 20 = 0$; $-1 + \sqrt{3}\,i$ is a root.
6. $x^4 - 4x^3 + 5x^2 - 2x - 2 = 0$; $1 - i$ is a root.
7. $x^4 - 2x^3 + 6x^2 + 22x + 13 = 0$; $2 + 3i$ is a root.

8. $x^4 + 1 = 0$; $\dfrac{1}{\sqrt{2}} + \dfrac{1}{\sqrt{2}}i$.

9. $x^6 - 1 = 0$; $\dfrac{1}{2} + \dfrac{\sqrt{3}}{2}i$, and $-\dfrac{1}{2} - \dfrac{\sqrt{3}}{2}i$ are roots.

10. $x^6 - x^5 - 8x^4 + 2x^3 + 21x^2 - 9x - 54 = 0$; $\sqrt{2} + i$ and $-\sqrt{2} - i$ are roots.
11. A cubic polynomial over the real numbers has zeros -1 and $2 + i$. What is the third zero?
12. A cubic polynomial over the real numbers has zeros 3 and $3i + 1$. What is the third zero?
13. A fourth degree polynomial over the real numbers has zeros $2i$ and $1 + 3i$. What are the other two zeros?
14. A fifth degree polynomial over the real numbers has zeros $2, -2, 3$, and $2 + 3i$. What is the fifth zero?

13.6 ◆ Rational Roots of $P(x) = 0$

The theorems we proved in Section 13.5 together with synthetic division provide a means of testing whether a number r is a solution of $P(x) = 0$ when $P(x)$ is a polynomial of degree n.

There is no general method for solving the equation $P(x) = 0$ if the degree of $P(x)$ is greater than four. However, if $P(x) = 0$ has roots which are rational numbers, there is a guide as to the possible values of the rational roots. Knowing the possible values of the rational roots, we can find the correct ones by the methods described in the previous sections of this chapter.

THEOREM 13.8. If $P(x) = a_0x^n + a_1x^{n-1} + \cdots + a_{n-1}x + a_n$ is a polynomial with integral coefficients, and if a rational number $\dfrac{p}{q}$, where p and q are integers without common divisors other than unity, is a root of $P(x) = 0$, then p is a factor of a_n, the constant term, and q is a factor of a_0, the coefficient of x^n.

PROOF. Since $\dfrac{p}{q}$ is a root of $P(x) = 0$, we have

$$(1) \qquad a_0\left(\frac{p}{q}\right)^n + a_1\left(\frac{p}{q}\right)^{n-1} + \cdots + a_{n-1}\left(\frac{p}{q}\right) + a_n = 0$$

If we multiply each member of Equation 1 by q^n we obtain

$$(2) \qquad a_0p^n + a_1p^{n-1}q + \cdots + a_{n-1}q^{n-1}p + a_nq^n = 0$$

If we subtract a_0p^n from each member of Equation 2 we have

$$(3) \qquad q(a_1p^{n-1} + a_2p^{n-2}q + \cdots + a_nq^{n-1}) = -a_0p^n$$

By hypothesis, all of $a_1, a_2, \ldots, a_n$ are integers. Therefore, since q is a factor of the left member of Equation 3, it is also a factor of the right member. Since p and q have no common factor other than unity (± 1), no factor of q is a factor of p^n. Therefore, q must be a factor of a_0. If a_nq^n is subtracted from both members of Equation 2 we have

$$p(a_0p^{n-1} + a_1p^{n-2} + \cdots + a_{n-1}q^{n-1}) = -a_nq^n$$

Using the same reasoning as before, it follows that p is a factor of a_n.

As a direct result of Theorem 13.8 we have:

THEOREM 13.8a. If $P(x) = a_0x^n + a_1x^{n-1} + \cdots + a_{n-1}x + a_n$ is a polynomial with integral coefficients, and if $a_0 = 1$, then any rational root of $P(x) = 0$ must be an integer and this integral root must be a factor of a_n.

The proof of this theorem is left to the reader.

As soon as a rational root, r_1, of $P(x) = 0$ has been found, the division of

$P(x)$ by $x - r_1$ should be performed, by synthetic division, to find a quotient $P_1(x)$. The remaining roots of $P(x) = 0$ will then be the roots of $P_1(x) = 0$, and the subsequent search for roots should be continued in the equation $P_1(x) = 0$. An equation such as $P_1(x) = 0$, found by reducing a given equation after a root has been found, is called a **depressed equation.**

EXAMPLE 1. Determine the rational roots of the equation $6x^4 - 7x^3 + 8x^2 - 7x + 2 = 0$.

SOLUTION. We apply Theorem 13.8 and determine that any rational root must be found among the numbers

$$\pm 1, \pm 2, \pm\tfrac{1}{2}, \pm\tfrac{1}{3}, \pm\tfrac{2}{3}, \pm\tfrac{1}{6}$$

We now use synthetic division to test each of these numbers, discarding those for which the remainder is not equal to zero. After a root has been found, the work is continued in the depressed equation. Thus, showing only the successful tries, we have:

$$
\begin{array}{r|rrrrr}
\tfrac{1}{2} & 6 & -7 & +8 & -7 & +2 \\
  &   & 3 & -2 & +3 & -2 \\
\hline
\tfrac{2}{3} & 6 & -4 & +6 & -4 & 0 \\
  &   & 4 & 0 & 4 & \\
\hline
  & 6 & 0 & +6 & 0 & \\
\end{array}
$$

The last depressed equation is $6x^2 + 6 = 0$, which reduces to

$$x^2 + 1 = 0$$

Since this equation is quadratic, it can be solved and all four roots of the given equation are determined. The solution set of the given equation is

$$\{\tfrac{1}{2}, \tfrac{2}{3}, i, -i\}$$

EXAMPLE 2. Determine the rational roots of $x^5 - 7x^4 + 9x^3 + 43x^2 - 166x + 168 = 0$.

SOLUTION. Since $a_0 = 1$, any rational root is an integral root by Theorem 13.8a. The possible roots, using Theorem 13.8a, are

$$\pm 1, \pm 2, \pm 3, \pm 4, \pm 6, \pm 7, \pm 8, \pm 12, \pm 14$$
$$\pm 21, \pm 24, \pm 28, \pm 42, \pm 56, \pm 84, \pm 168$$

We now use synthetic division to test each of these numbers, discarding those for which the remainder is not equal to zero. Thus, showing only the successful tries, we have:

$$
\begin{array}{r|rrrrr}
2|1 & -7 & +9 & +43 & -166 & +168 \\
  & 2 & -10 & -2 & +82 & -168 \\
\hline
4|1 & -5 & -1 & +41 & -84 & 0 \\
  & 4 & -4 & -20 & +84 & \\
\hline
-3|1 & -1 & -5 & +21 & 0 \\
  & -3 & +12 & -21 & \\
\hline
1 & -4 & +7 & 0 \\
\end{array}
$$

The last depressed equation is $x^2 - 4x + 7 = 0$. Since this equation is quadratic, it can be solved and all five roots of the given equation are determined. The solution set of the given equation is

$$\{-3, 2, 4, 2 + \sqrt{3}\,i, 2 - \sqrt{3}\,i\}$$

EXAMPLE 3. Determine the rational roots of $3x^4 - 8x^3 + 6x^2 - 1 = 0$.

SOLUTION. We apply Theorem 13.8 and determine that any rational roots must be found among the numbers

$$\pm 1, \pm\tfrac{1}{3}$$

We now use synthetic division to test each of these numbers, discarding those for which the remainder is not equal to zero. After a root has been found, the work is continued on the depressed equation. Thus, showing only the successful tries, we have:

$$
\begin{array}{r|rrrrr}
1|3 & -8 & 6 & 0 & -1 \\
  & 3 & -5 & 1 & 1 \\
\hline
1|3 & -5 & 1 & 1 & 0 \\
  & 3 & -2 & -1 & \\
\hline
1|3 & -2 & -1 & 0 \\
  & 3 & 1 & \\
\hline
-\tfrac{1}{3}|3 & 1 & 0 \\
  & -1 & \\
\hline
3 & 0 \\
\end{array}
$$

The rational roots are 1 and $-\tfrac{1}{3}$. The root 1 has multiplicity three. The solution set of the given equation is $\{1, 1, 1, -\tfrac{1}{3}\}$.

EXERCISES 13.5

Determine all the rational roots of the following equations. Where the depressed equation is of degree two, find all the roots (Exercises 1–20).

1. $x^3 + 8x^2 + 13x + 6 = 0$.
2. $x^3 - 5x^2 - 2x + 24 = 0$.
3. $x^3 - 10x^2 + 27x - 18 = 0$.
4. $x^4 + 4x^3 + 8x + 32 = 0$.
5. $6x^3 - 11x^2 + 6x - 1 = 0$.
6. $96x^3 - 16x^2 - 6x + 1 = 0$.
7. $24x^3 - 2x^2 - 5x + 1 = 0$.
8. $10x^3 + 19x^2 - 30x + 9 = 0$.
9. $6x^5 + x^4 - 14x^3 + 4x^2 + 5x - 2 = 0$.
10. $6x^4 - 11x^3 - x^2 - 4 = 0$.
11. $2x^3 + 3x^2 - 14x - 21 = 0$.
12. $4x^3 + 14x^2 - 6x - 21 = 0$.
13. $2x^3 - 11x^2 - 7x + 6 = 0$.
14. $4x^4 + 16x^3 + x^2 + 6x + 8 = 0$.
15. $3x^4 + 8x^3 - 39x^2 - 96x + 36 = 0$.
16. $4x^4 - 16x^3 - 13x^2 + 4x + 3 = 0$.
17. $x^4 - 1 = 0$.
18. $x^8 - 1 = 0$.
19. $24x^8 - 28x^7 - 210x^6 + 99x^5 + 451x^4 - 57x^3 - 225x^2 - 54x = 0$.
20. $x^4 + 8x^2 - 9 = 0$.
21. Prove Theorem 13.8a.

13.7 ◆ Irrational Roots of Polynomial Equations

We found that the complex roots of polynomial equations occur in conjugate pairs. We also found a method for finding rational roots of polynomial equations with integral coefficients. We now wish to find ways of identifying irrational roots of these polynomial equations.

We state the following theorem on continuity. We shall accept this theorem without proof.

THEOREM 13.9. If $P(x) = a_0x^n + a_1x^{n-1} + \cdots + a_{n-1}x + a_n$ is a polynomial over the field of real numbers and if k is a real number such that $P(x_1) < k < P(x_2)$, then there exists at least one real number c, $x_1 < c < x_2$, such that $P(c) = k$.

Theorem 13.9 asserts that there are no breaks or jumps in the values of $P(x)$, so that we know that $P(x)$ assumes all values between any two of its values. Curves that contain no gaps or that are not made up of separate or disjoint parts are called **continuous** curves. Theorem 13.9 tells us that graphs of polynomials are continuous curves.

THEOREM 13.10. If $P(x)$ is a polynomial over the field of real numbers and a and b are real numbers such that $P(a)$ and $P(b)$ have opposite signs, then the equation $P(x) = 0$ has at least one real root between $x = a$ and $x = b$ and in fact, an odd number of such roots if a root of multiplicity m is counted m times.

The proof of Theorem 13.10 is intuitively evident from geometric considerations. If the graph of $P(x)$ passes through two points on opposite sides of the x-axis, since the curve is continuous it must cross the x-axis an odd number of times between the points.

EXAMPLE 1. Locate the real roots of $8x^3 - 4x^2 - 18x + 9 = 0$.

SOLUTION. It can be verified that $P(0) = 9$ and $P(1) = -5$, $P(2) = 21$, $P(-1) = 15$, and $P(-2) = -35$. By Theorem 13.10 there are an odd number of roots between -2 and -1, between 0 and 1, and between 1 and 2. Since the degree of the equation is 3, it has only three roots, hence there is exactly one real root in each of the above intervals.

EXAMPLE 2. Verify that there is at least one real root of $x^4 - 6x^3 + 5x^2 + 14x - 4 = 0$ between -2 and -1.

SOLUTION. $P(-2) = 52$ and $P(-1) = -6$. Since $P(-2)$ and $P(-1)$ have opposite signs, by Theorem 13.10 there is at least one root between $x = -2$ and $x = -1$.

We use the fact that the graph of a polynomial is a continuous curve and Theorem 13.10 to find irrational roots of polynomial equations with integral coefficients.

EXAMPLE 3. Find the irrational root between 0 and 1 of $x^3 + x^2 - 10x + 4 = 0$.

SOLUTION. We first graph the polynomial function specified by $y = x^3 + x^2 - 10x + 4$. This graph is shown in Figure 13.3.

Since $P(0) = 4$ and $P(1) = -4$, we know that there is a root of $P(x) = 0$ between 0 and 1. We now approximate this root. Since the root we are seeking lies between 0 and 1, we approximate it, say, 0.5. We now test this value in the original polynomial:

$$
\begin{array}{r|rrrr}
0.5 & 1 & +1.0 & -10.0 & +4.0 \\
& & 0.5 & +0.8 & -4.6 \\
\hline
& 1 & +1.5 & -9.2 & -0.6 \quad P(0.5) = -0.6
\end{array}
$$

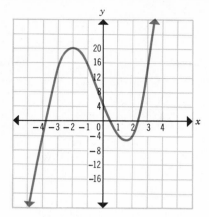

Figure 13.3

Since $P(0.5) < 0$ and $P(0) > 0$, it follows that the actual root must be less than 0.5. We now try $x = 0.45$, which gives $P(0.45) = -0.21$:

$$\begin{array}{r|rrrr}
0.45 & 1 & +1.00 & -10.00 & +4.00 \\
& & 0.45 & +0.65 & -4.21 \\
\hline
& 1 & +1.45 & -9.35 & -0.21
\end{array} \qquad P(0.45) = -0.21$$

Since $P(-0.45) < 0$ and $P(0) > 0$, it follows that the actual root must be less than 0.45. We try 0.43 which gives $P(0.43) = -0.04$:

$$\begin{array}{r|rrrr}
0.43 & 1 & +1.00 & -10.00 & +4.00 \\
& & 0.43 & +0.61 & -4.04 \\
\hline
& 1 & +1.43 & -9.39 & -0.04
\end{array} \qquad P(0.43) = -0.04$$

Since $P(0.43) < 0$ and $P(0) > 0$, it follows that the actual root must be less than 0.43. We try $x = 0.42$, which gives $P(0.42) = 0.05$.

$$\begin{array}{r|rrrr}
0.42 & 1 & +1.00 & -10.00 & +4.00 \\
& & 0.42 & +0.60 & -3.95 \\
\hline
& 1 & +1.42 & -9.40 & +0.05
\end{array} \qquad P(0.42) = 0.05$$

Since $P(0.43)$ and $P(0.42)$ are of opposite signs, the root must lie between 0.42 and 0.43. The portion of the graph which lies between $x = 0.42$ and $x = 0.43$ is now drawn with an enlarged scale as shown in Figure 13.4.

The curve crosses the x-axis at point P, whose x-coordinate is the desired value of x. This value is given approximately by the abscissa of point Q where the line segment joining A and B crosses the x-axis.

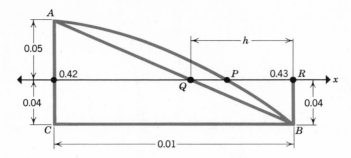

Figure 13.4

The triangle QBR is similar to triangle ABC. Since corresponding sides of similar triangles are proportional, we have

$$\frac{h}{0.01} = \frac{0.04}{0.09}$$

$$h \doteq 0.00444$$

$$\doteq 0.004$$

The corresponding value of x is 0.424. We find $P(0.426) = 0.016$:

$$
\begin{array}{r|rrrr}
0.424 & 1 & +1.000 & -10.000 & +4.000 \\
       &   & +0.424 & +\ 0.604 & -3.984 \\
\hline
       & 1 & 1.424 & -\ 9.396 & 0.016
\end{array}
$$

Since $P(0.424) > 0$, the exact root is more than 0.424 (correct to three decimal places). A continuation of the above process would give the root to as many decimal places as we desire.

The method used in the above example to approximate an irrational root of a polynomial equation is called the method of **linear interpolation**, since each interval considered in the curve is approximated by a straight line.

EXERCISES 13.6

Find the root of each of the following equations that lies in the interval indicated. Compute the roots to three decimal places.

1. $x^3 + x^2 - 2x - 1 = 0$; between 1 and 2.
2. $x^3 + 3x - 5 = 0$; between 1 and 2.
3. $x^3 - 2x - 5 = 0$; between 2 and 3.
4. $x^4 - 27x^2 + 48x - 24 = 0$; between 4 and 5.

5. $x^3 - 12x + 8 = 0$; between 3 and 4.
6. $x^4 - 2x^3 - 5x^2 + 10x - 3 = 0$; between 0 and 1.
7. $x^3 - 7x + 7 = 0$; between -4 and -3.
8. $x^4 - 4x + 1 = 0$; between 1 and 2.
9. $x^3 - 3x + 1 = 0$; between 1 and 2.
10. $x^3 - 7x + 7 = 0$; between 1 and 1.5.
11. $x^3 - 3x + 1 = 0$; between 0 and 1.
12. $x^3 - 9x^2 + 24x - 7 = 0$; between 0 and 1.
13. $x^3 - 4x^2 - x + 3 = 0$; between 4 and 5.
14. $x^5 - 10x - 60 = 0$; between 2 and 3.
15. $x^7 - 2x^5 - 10x^2 + 1 = 0$; between 0 and $\frac{1}{2}$.

CHAPTER REVIEW

1. Use synthetic division to determine the quotient and remainder in the division $(2x^5 - 3x^3 + 2x + 1) \div (x + 2)$.
2. Use synthetic division to find the specified values of $P(x) = 5x^6 - 3x^5 + 7x^3 - 2x^2 + 8x - 11$.
   (a) $P(2)$ (c) $P(-3)$
   (b) $P(-1)$ (d) $P(-2)$
3. Determine whether or not the following linear expressions are factors of $x^4 + 6x^3 + 13x^2 + 12x + 4$.
   (a) $x + 2$ (c) $x + 1$
   (b) $x - 1$ (d) $x - 2$
4. Draw the graph of $y = x^4 + 12x - 5$.
5. Find all the roots of $x^4 + 2x^3 - 7x^2 - 8x + 12 = 0$.
6. Find all the roots of $x^4 - 4x^2 + 8x - 4 = 0$ given that $1 + i$ is a root.
7. A third degree equation with real coefficients has roots $1 + i$ and -2. What is the third root?
8. A fourth degree equation with real coefficients $3 + 2i$ and $7 - 2i$. What are the other roots?
9. Find all the integral roots of $x^3 + x^2 - 3x + 9 = 0$.
10. Find all the rational roots of $3x^4 - 40x^3 + 130x^2 - 120x + 27 = 0$.
11. Find the real root between -3 and -2 of $x^3 + 2x + 20 = 0$. Compute the root to two decimal places.
12. Find the real root between 3 and 4 of $x^4 - 11727x + 40385 = 0$. Compute the root correct to three decimal places.
13. Draw the graph of $y = x^3 + 3x^2 - 6x - 3$.
14. Use synthetic division to show that $\frac{1}{3}$ is a root of $3x^3 - x^2 - 15x + 5 = 0$.
15. Find the solution set of $x^3 + 4x^2 - 47x - 210 = 0$ given that 7 and -5 are roots.

16. Find the solution set of $x^4 - 3x^3 + 3x^2 - 3x + 2 = 0$ given the roots 1 and 2.
17. Name the possible rational roots of $108x^3 - 270x^2 - 42x + 16 = 0$.
18. Name the possible integral roots of $2x^3 + 6x^2 - 3x + 12 = 0$.
19. Could $\{1 + i, 1 - i, 3 + i\}$ be the solution set of $x^3 - 3x^2 + 7x + 6 = 0$? Why?
20. Could 6 be an integral root of $x^4 + 27x^3 + 3x^2 - 13x + 101 = 0$? Why?

CHAPTER 14

Sequences and Series

14.1 ✦ Mathematical Induction

One of the most important methods of proof used in mathematics is that of **mathematical induction.** Let us suppose that we have noticed the following pattern:

$$1 = 1^2$$
$$1 + 3 = 2^2$$
$$1 + 3 + 5 = 3^2$$
$$1 + 3 + 5 + 7 = 4^2$$

In observing the above pattern we suspect that the sum of the first n odd positive integers may be equal to n^2 for all values of n. To prove that this is the case, that is, that the sum of the first n positive odd integers is n^2 for any value of n, we use the method of mathematical induction.

A proof by mathematical induction requires the following two parts.

PART I. Prove that if a relation were true for any positive integer value of n, say, $n = k$, then it must be true for the next larger value of n, that is, $n = k + 1$.

PART II. Show by direct substitution that the relation is true for $n = 1$.

A proof by mathematical induction is based on a property of the natural numbers called the property of finite induction. This property is stated below.

Property of Finite Induction. Every set M of natural numbers contains all the natural numbers if (1) $1 \in M$; and (2) if $a \in M$, then $a + 1 \in M$.

Part I of a proof by mathematical induction is the more difficult part of the proof. In this part we are not concerned with whether the given relation is true for any value of n, but only with the proposition: if the relation were true for $n = k$, then it must be true for $n = k + 1$.

EXAMPLE 1. Prove for any natural number n,

$$1 + 2 + 3 + \cdots + n = \tfrac{1}{2}n(n + 1)$$

SOLUTION. We prove this relation by using mathematical induction. We carry out Part II of the proof first. Substituting $n = 1$ in

(1) $$1 + 2 + 3 + \cdots + n = \tfrac{1}{2}n(n + 1)$$

we have

$$1 = \tfrac{1}{2}(1)(1 + 1) = \tfrac{1}{2}(1)(2) = 1$$

which is a true statement, and therefore Part II of the proof is complete. Now we wish to carry out Part I. We assume that the relation is valid for some natural number $n = k$. That is, we assume

(2) $$1 + 2 + 3 + \cdots + k = \tfrac{1}{2}k(k + 1)$$

is true. We wish to prove that this assumption implies the relation is true for $n = k + 1$. That is,

(3) $$1 + 2 + 3 + \cdots + k + (k + 1) = \tfrac{1}{2}(k + 1)(k + 2)$$

must be true. To do this we add $(k + 1)$ to both members of equation 2

$$1 + 2 + 3 + \cdots + k = \tfrac{1}{2}k(k + 1)$$

which we assumed to be true. We obtain

$$[1 + 2 + \cdots + k] + (k + 1) = \tfrac{1}{2}k(k + 1) + (k + 1)$$
$$= \frac{k(k + 1) + 2(k + 1)}{2}$$
$$= \tfrac{1}{2}(k + 1)(k + 2)$$

This proves that if Equation 2 is true, then Equation 3 must be true and completes the proof by mathematical induction.

EXAMPLE 2. Prove by mathematical induction for any natural number n,

$$1 + 3 + 5 + \cdots + (2n - 1) = n^2$$

SOLUTION. First, in Part II of the proof, for $n = 1$, we have $1 = 1^2$, which is true. For Part I, we assume that the relation is true for $n = k$, that is,

(1) $$1 + 3 + 5 + \cdots + (2k - 1) = k^2$$

To prove the relation for $n = k + 1$, that is, to prove

(2) $$1 + 3 + 5 + \cdots + [2(k + 1) - 1] = (k + 1)^2$$

we add $2(k + 1) - 1 = 2k + 1$ (the next odd number after $2k - 1$) to both members of Equation 1. We obtain

$$[1 + 3 + 5 + \cdots + (2k - 1)] + (2k + 1) = k^2 + (2k + 1)$$
$$= k^2 + 2k + 1$$
$$= (k + 1)^2$$

which completes the proof.

EXERCISES 14.1

Prove the following true for all natural numbers using mathematical induction (Exercises 1–20).

1. $1^3 + 2^3 + 3^3 + \cdots + n^3 = \dfrac{n^2(n + 1)^2}{4}$.

2. $2 + 4 + 6 + \cdots + 2n = n(n + 1)$.

3. $(1)(2) + (2)(3) + (3)(4) + \cdots + n(n + 1) = \dfrac{n(n + 1)(n + 2)}{3}$.

4. $(1)(2) + (3)(4) + (5)(6) + \cdots + (2n - 1)(2n) = \frac{1}{3}n(4n^2 + 3n - 1)$.

5. $1^2 + 3^2 + 5^2 + \cdots + (2n - 1)^2 = \dfrac{n(4n^2 - 1)}{3}$.

6. $1 + 3 + 6 + \cdots + \frac{1}{2}n(n + 1) = \frac{1}{6}n(n + 1)(n + 2)$.

7. $(1)(3) + (2)(4) + (3)(5) + \cdots + n(n + 2) = \frac{1}{6}n(n + 1)(2n + 7)$.

8. $1^2 + 2^2 + 3^2 + \cdots + n^2 = \frac{1}{6}n(n + 1)(2n + 1)$.

9. $\dfrac{1}{(1)(2)} + \dfrac{1}{(2)(3)} + \dfrac{1}{(3)(4)} + \cdots + \dfrac{1}{n(n + 1)} = \dfrac{n}{n + 1}$.

10. $(1)(2)(3) + (2)(3)(4) + (3)(4)(5) + \cdots + n(n+1)(n+2) = \frac{1}{4}n(n+1)(n+2)(n+3)$.

11. $(1)(6) + (4)(9) + (7)(12) + \cdots + (3n - 2)(3n + 3) = 3n(n^2 + 2n - 1)$.

12. $1^3 + 3^3 + 5^3 + \cdots + (2n - 1)^3 = n^2(2n^2 - 1)$.

13. $\dfrac{1}{(1)(3)} + \dfrac{1}{(3)(5)} + \dfrac{1}{(5)(7)} + \cdots + \dfrac{1}{(2n - 1)(2n + 1)} = \dfrac{n}{2n + 1}$.

14. $\dfrac{1}{(1)(2)(3)} + \dfrac{1}{(2)(3)(4)} + \dfrac{1}{(3)(4)(5)} + \cdots + \dfrac{1}{n(n+1)(n+2)} = \dfrac{n(n+3)}{4(n+1)(n+2)}$.

15. $3 + 3^2 + 3^3 + \cdots + 3^n = \frac{3}{2}(3^n - 1)$.

16. $2 + 2^2 + 2^3 + \cdots + 2^n = 2(2^n - 1)$.

17. $\dfrac{1}{2} + \dfrac{1}{2^2} + \dfrac{1}{2^3} + \cdots + \dfrac{1}{2^n} = 1 - \dfrac{1}{2^n}$.

18. $(1)(4) + (4)(7) + (7)(10) + \cdots + (3n - 2)(3n + 1) = n(3n^2 + 3n - 2)$.

19. $1 + (2)(2) + (3)(2^2) + (4)(2^3) + \cdots + (n)(2^{n-1}) = 1 + (n - 1)(2^n)$.

20. $(\cos x + i \sin x)^n = \cos nx + i \sin nx$.

14.2 ◆ Sequences

A **sequence** is an ordered set of numbers in one-to-one correspondence with the natural numbers:

$$a_1, a_2, a_3, \ldots, a_n, \ldots$$

Each number in the sequence is called a **term** of the sequence. The first term is denoted by a_1, the second by $a_2, \ldots$, the **nth term** or **general term,** by a_n, and so forth. Some examples of sequences are:

$$2, 4, 6, 8, 10, \ldots, 2n, \ldots$$

$$\frac{1}{3}, \frac{1}{9}, \frac{1}{27}, \ldots, \frac{1}{3n}, \ldots$$

$$1, 5, 9, 13, \ldots, 1 + 4(n - 1), \ldots$$

$$1, 4, 9, 16, \ldots, n^2, \ldots$$

A sequence with a finite number of terms is called a **finite sequence.** A sequence with an infinite number of terms is called an **infinite sequence.**

A sequence is a particular kind of function, one whose domain is the set of natural numbers. We could use the function notation $f(n)$ for the terms of a sequence as

$$f(n_1), f(n_2), \ldots, f(n_n), \ldots$$

but the notation used above is more common.

EXAMPLE 1. Find the first four terms of the sequence whose general term is

$$a_n = \frac{1}{n + 1}$$

SOLUTION. We find the terms in a sequence by replacing n in the formula for the general term of the sequence by the natural numbers in turn. Thus, for the given sequence,

$$a_1 = \frac{1}{1+1} = \frac{1}{2}$$

$$a_2 = \frac{1}{2+1} = \frac{1}{3}$$

$$a_3 = \frac{1}{3+1} = \frac{1}{4}$$

$$a_4 = \frac{1}{4+1} = \frac{1}{5}$$

When we are given the first few terms of a sequence, it may be possible to determine an expression for the general term of the sequence to which they belong. Sometimes this is quite easy; other times it is difficult. It should be emphasized that the general term of a sequence is not uniquely determined. For example, if the first three terms of a sequence are

$$3, 6, 9$$

the general term may be

$$a_n = 3n$$

or

$$a_n = 3n + (n-1)(n-2)(n-3)$$

Observe that using the above, we generate 3, 6, 9 for the first three terms of the sequence in each case.

We shall consider two special kinds of sequences in this book: arithmetic sequences called **arithmetic progressions** and geometric sequences called **geometric progressions.**

DEFINITION 14.1. An **arithmetic progression** is a sequence of numbers in which each, after the first, is obtained from the preceding one by adding a fixed number, d, called the **common difference.**

The sequence

$$8, 11, 14, 17, \ldots$$

is an arithmetic progression with common difference 3.

If a_1 denotes the first term of an arithmetic progression, we see that the terms of the progression are

$$a_1, a_1 + d, a_1 + 2d, \ldots, a_1 + (n-1)d, \ldots$$

The formula for the nth term, a_n, of the arithmetic progression is now clear and may be written

$$a_n = a_1 + (n - 1)d$$

EXAMPLE 2. Find the twelfth term of the arithmetic progression

$$-2, -5, -8, \ldots$$

SOLUTION. The first term is -2 and the common difference is -3. We are looking for the twelfth term, hence, $n = 12$. Using the formula, we have

$$a_n = a_1 + (n - 1)d$$
$$a_{12} = -2 + (12 - 1)(-3)$$
$$= -2 + (11)(-3)$$
$$= -2 - 33 = -35$$

DEFINITION 14.2. A **geometric progression** is a sequence of numbers each of which, after the first, is obtained from the preceding one by multiplying by a fixed number $r \neq 0$ called the **common ratio.**

The sequence

$$\tfrac{1}{2}, \tfrac{1}{4}, \tfrac{1}{8}, \tfrac{1}{16}, \ldots$$

is a geometric progression with common ratio $r = \tfrac{1}{2}$.

If a_1 denotes the first term of a geometric progression we see that the terms of the geometric progression are

$$a_1, a_1r, a_1r^2, a_1r^3, \ldots, a_1r^{n-1}, \ldots$$

The formula for the nth term, a_n, of the geometric progression is then

$$a_n = a_1r^{n-1}$$

EXAMPLE 3. Find the tenth term of the geometric progression

$$3, 6, 12, \ldots$$

SOLUTION. Here $a_1 = 3$, $n = 10$, and $r = 2$. Hence

$$a_n = a_1r^{n-1}$$
$$a_{10} = (3)(2)^{10-1}$$
$$= (3)(2)^9$$
$$= (3)(512)$$
$$= 1536$$

EXAMPLE 4. A bank pays an annual rate of interest of 5% which is compounded quarterly. If $100 is deposited, what will be the balance in the account after four years?

SOLUTION. With an annual rate of interest of 5% a depositor will receive $\dfrac{0.05}{4}$ of his balance at the end of each compounding period. This means that his balance at the end of each quarter will be $1 + \dfrac{0.05}{4} =$ 1.0125 times his balance at the beginning of each quarter. Thus a deposit of $100 will produce a balance of $100(1.0125) at the end of the first quarter, a balance of $100(1.0125)^2$ at the end of the second quarter, a balance of $100(1.0125)^3$ at the end of the third quarter, and so forth. Since there are 16 quarters in a 4-year period, the balance at the end of this time will be

$$\$100(1.0125)^{16}$$

Using logarithms, we find

$$
\begin{aligned}
\log_{10}(1.0125)^{16} &= 16\log_{10}1.0125 \\
&= 16(0.0054) \\
&= 0.0864 \\
(1.0125)^{16} &= \text{antilog}_{10}\,0.0864 \\
&= 1.22
\end{aligned}
$$

The final balance will be

$$\$100(1.22) = \$122$$

EXERCISES 14.2

Find the first four terms in the sequences with the general terms given below (Exercises 1–10).

1. $a_n = \tfrac{1}{2}n(n+1)$

2. $a_n = n^3$

3. $a_n = \dfrac{n}{n+2}$

4. $a_n = \tfrac{1}{6}n(n+1)(n+2)$

5. $a_n = (-1)^n 2^n$

6. $a_n = \dfrac{n}{2n+1}$

7. $a_n = (\tfrac{1}{2})^n$

8. $a_n = (-1)^{n-1}2^{n+1}$

9. $a_n = \dfrac{2^{1/3n}}{(n+1)!}$

10. $a_n = n[1 + (-1)^n]$

Find the indicated term in each of the following arithmetic progressions (Exercises 11–20).

11. $2, 4, 6, 8, \ldots, 50$th
12. $1, -3, -7, \ldots, 12$th
13. $4, 7, 10, \ldots, 15$th
14. $3, 3\frac{1}{2}, 4, \ldots, 25$th
15. $1.1, 1.3, 1.5, \ldots, 18$th

16. $-4, -2\frac{1}{2}, -1, \ldots, 40$th
17. $19, 17, 15, \ldots, 30$th
18. $3a + 4b, 2a - 3b, a - 10b, \ldots, 10$th
19. $-2\frac{2}{3}, -1\frac{1}{3}, 0, \ldots, 15$th
20. $1.20, 1.08, 0.96, \ldots, 50$th

Find a formula for the nth term of each of the following arithmetic progressions (Exercises 21–26).

21. $12, 7, 2, \ldots$
22. $11, 13\frac{1}{2}, 16, \ldots$
23. $1.001, 1.002, 1.003, \ldots$

24. $x, x + 2, x + 4, \ldots$
25. $z, z - 3, z - 6, \ldots$
26. $3a + 2b, 2a + 4b, a + 6b, \ldots$

Find the indicated term in each of the following geometric progressions (Exercises 27–36).

27. $50, 10, 2, \ldots, 10$th
28. $2, -6, 18, \ldots, 20$th
29. $18, -6, 2, \ldots, 12$th
30. $-\frac{1}{2}, \frac{3}{2}, -\frac{9}{2}, \ldots, 25$th
31. $1, \sin \theta, \sin^2 \theta, \ldots, 8$th

32. $\frac{3}{2}, 1, \frac{2}{3}, \ldots, 6$th
33. $1, 0.4, 0.16, \ldots, 5$th
34. $\frac{3}{4}, 3, 12, \ldots, 11$th
35. $\cos^2 \theta, 1, \sec^2 \theta, \ldots, 17$th
36. $\tan \theta, 1, \cot \theta, \ldots, 30$th

37. If the tenth term of an arithmetic progression is 32 and the eighteenth term is 48, what is the fifteenth term?
38. Which term of the arithmetic progression $-1, -6, -11, \ldots$ is -176?
39. If the fifth term of a geometric progression is 36 and the common ratio is 2, find the first term.
40. The sixth term of a geometric progression is $\tan^3 \dfrac{\theta}{2}$ and the twelfth term is $\sin^3 \theta$. Find the third term.
41. Ideal Bank pays an annual rate of interest of 100%. A deposit of $100 is made. What will be the balance at the end of the year if the interest is compounded semiannually? quarterly? monthly? weekly? daily? every hour? every minute? every second? Compare these results with the discussion of Napierian logarithms in Chapter 5, Section 5.7.
42. At the end of one year the trade-in value of a certain automobile is $500 less than the original cost. Each year thereafter the trade-in value decreases by $200. If the original cost of the automobile was $4000, what is the trade-in value after 8 years?
43. A missile fired vertically upward rises 15,840 feet the first second, 15,808 feet the following second, and 15,776 feet the third second. How many feet does it rise the 30th second?

14.3 ◆ The Summation Notation

A highly useful symbol in mathematics is the uppercase Greek letter sigma, Σ, used in **summation notation**. Let

$$a_1, a_2, \ldots, a_n$$

be a finite sequence. The sum of the n terms of this sequence is compactly denoted by the notation

$$\sum_{i=1}^{n} a_i$$

Let us consider the finite sequence whose general term is n^2. Then

$$\sum_{i=1}^{5} i^2$$

means the sum

$$1^2 + 2^2 + 3^2 + 4^2 + 5^2 = 1 + 4 + 9 + 16 + 25 = 55$$

Observe that

$$\sum_{j=1}^{5} j^2$$

also represents the sum $1^2 + 2^2 + 3^2 + 4^2 + 5^2 = 55$ and thus is the same number as $\sum_{i=1}^{5} i^2$; the use of the letter i or j is immaterial.

EXAMPLE 1. Write $\sum_{i=1}^{4} (2i - 1)$ in expanded form and calculate the sum.

SOLUTION

$$\sum_{i=1}^{4} (2i - 1) = (2 \cdot 1 - 1) + (2 \cdot 2 - 1) + (2 \cdot 3 - 1) + (2 \cdot 4 - 1)$$
$$= 1 + 3 + 5 + 7 = 16.$$

EXAMPLE 2. Calculate $\sum_{j=2}^{4} j^{-1}$

SOLUTION

$$\sum_{j=2}^{4} j^{-1} = 2^{-1} + 3^{-1} + 4^{-1}$$
$$= \tfrac{1}{2} + \tfrac{1}{3} + \tfrac{1}{4} = \tfrac{13}{12}$$

14.4 ◆ Series

Whenever we have a finite sequence

$$a_1, a_2, \ldots, a_n$$

the sum of all the terms of this sequence:

$$\sum_{i=1}^{n} a_i = a_1 + a_2 + \cdots + a_n$$

is called a **finite series**.

With the arithmetic progression of n terms

$$a_1, a_1 + d, a_1 + 2d, \ldots, a_1 + (n-1)d$$

is associated the finite series, called the **arithmetic series**:

$$\sum_{i=1}^{n} [a_1 + (i-1)d] = a_1 + [a_1 + d] + \cdots + [a_1 + (n-1)d]$$

Let us denote this series by S_n. Then

(1) $S_n = a_1 + (a_1 + d) + (a_1 + 2d)$
$$+ \cdots + [a_1 + (n-2)d] + [a_1 + (n-1)d]$$

If the terms of the right member of Equation 1 are written in reverse order, we have

(2) $S_n = [a_1 + (n-1)d] + [a_1 + (n-2)d] + \cdots + (a_1 + d) + a_1$

Adding Equations 1 and 2 member by member, the sums of the corresponding terms of the right members are the same and we have

$$2S_n = n[2a_1 + (n-1)d]$$

or

$$S_n = \frac{n}{2}[2a_1 + (n-1)d]$$

Notice that $2a_1 + (n-1)d$ may be written $a_1 + [a_1 + (n-1)d] = a_1 + a_n$, hence we may write:

$$S_n = \frac{n}{2}(a_1 + a_n)$$

We could also have used the result of Example 1 of Section 14.1 to get

$$S_n = a_1 + (a_1 + d) + (a_1 + 2d) + \cdots + [a_1 + (n-2)d] + [a_1 + (n-1)d]$$
$$= n \cdot a_1 + [1 + 2 + \cdots + (n-2) + (n-1)]d$$
$$= n \cdot a_1 + \frac{(n-1)n}{2}d$$
$$= \frac{n}{2}[2a_1 + (n-1)d]$$

EXAMPLE 1. Find the sum of the first five terms of the arithmetic series

$$3 + 7 + 11 + \cdots$$

SOLUTION. Here $a_1 = 3$, $d = 4$, and $n = 5$; hence,

$$S_5 = \tfrac{5}{2}[(2)(3) + (5 - 1)4]$$
$$= \tfrac{5}{2}[6 + 16]$$
$$= 55$$

Now let us consider the finite geometric progression with n terms

$$a_1, a_1r, a_1r^2, \ldots, a_1r^{n-1}$$

The series associated with this sequence is

$$\sum_{i=1}^{n} a_1r^{i-1} = a_1 + a_1r + a_1r^2 + \cdots + a_1r^{n-1}$$

Let us denote this series by S_n. Then

(3) $$S_n = a_1 + a_1r + a_1r^2 + \cdots + a_1r^{n-1}$$

Multiplying both members of Equation 3 by r, we obtain

(4) $$rS_n = a_1r + a_1r^2 + a_1r^3 + \cdots + a_1r^n$$

When we subtract the members of Equation 4 from the members of Equation 3, all terms in the right-hand member vanish except the first term on the right in Equation 3 and the last term on the right in Equation 4. We obtain

(5) $$S_n - rS_n = a_1 - a_1r^n$$

(6) $$S_n(1 - r) = a_1(1 - r^n)$$

(7) $$S_n = \frac{a_1(1 - r^n)}{1 - r}$$

Since $a_n = a_1r^{n-1}$, we see that

$$S_n = \frac{a_1 - a_1r^n}{1 - r}$$

$$= \frac{a_1 - (a_1r^{n-1})r}{1 - r}$$

$$S_n = \frac{a_1 - a_nr}{1 - r}$$

EXAMPLE 2. Find the sum of the first nine terms of the geometric progression

$$-\tfrac{1}{3}, \tfrac{2}{3}, -\tfrac{4}{3}, \cdots$$

SOLUTION. Here $a_1 = -\tfrac{1}{3}$, $r = -2$, and $n = 9$. Using the formula, we have

$$S_n = \frac{a_1(1 - r^n)}{1 - r}$$

$$S_9 = \frac{-\tfrac{1}{3}[1 - (-2)^9]}{1 - (-2)}$$

$$= \frac{-\tfrac{1}{3}(1 + 512)}{3}$$

$$= -\frac{513}{9}$$

$$= -\frac{171}{3}$$

EXAMPLE 3. Find the sum of $\displaystyle\sum_{i=1}^{12} (7i - 1)$

SOLUTION. We write the first two or three terms in expanded form:

$$6 + 13 + 20 + \cdots$$

This is an arithmetic series. The first term is 6 and the common difference is 7. We are looking for the sum of the first twelve terms, hence, $n = 12$. Using the formula,

$$S_n = \frac{n}{2}[2a + (n - 1)d]$$

$$S_{12} = \frac{12}{2}[(2)(6) + (12 - 1)7]$$

$$= 6[12 + (11)(7)]$$
$$= 6[12 + 77]$$
$$= 6 \cdot 89$$
$$= 534$$

EXERCISES 14.3

Write in expanded form:

1. $\displaystyle\sum_{i=1}^{5} (3i + 1)$ 2. $\displaystyle\sum_{i=1}^{10} (-1)^i$

3. $\displaystyle\sum_{i=1}^{5} (\tfrac{1}{3})^{-i}$ 5. $\displaystyle\sum_{i=1}^{n} \frac{n}{n+1}$

4. $\displaystyle\sum_{i=1}^{7} \frac{i^2}{i+2}$ 6. $\displaystyle\sum_{i=1}^{k} (-1)^i i^2$

Find the sum of the specified number of terms of the following arithmetic and geometric progressions (Exercises 7–20).

7. $-5, -3, -1, \ldots, 20$ terms   16. $\sin^2\theta, 1, \cos^2\theta, \ldots, 6$ terms

8. $4, 2\tfrac{1}{2}, 1, \ldots, 25$ terms   17. $\displaystyle\sum_{i=1}^{12} (3i + 1)$

9. $40, 20, 10, \ldots, 15$ terms

10. $-18, -6, -2, \ldots, 30$ terms   18. $\displaystyle\sum_{i=1}^{20} (2i - 2)$

11. $1, -\tfrac{1}{5}, \tfrac{1}{25}, \ldots, 12$ terms

12. $27, -18, 12, \ldots, 18$ terms   19. $\displaystyle\sum_{i=1}^{15} (\tfrac{1}{2}i - 3)$

13. $\sin^2\theta, 1, \csc^2\theta, \ldots, 8$ terms

14. $\cos^6\theta, \cos^4\theta, \cos^2\theta, \ldots, 12$ terms   20. $\displaystyle\sum_{i=1}^{400} i$

15. $3, 3\tfrac{1}{4}, 3\tfrac{1}{2}, \ldots, 16$ terms

21. Find the sum of all integral multiples of 3 between 11 and 21.
22. Find the sum of all integral multiples of 5 between 32 and 67.
23. Show that the sum of the first n positive integers is equal to $\tfrac{1}{2}n(n + 1)$.
24. Show that the sum of the first n even positive integers is equal to $n(n + 1)$.
25. Show that the sum of the first n terms of the geometric progression $1, \tfrac{2}{3}, \tfrac{4}{9}, \ldots$ is given by the formula $S_n = 3[1 - (\tfrac{2}{3})^n]$.
26. A man accepts a position at a salary of $5000 for the first year with an increase of $200 per year each year thereafter. How many years will he have to work for his total earnings to equal $104,000?
27. A ball is dropped from a height of 9 feet. Each time it strikes the ground it rebounds to approximately $\tfrac{1}{3}$ of the height from which it last fell. What distance has it traveled at the instant it strikes the ground for the seventh time?
28. A lathe makes a total of 211 revolutions in the first 5 seconds after the motor is turned off. In any 1 second, its speed is two-thirds of its speed during the preceding second. What was its speed in revolutions per second at the time the motor was turned off?
29. How many blocks will be in a pile one block in thickness if there are 35 blocks in the bottom row, 33 in the second row, and so forth, and one in the top row?
30. A pyramid is formed from square blocks arranged to form tiers. In each tier the blocks are arranged to form squares in such a manner that the base of the pyramid measures six blocks on each side, the second tier measures four blocks on each side, and the top tier two blocks on each side. (a) How many blocks are in the pyramid? (b) How many blocks are in the pyramid if the base measures nine blocks on each side and the top tier has one block?

31. When Mr. Hurd's salary reached $12,000 he put $1000 in the bank. Every year thereafter he increased his deposit by $120. How much will he have accumulated in ten years?

32. A carillon in a campus bell tower chimes as many times as the hour. How many times does it chime between 8:00 A.M. and 4:00 P.M.?

33. If the taxi rate in a certain town is 50¢ for the first mile and 30¢ for each additional mile, what is the fare for a 15-mile trip?

14.5 ◆ Limits of Sequences and Sums of Series

Some infinite sequences have the property that as n increases, the term a_n gets very close to some real number L, called the **limit** of the sequence. Another way of saying this is that the numerical difference, $|a_n - L|$, is very small when n is very large. As an illustration consider the infinite sequence

$$\frac{3}{2}, \frac{4}{3}, \frac{5}{4}, \ldots, 1 + \frac{1}{n}, \ldots$$

It appears that as n increases, the terms approach 1. Let us consider $|a_n - 1|$ for any positive integer n:

$$|a_n - 1| = \left|\left(1 + \frac{1}{n}\right) - 1\right| = \left|\frac{1}{n}\right| = \frac{1}{n}$$

The number $\dfrac{1}{n}$ and, hence, $|a_n - 1|$ can be made very, very close to zero by choosing n sufficiently large. That is, we can make $|a_n - 1|$ arbitrarily close to zero by choosing sufficiently large values for n. We say that the sequence

$$\frac{3}{2}, \frac{4}{3}, \frac{5}{4}, \ldots, 1 + \frac{1}{n}, \ldots$$

has limit 1.

If an infinite sequence with nth term a_n has a limit L, we write

$$\lim_{n \to \infty} a_n = L$$

Not every infinite sequence has a limit. For example, consider the infinite sequence

$$1, 3, 5, 7, \ldots, (2n - 1), \ldots$$

Here the number $|(2n - 1) - L|$ can be made arbitrarily large for every real number L, and consequently the sequence does not have a limit.

EXAMPLE 1. Find the limit of the infinite sequence

$$3, \frac{5}{2}, \frac{7}{3}, \frac{9}{4}, \ldots, \frac{2n+1}{n}, \ldots$$

SOLUTION. Since

$$\frac{2n+1}{n} = 2 + \frac{1}{n}$$

approaches 2 as n increases (because as n gets larger and larger, $\frac{1}{n}$ approaches 0), it appears that the limit of the sequence is 2. We see that

$$\left| \frac{2n+1}{n} - 2 \right| = \left| \frac{2n+1-2n}{n} \right| = \left| \frac{1}{n} \right| = \frac{1}{n}$$

The number $\frac{1}{n}$ can be made arbitrarily close to 0 by choosing n sufficiently large. Hence the limit of the given sequence is 2.

With every infinite sequence is associated an **infinite series.** With the infinite sequence

$$a_1, a_2, a_3, \ldots, a_n, \ldots$$

is associated the infinite series

$$a_1 + a_2 + a_3 + \cdots + a_n + \cdots$$

To indicate an infinite series we use the notation

$$\sum_{i=1}^{\infty} a_i$$

Let us consider the infinite series

$$\sum_{i=1}^{\infty} a_i = a_1 + a_2 + \cdots + a_n + \cdots$$

Now let us consider the infinite sequence of numbers $s_1, s_2, \ldots, s_n, \ldots$, called **partial sums,** formed as follows:

$$s_1 = a_1$$
$$s_2 = a_1 + a_2$$
$$s_3 = a_1 + a_2 + a_3$$
$$\vdots$$
$$s_n = a_1 + a_2 + \cdots + a_n$$

and so forth.

If this infinite sequence, $s_1, s_2, \ldots, s_n, \ldots$, of partial sums has a limit, S, we say that the infinite series $\sum\limits_{i=1}^{\infty} a_i$ has a sum S.

EXAMPLE 2. Find the sum of the infinite series

$$1 + \frac{1}{3} + \frac{1}{9} + \cdots + \frac{1}{3^{n-1}} + \cdots$$

SOLUTION. The given infinite series is determined by the geometric progression with first term 1 and common ratio $\frac{1}{3}$. By adding successive terms we obtain the partial sums

$$S_1 = 1$$
$$S_2 = 1 + \tfrac{1}{3} = \tfrac{4}{3} = 1\tfrac{1}{3}$$
$$S_3 = 1 + \tfrac{1}{3} + \tfrac{1}{9} = \tfrac{13}{9} = 1\tfrac{4}{9}$$
$$S_4 = 1 + \tfrac{1}{3} + \tfrac{1}{9} + \tfrac{1}{27} = \tfrac{40}{27} = 1\tfrac{13}{27}$$
$$S_5 = 1 + \tfrac{1}{3} + \tfrac{1}{9} + \tfrac{1}{27} + \tfrac{1}{81} = 1\tfrac{40}{81}$$
$$S_6 = 1 + \tfrac{1}{3} + \tfrac{1}{9} + \tfrac{1}{27} + \tfrac{1}{81} + \tfrac{1}{243} = 1\tfrac{121}{243}$$

and so forth.

It appears that if more and more terms are added, S_n approaches $1\frac{1}{2}$. As a matter of fact, we can find the partial sum S_n:

$$S_n = \frac{a_1 - a_1 r^n}{1 - r}$$
$$= \frac{1 - (\tfrac{1}{3})^n}{1 - \tfrac{1}{3}}$$
$$= \tfrac{3}{2}[1 - (\tfrac{1}{3})^n]$$
$$= \tfrac{3}{2} - \tfrac{1}{2}(\tfrac{1}{3})^{n-1}$$

Since $(\tfrac{1}{3})^{n-1}$ can be made as close to zero as we please by choosing sufficiently large values of n, we have the limit of the sequence of partial sums $s_1, s_2, \ldots, s_n, \ldots$, is $1\frac{1}{2}$ and the sum of the series is also $1\frac{1}{2}$.

The method used to find the sum of the infinite series in Example 2 can be extended to general geometric series. Consider the geometric progression

$$a_1, a_1 r, a_1 r^2, \ldots, a_1 r^{n-1}, \ldots$$

If $r \neq 1$ we have

$$S_n = \frac{a_1 - a_1 r^n}{1 - r}$$
$$= \frac{a_1}{1 - r} - \frac{a_1}{1 - r} \cdot r^n$$

If $|r| < 1$, it can be shown that r^n approaches zero as n increases and hence

$$\lim_{n \to \infty} S_n = \frac{a_1}{1 - r}$$

Hence, if $|r| < 1$, the infinite geometric series

$$a_1 + a_1 r + a_1 r^2 + \cdots + a_1 r^{n-1} + \cdots$$

has a sum

$$S = \frac{a_1}{1 - r}$$

EXAMPLE 3. Find the sum of the infinite geometric series

$$\frac{1}{4} + \frac{1}{16} + \frac{1}{64} + \cdots + \frac{1}{4^n} + \cdots$$

SOLUTION. The first term of the given geometric series is 1 and the common ratio is $\frac{1}{4}$. Since $|\frac{1}{4}| < 1$, to find the sum we use the formula

$$S = \frac{a_1}{1 - r}$$

$$= \frac{\frac{1}{4}}{1 - \frac{1}{4}}$$

$$= \frac{\frac{1}{4}}{\frac{3}{4}}$$

$$= \frac{1}{3}$$

EXAMPLE 3. Find the rational number which names the repeating decimal

$$3.1\overline{27}$$

SOLUTION. From the repeating decimal

$$3.1\overline{27} = 3.1272727 \ldots$$

we obtain the infinite series

$$3.1 + 0.027 + 0.00027 + 0.0000027 + \cdots$$

The portion

$$0.027 + 0.00027 + 0.0000027 + \cdots$$

of this series is an infinite geometric series with $a_1 = 0.027$ and $r = 0.01$; hence its sum is

$$S = \frac{0.027}{1 - .01} = \frac{0.027}{0.99} = \frac{27}{990} = \frac{3}{110}$$

Thus the desired number is

$$3.127 = 3\tfrac{1}{10} + \tfrac{3}{110}$$
$$= \tfrac{31}{10} + \tfrac{3}{110}$$
$$= \tfrac{341}{110} + \tfrac{3}{110}$$
$$= \tfrac{344}{110}$$
$$= \tfrac{172}{55}$$

EXERCISES 14.4

Find the sum of each geometric series in Exercises 1–10.

1. $3 + \tfrac{1}{2} + \tfrac{1}{12} + \cdots$
2. $10 - 1 + \tfrac{1}{10} - \cdots$
3. $4 - \tfrac{1}{2} + \tfrac{1}{16} - \cdots$
4. $3 + \tfrac{3}{2} + \tfrac{3}{4} + \cdots$
5. $1 - \tfrac{1}{5} + \tfrac{1}{25} - \cdots$
6. $1 - \tfrac{1}{3} + \tfrac{1}{6} - \cdots$
7. $\tfrac{1}{2} + \tfrac{1}{4} + \tfrac{1}{8} + \cdots$
8. $\tfrac{1}{3} - \tfrac{1}{9} + \tfrac{1}{27} - \cdots$
9. $\tfrac{1}{10} + \tfrac{1}{100} + \tfrac{1}{1000} + \cdots$
10. $\tfrac{2}{3} + \tfrac{1}{6} + \tfrac{1}{24} + \cdots$

Find the rational number which names each of the following repeating decimals.

11. $1.\overline{3}$
12. $2.\overline{45}$
13. $6.5\overline{11}$
14. $0.41\overline{2}$
15. $8.\overline{345}$
16. $2.1\overline{66}$
17. $0.\overline{53}$
18. $0.1\overline{42}$
19. $1.23\overline{123}$
20. $14.56\overline{7}$

Find the limit of each of the following infinite sequences.

21. $\dfrac{1}{2}, \dfrac{1}{4}, \dfrac{1}{8}, \ldots, \dfrac{1}{2^n}, \ldots$

22. $\dfrac{1}{3}, \dfrac{1}{9}, \dfrac{1}{27}, \ldots, \dfrac{1}{3^n}, \ldots$

23. $\dfrac{1}{10}, \dfrac{1}{100}, \dfrac{1}{1000}, \ldots, \dfrac{1}{10^n}, \ldots$

24. $\dfrac{1}{2}, \dfrac{2}{3}, \dfrac{3}{4}, \dfrac{4}{5}, \dfrac{5}{6}, \ldots, \dfrac{n}{n+1}, \ldots$

25. $\cos 1, \cos \dfrac{1}{2}, \cos \dfrac{1}{3}, \ldots, \cos \dfrac{1}{n}, \ldots$

14.6 ✦ The Binomial Expansion

By direct multiplication we can obtain the following:

$$(a + b)^1 = a + b$$
$$(a + b)^2 = a^2 + 2ab + b^2$$
$$(a + b)^3 = a^3 + 3a^2b + 3ab^2 + b^3$$
$$(a + b)^4 = a^4 + 4a^3b + 6a^2b^2 + 4ab^3 + b^4$$
$$(a + b)^5 = a^5 + 5a^4b + 10a^3b^2 + 10a^2b^3 + 5ab^4 + b^5$$

Notice that in all the above cases:

(a) The first term is a^n and the last term is b^n.

(b) The second term is $na^{n-1}b$.

(c) In each successive term after the first the exponent of a is one less and that of b is one more than in the term immediately preceding, so that the sum of the exponents of a and b is equal to n in every term.

(d) There are $n + 1$ terms in the expansion $(a + b)^n$.

(e) The coefficients of the terms form the following pattern known as **Pascal's triangle,** shown below:

$$
\begin{array}{ccccccccccc}
 & & & & & 1 & & & & & \\
 & & & & 1 & & 1 & & & & \\
 & & & 1 & & 2 & & 1 & & & \\
 & & 1 & & 3 & & 3 & & 1 & & \\
 & 1 & & 4 & & 6 & & 4 & & 1 & \\
1 & & 5 & & 10 & & 10 & & 5 & & 1
\end{array}
$$

We observe that if n is the exponent and r is the number of the term in the expansion, the coefficient of $a^{n-r+1}b^{r-1}$ is

$$\frac{n(n - 1)(n - 2) \cdot \cdots \cdot (n - r + 2)}{1 \cdot 2 \cdot 3 \cdot \cdots \cdot (r - 1)}$$

We also note that each number in Pascal's triangle, apart from those that bound the triangle, is the sum of the numbers immediately to its left and its right in the row above. Thus the third number, 10, in the sixth row is the sum of 4 and 6 immediately to its left and right in the fifth row.

From the above observations, we conclude

$$(a + b)^n = a^n + na^{n-1}b + \frac{n(n - 1)}{1 \cdot 2} a^{n-1}b^2 + \frac{n(n - 1)(n - 2)}{1 \cdot 2 \cdot 3} a^{n-3}b^3$$

$$+ \cdots + \frac{n(n - 1)(n - 2) \cdots (n - r + 2)}{1 \cdot 2 \cdot 3 \cdots (r - 1)} a^{n-r+1}b^{r-1} + \cdots + b^n$$

This formula is called the **binomial expansion** or the **binomial theorem.**

To facilitate the writing of the binomial theorem we use a special symbol $n!$ (read: n factorial) which is defined by

$$0! = 1$$
$$n! = n(n - 1)(n - 2) \cdots (2)(1)$$

Thus

$$6! = (6)(5)(4)(3)(2)(1) = 720$$
$$3! = (3)(2)(1) = 6$$

We may now write

$$(a + b)^n = a^n + na^{n-1}b + \frac{n(n - 1)}{2!} a^{n-2}b^2 + \cdots$$

$$+ \frac{n(n - 1) \cdots (n - r + 2)}{(r - 1)!} a^{n-r+1}b^{r-1}$$

$$+ \frac{n(n - 1) \cdots (n - r + 1)}{r!} a^{n-r}b^r + \cdots + nab^{n-1} + b^n$$

We use mathematical induction to prove the binomial theorem. For $n = 1$ we have

$$(a + b)^1 = a^1 + b^1$$

and Part II of the induction process is verified.

Now we assume that the formula is true for $n = k$. That is,

$$(a + b)^k = a^k + ka^{k-1}b + \frac{k(k - 1)}{2!} a^{k-2}b^2 + \cdots$$

$$+ \frac{k(k - 1) \cdots (k - r + 2)}{(r - 1)!} a^{k-r+1}b^{r-1}$$

$$+ \frac{k(k - 1) \cdots (k - r + 1)}{r!} a^{k-r}b^r + \cdots + kab^{k-1} + b^k$$

To show that the formula is true for $n = k + 1$ we multiply each member of the equation above by $(a + b)$ and obtain

$$(1) \quad (a + b)^{k+1} = a^{k+1} + ka^kb + \frac{k(k - 1)}{2!} a^{k-1}b^2 + \cdots$$

$$+ \frac{k(k - 1) \cdots (k - r + 2)}{(r - 1)!} a^{k-r+2}b^{r-1}$$

$$+ \frac{k(k - 1) \cdots (k - r + 1)}{r!} a^{k-r+1}b^r + \cdots + ka^2b^{k-1} + ab^k$$

$$+ a^k b + ka^{k-1}b^2 + \frac{k(k-1)}{2!} a^{k-2}b^3 + \cdots$$

$$+ \frac{k(k-1) \cdots (k-r+2)}{(r-1)!} a^{k-r+1}b^r$$

$$+ \frac{k(k-1) \cdots (k-r+1)}{r!} a^{k-r}b^{r+1} + \cdots + kab^k + b^{k+1}$$

When we combine like terms of equation 1 above, we see that the coefficient of $a^k b$ is $(k+1)$. The coefficient of $a^{k-1}b^2$ is

$$\frac{k(k-1)}{2!} + k = k\left(\frac{k-1}{2} + 1\right) = k\left(\frac{k+1}{2}\right) = \frac{(k+1)k}{2!}.$$

The coefficient of $a^{k-r+1}b^k$ is

$$\frac{k(k-1) \cdots (k-r+1)}{r!} + \frac{k(k-1) \cdots (k-r+2)}{(r-1)!}$$

$$= \frac{k(k-1) \cdots (k-r+2)(k-r+1)}{r(r-1)!} + \frac{k(k-1) \cdots (k-r+2)}{(r-1)!}$$

$$= \frac{k(k-1) \cdots (k-r+2)}{(r-1)!}\left(\frac{k-r+1}{r} + 1\right)$$

$$= \frac{k(k-1) \cdots (k-r+2)}{(r-1)!}\left(\frac{k+1}{r}\right)$$

$$= \frac{(k+1)(k)(k-1) \cdots (k-r+2)}{r!}$$

The coefficient of ab^k is $(k+1)$. We then have

$$(a+b)^{k+1} = a^{k+1} + (k+1)a^k b + \frac{(k+1)k}{2!} a^{k-1}b^2 + \cdots$$

$$+ \frac{(k+1)(k)(k-1) \cdots (k-r+2)}{r!} a^{k-r+1}b^r$$

$$+ \cdots + (k+1)ab^k + b^{k+1}$$

This is $(a+b)^n$ for the case $n = k+1$. The proof is therefore complete.

EXAMPLE 1. Expand $(2x - y)^5$ and simplify.

SOLUTION

$$(2x - y)^5 = (2x)^5 + 5(2x)^4(-y) + \frac{(5)(4)}{2!} (2x)^3(-y)^2$$

$$+ \frac{(5)(4)(3)}{3!} (2x)^2(-y)^3 + \frac{(5)(4)(3)(2)}{4!} (2x)(-y)^4 + (-y)^5$$

$$= 32x^5 - 80x^4y + 80x^3y^2 - 40x^2y^3 + 10xy^4 - y^5$$

EXAMPLE 2. Find the term involving x^5 in the expansion $(x + 3y)^7$.

SOLUTION. The term involving x^5 would be the term of the expansion such that $x^{7-r+1} = x^5$; hence,

$$7 - r + 1 = 5 \quad \text{and} \quad r = 3$$

We are looking for the third term of the expansion. Then

$$\frac{(7)(6)}{(3-1)!} x^{7-3+1}(3y)^{3-1} = \frac{42}{2} x^5(3y)^2$$

$$= (21)(9)x^5y^2$$

$$= 189x^5y^2$$

EXAMPLE 3. Find the term that does not involve x in the expansion of

$$\left(x^2 + \frac{1}{x^3}\right)^{10}$$

SOLUTION. We are looking for the term that does not contain an x, that is, the term in which the exponent of x is 0 since $x^0 = 1$. Then

$$(x^2)^{10-r+1}\left(\frac{1}{x^3}\right)^{r-1} = x^0$$

But

$$(x^2)^{10-r+1}\left(\frac{1}{x^3}\right)^{r-1} = (x^2)^{11-r}\left(\frac{1}{x^3}\right)^{r-1}$$

$$= \frac{x^{22-2r}}{x^{3r-3}}$$

$$= x^{25-5r}$$

But $x^{25-5r} = x^0$, hence

$$25 - 5r = 0$$

$$r = 5$$

We are looking for the fifth term which is

$$\frac{10 \cdot 9 \cdot 8 \cdot 7}{(5-1)!} (x^2)^{10-5+1}\left(\frac{1}{x^3}\right)^{5-1} = 210x^{12} \cdot \frac{1}{x^{12}}$$

$$= 210x^0$$

$$= 210$$

EXAMPLE 4. Find the fifth term of $\left(x - \frac{1}{x}\right)^{10}$.

SOLUTION. The rth term of the binomial expansion is

$$\frac{n(n-1)(n-2)\cdots(n-r+2)}{(r-1)!}a^{n-r+1}b^{r-1}$$

In this case $n = 10$, $r = 5$, and

$$n - r + 1 = 10 - 5 + 1 = 6$$
$$n - r + 2 = 10 - 5 + 2 = 7$$

The fifth term is

$$\frac{10\cdot 9\cdot 8\cdot 7}{1\cdot 2\cdot 3\cdot 4}x^7\cdot\left(-\frac{1}{x}\right)^3 = -210x^4$$

EXERCISES 14.5

Expand each of the following by use of the binomial theorem. Simplify (Exercises 1–10).

1. $(x + y)^8$

2. $(a - b)^9$

3. $(x + 2)^5$

4. $(a - 2b)^4$

5. $(2a^2 - 3b^3)^3$

6. $\left(\dfrac{1}{x} - \dfrac{1}{y}\right)^6$

7. $\left(\dfrac{1}{x} + x\right)$

8. $\left(3x - \dfrac{1}{3}y\right)^5$

9. $(5x + y)^7$

10. $\left(3a^2 - \dfrac{2}{a}\right)^5$

Write the first five terms of each of the following expansions (Exercises 11–18).

11. $(a - 3b)^8$

12. $\left(x + \dfrac{1}{2}y\right)^{12}$

13. $(3x - y)^{15}$

14. $\left(x - \dfrac{1}{x}\right)^{20}$

15. $\left(x - \dfrac{2}{x}\right)^{18}$

16. $(2x - \sqrt{y})^{25}$

17. $\left(x - \dfrac{1}{x^2}\right)^{14}$

18. $(1 + 2\sqrt{x})^{13}$

Find the indicated term in each of the following expansions without writing the preceding terms (Exercises 19–25).

19. $(x + y)^{25}$, eighth term.
20. $(y^2 - 3)^{10}$, fifth term.

21. $\left(x - \dfrac{2}{x}\right)^{15}$, seventh term.

22. $\left(x^2 - \dfrac{1}{x^3}\right)^{20}$, ninth term.

23. $(2a - 4b^2)^{12}$, sixth term.

24. $\left(x + \dfrac{3}{x}\right)^{12}$, middle term.

25. $(2x + 3y)^{30}$, tenth term.

26. Find the term that does not involve y in the expansion of $\left(\dfrac{3}{y^2} + \dfrac{y^4}{9}\right)^{18}$.

27. Find the term involving x^5 in the expansion of $(2x + 3y)^7$.

28. Find the term that involves x^{-3} in $\left(x + \dfrac{y}{x}\right)^{11}$.

14.7 ◦ The Binomial Series

Using the binomial theorem, we have the expansion of $(1 + x)^n$ for n a positive integer:

$$(1 + x)^n = 1 + nx + \frac{n(n - 1)}{2!} x^2 + \frac{n(n - 1)(n - 2)}{3!} x^3$$

$$+ \frac{n(n - 1)(n - 2)(n - 3)}{4!} x^4 + \cdots + x^n$$

The expansion has $n + 1$ terms.

If n is any real number that is not a positive integer, the corresponding expansion can be written in a purely formal manner without proof and then investigated. Thus, for the case $n = \frac{1}{2}$ we may write

$$(1 + x)^{1/2} = 1 + \frac{1}{2}x + \frac{\frac{1}{2}(-\frac{1}{2})}{2!} x^2 + \frac{\frac{1}{2}(-\frac{1}{2})(-\frac{3}{2})}{3!} x^3 + \frac{\frac{1}{2}(-\frac{1}{2})(-\frac{3}{2})(-\frac{5}{2})}{4!} x^4 + \cdots$$

$$= 1 + \frac{1}{2}x - \frac{1}{2^2 \cdot 2!} x^2 + \frac{1 \cdot 3}{2^3 \cdot 3!} x^3 - \frac{1 \cdot 3 \cdot 5}{2^4 \cdot 4!} x^4 + \cdots$$

We observe that this expansion does not terminate as it does when n is a positive integer—the terms continue indefinitely. This happens whenever n is not a positive integer or zero. The result is an infinite series called a **binomial series**.

We now ask, does this series have a sum? That is, does the sum of the first n terms approach a limit as n becomes sufficiently large? The answer is yes if $|x| < 1$, but the proof will not be given here.

We see that if n is not a positive integer and $|x| < 1$, then a finite number of terms of the expansion of $(1 + x)^n$ may be used to approximate the value of $(1 + x)^n$; in general, the greater the number of terms used the better the approximation will be.

EXAMPLE 1. Compute $\sqrt{98} = (98)^{1/2}$ correct to four decimal places.

SOLUTION

$$
\begin{aligned}
\sqrt{98} &= \sqrt{100 - 2} \\
&= 10\sqrt{1 - 0.02} \\
&= 10(1 - 0.02)^{1/2} \\
&= 10\left[1 + \frac{1}{2}(-0.02) + \frac{\frac{1}{2}(-\frac{1}{2})(-0.02)^2}{2!} \right. \\
&\qquad\qquad\qquad \left. + \frac{\frac{1}{2}(-\frac{1}{2})(-\frac{3}{2})(-0.02)^3}{3!} + \cdots \right] \\
&= 10[1 - 0.01 - \tfrac{1}{8}(0.0004) - \tfrac{1}{16}(0.000008) + \cdots] \\
&= 10[1 - 0.01 - 0.00005 - 0.0000005 + \cdots] \\
&= 10(1 - 0.01005005) \\
&= 10(0.98994995) \\
&= 9.8994995 \doteq 9.899
\end{aligned}
$$

EXAMPLE 2. Find the value of $(1.02)^{-1}$ correct to four decimal places.

SOLUTION. The binomial series for $n = -1$ is given by

$$
\begin{aligned}
(1 + x)^{-1} &= 1 + \frac{-1}{1!}x + \frac{(-1)(-2)}{2!}x^2 + \frac{(-1)(-2)(-3)}{3!}x^3 \\
&\qquad + \frac{(-1)(-2)\cdots(-1 - r + 2)}{(r - 1)!}x^{r-1} + \cdots \\
&= 1 - x + x^2 - x^3 + x^4 - x^5 + \cdots
\end{aligned}
$$

We may now replace x by 0.02 to obtain

$$
\begin{aligned}
(1.02)^{-1} &= 1 - 0.02 + 0.0004 - 0.000008 - \cdots \\
&= 9.804, \text{ correct to four decimal places.}
\end{aligned}
$$

EXERCISES 14.6

Write the first five terms of each of the following expansions. Simplify.

1. $\sqrt{1 + x}$ 2. $\sqrt[3]{1 - x}$

3. $\sqrt[3]{1 + x}$ 6. $\dfrac{1}{\sqrt{1 - x}}$

4. $(1 + x)^{-3}$ 7. $\dfrac{1}{(1 - x)^2}$

5. $\dfrac{1}{\sqrt{1 + x}}$ 8. $\dfrac{1}{(1 - x)^3}$

Use the binomial series to find the value of each of the following correct to four decimal places.

9. $\sqrt{50}$ 13. $\sqrt[3]{25}$

10. $\sqrt[3]{130}$ 14. $(1.04)^{-5}$

11. $\sqrt[4]{1.02}$ 15. $(1.03)^{-6}$

12. $\sqrt[3]{29}$ 16. $(1.05)^{-10}$

17. Divide 1 by $1 - x^2$ and compare the result with that obtained by expanding $(1 - x^2)^{-1}$ by the binomial formula.

CHAPTER REVIEW

1. Using mathematical induction, prove that for all natural numbers n,

$$\frac{1}{1 \cdot 2} + \frac{1}{2 \cdot 3} + \frac{1}{3 \cdot 4} + \cdots + \frac{1}{n(n + 1)} = \frac{n}{n + 1}$$

2. Using mathematical induction, prove that for all natural numbers n,

$$1 \cdot 4 + 2 \cdot 9 + 3 \cdot 16 + \cdots + n(n + 1)^2 = \tfrac{1}{12}n(n + 1)(n + 2)(3n + 5)$$

3. Find the first four terms in the sequences with the general terms given below.
   (a) $a_n = (-1)^n 3n$ (c) $a_n = (\tfrac{1}{4})^n$
   (b) $a_n = \tfrac{1}{2}(n + 1)(n + 2)$ (d) $a_n = (-1)^{n-1}(\tfrac{1}{2})^{n+1}$

4. Find the indicated terms in each of the following arithmetic progressions.
   (a) $3, -1, -5, \ldots$, tenth
   (b) $4, 4\tfrac{1}{2}, 5, \ldots$, twentieth
   (c) $1.2, 1.6, 2.0, 2.4, \ldots$, twenty-fifth
   (d) $2.1, 2.7, 3.3, \ldots$, fiftieth

5. Find the indicated term in each of the following geometric progressions.
   (a) $12, 6, 3, \ldots$, tenth
   (b) $4, 12, 36, \ldots$, fifteenth
   (c) $\tfrac{1}{2}, \tfrac{1}{4}, \tfrac{1}{8}, \ldots$, twentieth
   (d) $\sin \theta, 1, \csc \theta, \ldots$, twelfth

6. The seventh term of an arithmetic progression is 12 and the fifteenth term is -4. Find the eleventh term.

7. The fifth term of a geometric progression is $\frac{1}{32}$ and the tenth term is $\frac{1}{1024}$. Find the eighth term.

8. Find the sum of the following arithmetic series.

   (a) $\sum_{i=1}^{10} 4i$ (c) $\sum_{i=1}^{12} (2 - i)$

   (b) $\sum_{i=1}^{15} \frac{1}{2}(i + 1)$ (d) $\sum_{i=1}^{20} (4 - 2i)$

9. Find the sum of the following geometric series.

   (a) $\sum_{i=1}^{6} 5(2)^{i-1}$ (c) $\sum_{i=1}^{5} 3(-\frac{1}{3})^{i-1}$

   (b) $\sum_{i=1}^{8} 2^{i-1}$ (d) $\sum_{i=1}^{4} (-\frac{1}{2})(-\frac{1}{4})^{i-1}$

10. Find the sums of the infinite geometric series.
    (a) $9, -6, 4, \ldots$
    (b) $\sqrt{6}, \sqrt{2}, \sqrt{\frac{2}{3}}, \ldots$
    (c) $0.8, 0.08, 0.008, \ldots$

11. Find the rational number that approximates each of the following repeating decimals.
    (a) $0.\overline{5}$ (c) $0.\overline{13}$
    (b) $0.\overline{8}$ (d) $1.4\overline{26}$

12. Expand each of the following binomials expressing the result in simplest form.
    (a) $(x^2 + 1)^9$
    (b) $(3y - 1)^7$

13. Find the first three terms of each expansion.
    (a) $(x^2 - 3)^9$
    (b) $(\cos x - 1)^{12}$

14. Find the indicated term in each expansion.
    (a) $(2 + x)^8$; fifth
    (b) $(x^3 - 2)^7$; fourth
    (c) $\left(x - \dfrac{1}{x}\right)^{12}$; seventh
    (d) $(1 + \tan x)^{20}$; middle

15. Write the first four terms of each of the following expansions. Simplify.
    (a) $\sqrt{1 - x}$ (c) $\dfrac{1}{(1 + x)^3}$
    (b) $(1 - x)^{-3}$ (d) $\sqrt[3]{1 - x}$

16. A man is driving along a highway at 60 miles per hour (88 feet/second). He applies the brake and comes to a complete stop in 22 seconds. If the speed at which he travels in successive seconds are in arithmetic progression, how fast did he travel the tenth second after braking?

17. Betty has a 1050 page novel to read for a literature class. She reads 10 pages the first day, 20 the second, 30 the third, and so on. How long will it take her to read the novel?

18. View lots in a town bordering the ocean are becoming more and more scarce, and hence are increasing in value. The City Planning Commission finds that view lots are rising in value by 10% annually. If a view lot is purchased for $7500, what will it be worth in 10 years?

19. If there were no intermarriages, how many ancestors have you had in the twelve generations preceding you?

20. One side of an equilateral triangle is 12 inches. The midpoints of its sides are joined to form an inscribed equilateral triangle, and the process is continued. Find the sum of the perimeters of the triangles if the process is continued indefinitely.

CHAPTER 15

Permutations, Combinations, and Probability

15.1 ✦ Fundamental Counting Principle

Let us suppose that four students are participating in a mathematics contest in which first, second, third, and fourth places will be determined. We label the students a, b, c, and d. Let us first find out the number of different ways the first two places can be filled. To do this let us use the **tree diagram** shown in Figure 15.1 (see page 396).

In the first place there are four possibilities, a, b, c, or d. After the first place has been filled, there are three possible choices for the second place. There are twelve different ways of filling the first and second places. These outcomes are noted in the tree diagram by the twelve different paths, called **branches,** from the first place to the second place. We note that the number of possible outcomes is the product of the number of choices for the first place and the number of choices for the second place (after the first place has been determined).

Now let us determine the number of different ways in which the first, second, third, and fourth places can be determined. Again we may use a tree diagram. We begin as shown in Figure 15.2. When the first place has been filled, we have three choices for the second place. When the second place has been filled, there are two choices for the third place. After the third place has been filled, there is one choice for the fourth place. We see that we have

$$4 \times 3 \times 2 \times 1 = 4! = 24$$

different ways of filling the four places.

The example above illustrates the **fundamental counting principle.**

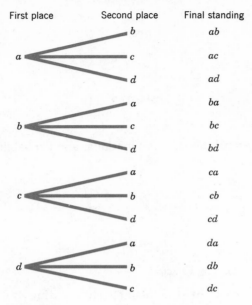

Figure 15.1

DEFINITION 15.1 (Fundamental Counting Principle). Given a sequence $E_1, E_2, \ldots, E_n$ of n events, if for each i, E_i can occur m_i ways, then the total number of distinct ways the events may take place is $m_1 m_2 \ldots m_n$.

EXAMPLE 1. How many numerals of two different digits can be formed with the digits 1, 2, 3, and 4?

SOLUTION. We wish to determine numerals of the form $x_1 x_2$ using the digits 1, 2, 3, and 4. The first choice may be made in any one of four

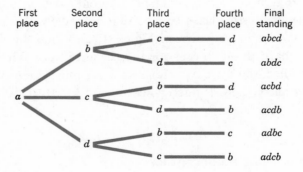

Figure 15.2

ways and after it, the second choice can be made in any one of three ways (since the digits must be different). Hence we have

$$4 \times 3 = 12$$

EXAMPLE 2. A club consists of 10 men and 6 women. In how many ways can a president, vice-president, secretary, and treasurer be chosen if the secretary must be a woman and the treasurer must be a man?

SOLUTION. Let E_1 be the choice of secretary, E_2 the choice of treasurer, E_3 the choice of president, and E_4 the choice of vice-president. Then E_1 can occur in 6 ways, E_2 can occur in 10 ways, E_3 can occur in 14 ways (why?), and E_4 can occur in 13 ways (why?). The total number of possibilities is

$$6 \times 10 \times 14 \times 13 = 10,920$$

EXERCISES 15.1

1. In how many ways can four books be arranged on a shelf?
2. In how many ways can six people be arranged in a line to have their photograph taken?
3. How many different four-digit numerals can be formed from the digits 1, 2, 3, 4, 5, and 6 if no digit is repeated in a numeral?
4. In how many ways can the eight officers of a club be seated in a row of eight chairs?
5. How many different four-letter arrangements are there of the letters of the word LUCK?
6. How many different signals can be made from six different flags if each signal is to consist of six flags hung in a vertical line?
7. How many teams for a game of bridge can be arranged if each team is to be composed of a man and a woman from a group of five women and four men?
8. How many different automobile license plates can be made using one letter followed by four digits?
9. An automobile manufacturer builds four different models of cars; each model is available in any one of six exterior colors, any one of five interior colors, and with or without white-wall tires. How many different appearing automobiles does the manufacturer make?
10. A man has 6 pairs of slacks and 4 jackets. How many different outfits can he wear?
11. How many arrangements of 3 different novels and 5 different nonfiction books can be made on a shelf with space for eight books if the books of the same kind are to be kept together?

12. How many positive integers less than 300 can be formed from the digits 1, 2, 3, 4, and 5 if repetitions of digits is not allowed?
13. How many different signals can be made from 5 flags if each signal consists of 3 flags hung in a vertical line?
14. A penny, a nickel, and a quarter are tossed together. In how many different ways may they fall?
15. Two dice are tossed. In how many different ways may they fall?

15.2 ◆ Permutations

Let S be a set of n distinct objects. Generally, when we speak of sets of objects we are not concerned with the order or arrangement of the objects. Thus $\{a, b, c\}$ is the same set as $\{c, a, b\}$. Sometimes, however, the arrangement of the elements of a set is important. If the elements of a set S are arranged according to some rule we call the set an **ordered set**. Any ordered set of elements is called a **permutation** of those elements. Thus 5,4 and 4,5 are different permutations of the elements of the set $\{4, 5\}$.

DEFINITION 15.2. A **permutation** of r elements of a set S is an arrangement, without repetitions, of r elements.

In many problems it is necessary to know the number of permutations of n different objects taken r at a time. We use the symbol $P(n, r)$ to denote the number of different permutations of r elements that can be obtained from a set containing n elements.

A formula can be developed for $P(n, r)$ by using the fundamental counting principle. There are n choices for the first position. After the first position has been filled, there are $n - 1$ choices for the second position; hence the first two positions may be filled in $n(n - 1)$ ways. For the third position there are $n - 2$ choices, so that the first three positions can be filled in $n(n - 1)(n - 2)$ ways. Continuing in this manner, we see that the rth position can be filled in $n - r + 1$ ways, and we obtain the formula

$$(1) \qquad P(n, r) = n(n - 1)(n - 2) \cdots (n - r + 1)$$

As a special case, when $r = n$ we obtain the number of different permutations of n elements taken n at a time. Since in this case $n - r + 1 = n - n + 1$, we have

$$(2) \qquad P(n, n) = n(n - 1)(n - 2) \cdots (2)(1)$$

Recalling that

$$n! = n(n - 1)(n - 2) \cdots (2)(1) \qquad \text{if } n \text{ is a positive integer}$$
$$0! = 1$$

we may write Equation 2 as

$$P(n, n) = n!$$

An alternate form of Equation 1 may be obtained by multiplying the right member by $\dfrac{(n - r)!}{(n - r)!}$. This gives

$$P(n, r) = \frac{n(n - 1)(n - 2) \cdots (n - r + 1)(n - r)!}{(n - r)!}$$

or

(3) $$P(n, r) = \frac{n!}{(n - r)!}$$

EXAMPLE 1. How many serial numbers of six different digits can be formed from the digits 0, 1, 2, 3, 4, 5, 6, 7, 8, and 9?

SOLUTION. There are ten digits to choose from and six different ones may be chosen at a time. Hence, $n = 10$, $r = 6$, and we have

$$P(10, 6) = (10)(9)(8)(7)(6)(5) = 151{,}200$$

EXAMPLE 2. A baseball team consists of nine players. Find the number of different batting orders if the pitcher must bat last.

SOLUTION. Since the pitcher must bat last, there are only eight players to be arranged. Hence we are looking for the number of permutations of 8 objects taken 8 at a time. Then

$$P(n, n) = 8! = 40{,}320$$

15.3 ✦ Permutations of *n* Objects Some of Which Are Alike

Now let us consider the problem of finding the number of permutations of *n* objects taken *n* at a time if some of the objects are identical. As an example, let us take the number of permutations of the letters of the word ELEVEN.

If we assign subscripts to the *E*'s so that they are distinguishable, we have

$$E_1 \, L \, E_2 \, V \, E_3 \, N$$

The number of permutations of these six letters is 6!. If the letters other than E_1, E_2, and E_3 are retained in the position they occupy in the permutation above, E_1, E_2, and E_3 may be permuted among themselves 3! ways. Thus, if *P* is the number of distinguishable permutations of the letters of the word eleven, and, if for each of these there are 3! ways in which the *E*'s can be permuted without otherwise changing the order of the other letters, then

$$3! \cdot P = 6!$$

and hence

$$P = \frac{6!}{3!}$$

As another example, consider the letters of the word TENNESSEE. There would exist 9! distinguishable permutations of the letters in this word if all the letters were distinct. However, the letters N and S each appear twice and the letter E appears four times. Using the same reasoning as we did before, we have

$$2! \cdot 2! \cdot 4! \cdot P = 9!$$

from which we obtain

$$P = \frac{9!}{2!2!4!}$$

The examples above illustrate the following theorem.

THEOREM 15.1. The number P of distinct permutations of n objects taken n at a time of which n_1 are alike, n_2 are alike of another kind, $\ldots, n_k$ are alike of still another kind, and if $n_1 + n_2 + \cdots + n_k = n$ is

$$P = \frac{n!}{n_1!n_2!\ldots n_k!}$$

EXERCISES 15.2

1. Evaluate.
   (a) $P(8; 5)$ (d) $P(10; 3)$
   (b) $P(11; 6)$ (e) $P(19; 7)$
   (c) $P(15; 9)$ (f) $P(25; 11)$
2. Evaluate.
   (a) $P(20; 10)$ (d) $P(15; 7)$
   (b) $P(16; 8)$ (e) $P(17; 2)$
   (c) $P(27; 11)$ (f) $P(19; 11)$
3. How many different signals can be made from 6 different flags if each signal consists of 3 flags hung in a vertical line?
4. In how many different ways can the four positions of chairman, vice-chairman, group leader, and section leader be filled from a group of 20 persons if only one person can hold any one of the positions?

5. How many numerals of four different digits can be formed from the digits 1, 2, 3, 4, 5, 6, 7, 8, and 9?

6. How many even numbers named by four-digit numerals can be formed from the digits 3, 4, 5, 6, 7, 8, and 9?

7. In how many ways can the group photograph be taken of 10 persons if the people are to stand in a line?

8. A telephone dial has 10 holes. How many different signals, each consisting of 7 impulses in succession, can be made if:
   (a) Repetition of an impulse is not allowed?
   (b) Repetitions are allowed?

9. If a penny, a nickel, and a dime are tossed together, in how many different ways may they fall?

10. Find the number of permutations of the letters of the word CALIFORNIA.

11. Find the number of permutations of the letters of the word MISSOURI.

12. Find the number of permutations of the letters of the word MISSISSIPPI.

13. Find the number of permutations of the letters of the word KEALAKEKUA.

14. In how many different ways can 12 beads of different colors be strung on a straight wire?

15. How many positive integers less than 500 can be named by the digits 1, 3, 4, and 7 if repetitions are *not* allowed?

16. In how many ways can a game of doubles in tennis be arranged if each team is to be composed of a man and a woman from a group of 5 men and 4 women?

17. A map of five states is to be colored using a different color for each state. How many different results are possible if there are eight different colors available?

18. Find the solution sets of the following:
   (a) $P(3; k) = 3!$
   (b) $P(5; k) = 20$
   (c) $P(n; 5) = 10 \, P(n - 1; 4)$

19. In how many ways may 6 people be seated in 9 chairs?

20. In how many ways may 8 people be seated in 12 chairs?

15.4 ◆ Combinations

When working with permutations our concern is with ordered sets or arrangements of elements in sets. Now let us consider the elements rather than the arrangement of the elements in a set.

DEFINITION 15.3. A **combination** of r elements of a set S is a subset of S containing r distinct elements.

Suppose we have the set of letters a, b, c, and d. How many different combinations of three letters can be selected from the four letters? There are four. They are

$$abc, \; abd, \; acd, \; bcd$$

Taken two at a time there are six different combinations. They are

$$ab, \; ac, \; ad, \; bc, \; bd, \; cd$$

We shall now prove that the number of combinations of n distinct objects taken r at a time, denoted by $_nC_r$ or $\binom{n}{r}$, is given by the formula

$$_nC_r = \binom{n}{r} = \frac{n!}{r!(n-r)!}$$

The number of permutations of n objects taken r at a time is

$$P(n, r) = \frac{n!}{(n-r)!}$$

Each of the combinations that can be formed furnished $r!$ of these, since the number of permutations of the r objects in each combination is $r!$. Then

$$r! \cdot _nC_r = \frac{n!}{(n-r)!}$$

from which we obtain

$$_nC_r = \binom{n}{r} = \frac{n!}{r!(n-r)!}$$

The above discussion proves theorem 15.2:

THEOREM 15.2. **The number of combinations of n distinct objects taken r at a time, denoted by $\binom{n}{r}$ is $\dfrac{n!}{r!(n-r)!}$.**

EXAMPLE 1. How many committees of 4 people may be chosen from a group of 9 people?

SOLUTION. There is no question of order in a committee, hence we are looking for the number of combinations of 9 things taken 4 at a time:

$$\begin{aligned}
_9C_4 = \binom{9}{4} &= \frac{9!}{4!5!} \\
&= \frac{9 \cdot 8 \cdot 7 \cdot 6}{1 \cdot 2 \cdot 3 \cdot 4} \\
&= 126
\end{aligned}$$

EXAMPLE 2. There are 20 students, 15 men and 5 women, in a mathematics class. In how many different ways can a committee of 3 men and 2 women be chosen?

SOLUTION. The number of combinations of 3 men chosen from 15 men is

$$_{15}C_3 = \frac{15!}{3!12!} = 455$$

The number of combinations of 2 women chosen from 5 women is

$$_5C_2 = \frac{5!}{2!3!} = 10$$

From the fundamental counting principle, the total number of committees that can be chosen is

$$(455)(10) = 4550$$

EXERCISES 15.3

1. Evaluate

   (a) $\binom{7}{4}$ (d) $\binom{15}{7}$

   (b) $\binom{9}{6}$ (e) $\binom{20}{9}$

   (c) $\binom{12}{3}$ (f) $\binom{18}{7}$

2. Evaluate

   (a) $\binom{24}{9}$ (c) $\binom{30}{6}$

   (b) $\binom{35}{7}$ (d) $\binom{38}{30}$

3. Determine n such that $\binom{n}{3} = 10$.

4. Determine n such that $\binom{n}{17} = 1140$.

5. Find the least value of n such that $\binom{n}{4} \geq 40$.

6. Simplify:

   (a) $\binom{3}{0} + \binom{3}{1} + \binom{3}{2}$

   (b) $\binom{5}{0} + \binom{5}{1} + \binom{5}{2} + \cdots + \binom{5}{5}$

(c) $\binom{12}{12} + \binom{12}{11} + \cdots + \binom{12}{0}$

7. How many committees of 4 members can be formed from a set of 20 possible members?

8. In a class of 20 students, how many combinations of 8 students may be selected to go to the board?

9. A student must do 10 of 15 problems on a final examination. In how many different ways can the choice be made?

10. In an examination a student is to choose any 10 questions from a set of 12. In how many ways can the choice be made if it is required that Questions 1 and 2 be answered?

11. How many straight lines are determined by 10 points no three of which lie on the same straight line?

12. From a standard deck of 52 playing cards, how many different 5-card hands can be formed?

13. A bag contains 7 red, 6 blue, and 3 green marbles. In how many ways can we select each of the following?
    (a) Two red marbles.
    (b) Two marbles of the same color.
    (c) One marble of each color.

14. In how many ways can a group of 9 boys be divided into two teams with 5 on one side and 4 on the other?

15. A legislative body of 17 people passes a law by a vote of 9 to 8. In how many ways could the vote have resulted?

16. In how many ways can a committee of 4 Republicans and 3 Democrats be selected from a group of 8 Republicans and 6 Democrats?

17. In Exercise 16, how many committees of 7 can be chosen if there are at least 4 Republicans on the committee?

15.5 ◆ Sample Spaces and Events

The study of probability was born in the gaming rooms during the sixteenth century. Since that time it has grown to be an important science with applications not only in gambling, but in the social and physical sciences, in business, industry, insurance, and agriculture.

Activities such as tossing a coin, rolling a pair of dice, or drawing a card from a hat are called **experiments.** With any kind of experiment there is associated a set of possible results. For example, when a coin is tossed, the possible results are that it will fall with heads showing or it will fall with tails showing. When rolling a die, a face showing 1, 2, 3, 4, 5, or 6 dots will appear when the die comes to rest. We see, then, that there are two possibilities, heads or tails, when a coin is tossed, and six possibilities, 1, 2, 3, 4, 5, or 6, when a die is rolled. In each case these are the only possible results.

DEFINITION 15.4. A **sample space** of an experiment is a set of all possible results of the experiment.

DEFINITION 15.5. Each element of a sample space of an experiment is called an **outcome** of the experiment.

Observe that in Definition 15.4 we said *a* sample space rather than *the* sample space. There may be many more than one possible sample space for an experiment.

Consider the experiment in which a Christmas tree bulb is drawn from a box. Suppose there are some good bulbs and some bad bulbs in the box. Suppose also that some bulbs are red, some green, and the rest white. We now draw a bulb, without looking, from the box. The sample space may be

$$\{G, B\}$$

if we are interested only in whether the bulb is good (G) or bad (B).

If we are interested in whether the bulb is green (g), red (r), or white (w), the sample space is

$$\{g, r, w\}$$

We may be interested in both the color and whether the bulb is good or bad. In this case the sample space is the set of ordered pairs:

$$\{(G, g), (G, r), (G, w), (B, g), (B, r), (B, w)\}$$

where (G, r) means a good bulb which is red.

In general, when choosing a sample space, it is a safe guide to include as much detail as possible in the description of the outcomes of the experiment.

From the elements of a set it is possible to make many subsets. We call each subset of a sample space an **event.**

DEFINITION 15.6. Any subset of a sample space is called an **event** of the sample space.

EXAMPLE 1. Suppose we toss a die. Let a sample space be

$$S = \{1, 2, 3, 4, 5, 6\}$$

List the elements of the event, E, "a number greater than 4 showing."

SOLUTION. The event, E, is the subset

$$\{5, 6\}$$

of the sample space S.

EXAMPLE 2. Two coins are tossed. List the elements of a sample space. List the elements of the event, E, "one head and one tail showing."

SOLUTION. The sample space is the set of ordered pairs (H, H), (H, T), (T, H), (T, T), where (H, H) denotes heads on both coins, (H, T) denotes heads on the first coin and tails on the second, and so forth. Then

$$S = \{(H, H), (H, T), (T, H), (T, T)\}$$

The event E is

$$E = \{(H, T), (T, H)\}$$

EXERCISES 15.4

Specify a sample space for each experiment described in Exercises 1–6.

1. Tossing two coins.
2. Tossing a pair of dice.
3. Picking a card from an ordinary deck of 52 playing cards.
4. Tossing three coins.
5. Picking two balls from a bag containing seven balls labeled a, b, c, d, e, f, and g.
6. Picking three girls from a group of five girls.
7. A committee of 3 students is to be chosen from a class of 25 students. How many elements will be in the sample space?
8. A box contains 25 electronic tubes. A repairman selects 4 at random. How many elements will be in the sample space?
9. A pair of dice is tossed. List the elements in the following events.
   (a) The sum of the two numbers showing is 8.
   (b) The sum of the two numbers showing is 7.
   (c) A number greater than 4 is showing on one die.
   (d) The number showing on one die is 6.
10. A card is selected from a standard deck of 52 playing cards. What is the subset defining each of the following events?
    (a) a red card   (d) a spade
    (b) an ace   (e) a king of hearts
    (c) a jack, queen, or king (f) a red jack
11. Three coins are tossed. What is the subset defining each of the following events?
    (a) all tails   (c) at least one head
    (b) two heads   (d) at least one tail

12. A committee of three persons is to be picked from a set of 7 persons, 3 men and 4 women. How many elements are in the subset defining each of the following events?
    (a) two men
    (b) two women
    (c) all women
    (d) at least one man
13. Four coins are tossed. What is the subset defining each of the following events?
    (a) all heads
    (b) an equal number of heads and tails
    (c) an even number of tails
    (d) at least one tail
14. A committee of three men is picked from a group of six men designated by A, B, C, D, E, and F. What is the subset defining each of the following events?
    (a) A is on the committee
    (b) A or B is on the committee
    (c) A and B are on the committee
    (d) A is not on the committee

15.6 ✦ Probability of an Event

When an experiment is performed, we determine a sample space, S, containing all the possible outcomes denoted by o_i, $i = 1, 2, \ldots n$:

$$S = \{o_1, o_2, \ldots, o_n\}$$

If an event E is a set containing exactly one outcome of S, then E is called a **simple event** of S. Any event which is not a simple event is called a **compound event**. Every compound event can be expressed as the union of simple events. For example, if a die is tossed a sample space is

$$S = \{1, 2, 3, 4, 5, 6\}$$

The event, E, "an even number showing," is

$$E = \{2, 4, 6\}$$

The event E is the union of the simple events $\{2\}$, $\{4\}$, $\{6\}$.

$$E = \{2, 4, 6\} = \{2\} \cup \{4\} \cup \{6\}$$

To each simple event, $\{o_i\}$ of $S = \{o_1, o_2, \ldots, o_n\}$ we assign a number denoted by $P(\{o_i\})$ called the **probability** of the event $\{o_i\}$. These numbers (probabilities) may be assigned arbitrarily, but they must satisfy two conditions:

(1) $P(\{o_i\}) \geq 0$, $i = 1, 2, \ldots, n$
(2) $P(\{o_1\}) + P(\{o_2\}) + \cdots + P(\{o_n\}) = 1$

We see from the above that the probability of each simple event is a non-negative number p, $0 \leq p \leq 1$.

If each of n outcomes of an experiment is equally likely to occur (we do not try to define "equally likely" but expect the reader to have an intuitive idea of this notion), we shall agree to assign probability $\frac{1}{n}$ to each of the simple events of the sample space S. With this agreement, if E is any event of S, it can be proved that

$$P(E) = \frac{\text{number of outcomes in } E}{\text{number of outcomes in } S}$$

In this chapter, unless stated otherwise, all outcomes of an experiment will be assumed to be equally likely. If we denote the number of outcomes in event E by $n(E)$ and the number of outcomes in the sample space S by $n(S)$, we have

$$P(E) = \frac{n(E)}{n(S)}$$

We shall accept this statement as the definition of the probability of event E of sample space S.

EXAMPLE 1. Consider the experiment of tossing a die with sample space

$$S = \{1, 2, 3, 4, 5, 6\}$$

What is the probability of the event, E, "an odd number showing?"

SOLUTION. The event E is

$$E = \{1, 3, 5\}$$

We see that $n(E) = 3$ and $n(S) = 6$; hence,

$$P(E) = \tfrac{3}{6} = \tfrac{1}{2}$$

EXAMPLE 2. Two coins are tossed. What is the probability of the event, E, "two heads?"

SOLUTION. A sample space for this experiment is

$$S = \{(H, H), (H, T), (T, H), (T, T)\}$$

The event, E, "two heads," is

$$E = \{(H, H)\}$$

We see that $n(E) = 1$ and $n(S) = 4$; hence,

$$P(E) = \tfrac{1}{4}$$

EXAMPLE 3. An urn contains 3 black balls and 4 white balls. We select a ball at **random** (this means one ball is as likely to be selected as another). What is the probability of drawing a black ball?

SOLUTION. We set up a sample space numbering the balls 1, 2, 3, 4, 5, 6, and 7. We assign numbers 1, 2, and 3 to the black balls and 4, 5, 6, and 7 to the white balls. Then

$$S = \{1, 2, 3, 4, 5, 6, 7\}$$

The event, E, "drawing a black ball," is

$$E = \{1, 2, 3\}$$

Since $n(S) = 7$ and $n(E) = 3$ we have

$$P(E) = \tfrac{3}{7}$$

EXAMPLE 4. Four defective switches are put in a carton containing 32 good switches. The defective switches look exactly like the good switches. What is the probability that if you select two switches at random from the box, both will be good?

SOLUTION. It is unnecessary in this problem to write out a sample space for the experiment. To find the probability we need only know the number of outcomes in the sample space and the number of elements in the event in which we are interested. The number of elements in the sample space is the number of combinations of 36 objects taken 2 at a time. Then

$$n(S) = {}_{36}C_2 = \frac{36!}{2!34!}$$

The number of elements in the event, E, "drawing 2 good switches," is the number of combinations of 32 objects taken 2 at a time. Then

$$n(E) = {}_{32}C_2 = \frac{32!}{2!30!}$$

Hence

$$P(E) = \frac{\dfrac{32!}{2!30!}}{\dfrac{36!}{2!34!}} = \frac{248}{315}$$

EXERCISES 15.5

1. A card is drawn from a standard deck of 52 playing cards. What is the probability of each of the following events?
   (a) a spade
   (b) a red card
   (c) an ace
   (d) the queen of hearts
   (e) a jack, queen, or king
   (f) a black jack
2. Two dice are thrown. What is the probability of each of the following events?
   (a) The sum of the numbers showing is 7.
   (b) The numbers showing are both odd.
   (c) The sum of the numbers showing is divisible by 3.
   (d) The sum of the numbers showing is 8.
3. Two letters are chosen at random from the set of letters in the word HORSE. What is the probability of each of the following events?
   (a) Both letters are vowels.
   (b) At least one letter is a vowel.
   (c) At least one letter is r.
4. A ball is drawn at random from a bag containing 3 red balls, 4 green balls, and 1 white ball. What is the probability that the ball is:
   (a) red?
   (b) green?
   (c) white?
   (d) red or white?
   (e) green or red?
5. Two coins are tossed. Find the probability of each of the following events.
   (a) two heads
   (b) one head and one tail
   (c) at least one head
6. Four coins are tossed. Find the probability of each of the following events.
   (a) an even number of heads
   (b) at least one tail
   (c) exactly three tails
   (d) at most two tails
7. Three letters are chosen from the letters of the word PEACH. Find the probability of each of the following events.
   (a) One of the letters selected is h.
   (b) At least one of the letters selected is a vowel
   (c) One of the letters selected is p
   (d) The letter p is not selected.
8. Consider the set of families with four children each of a different age. Find the probability of each of the following events.
   (a) Two girls.
   (b) The oldest child is a boy.
   (c) The youngest child is a boy.

(d) There are four boys.

(e) The oldest child is a boy and the youngest child is a girl.

9. A committee of three is selected from a group of six men designated as a, b, c, d, e, and f. Find the probability of each of the following events.

(a) a is selected.   (d) a or b is selected.

(b) f is selected.   (e) a is not selected.

(c) a and b are selected.

10. A box contains 20 slips of paper numbered consecutively from 1 through 20. A slip is selected at random from the box. Find the probability of each of the following events.

(a) drawing 5

(b) drawing a number less than 10

(c) drawing a number greater than 7

(d) drawing a number divisible by 4

(e) drawing a number that is odd

11. From a box containing 5 red, 3 black and 7 white balls, two balls are drawn at random. What is the probability of each of the following events?

(a) One is red and one is white.   (c) Both are red.

(b) Both are black.   (d) One is white.

12. Mr. Hurd has accidentally mixed 4 burned-out Christmas tree bulbs with 20 good ones. He cannot tell by looking which bulb is good. What is the probability that he will select a good bulb if he picks one at random?

13. In Exercise 12, suppose two bulbs are selected. What is the probability that:

(a) Both will be good?

(b) Both will be bad?

(c) One will be good and one bad?

14. An employer wishes to fill two positions from a group of 13 employees, 5 male and 8 female. If he selects without knowing the identity of the employee (say, by assigning a number to each and picking two names from a bag), what is the probability that one will be a male and the other a female?

15. A bag contains 100 jelly beans: 50 are red, 15 are green, 10 are black, and the rest are white. Two jelly beans are drawn at random from the bag. What is the probability of each of the following events?

(a) Both are red.

(b) Both are white.

(c) Both are green.

(d) Both are black.

(e) One is red and one is black.

15.7 ✦ Probability of More than One Event

In some probability problems it is necessary to consider two or more events taking place. Two or more events are called **mutually exclusive** events if they have no elements in common. Thus, if E_1 and E_2 are two mutually exclusive events, then $E_1 \cap E_2 = \phi$. For example, in the experiment of tossing a die, the events E_1, "an even number showing," and E_2, "an odd number showing," are mutually exclusive. If two events are mutually exclusive they *cannot* occur simultaneously.

THEOREM 15.3. If E_1 and E_2 are two mutually exclusive events of S, the probability of the event E_1 or E_2, which we denote by $E_1 \cup E_2$, is

$$P(E_1 \cup E_2) = P(E_1) + P(E_2)$$

PROOF. Since E_1 and E_2 are mutually exclusive, they have no elements in common and

$$n(E_1 \cup E_2) = n(E_1) + n(E_2)$$

But

$$\begin{aligned} P(E_1 \cup E_2) &= \frac{n(E_1 \cup E_2)}{n(S)} \\ &= \frac{n(E_1) + n(E_2)}{n(S)} \\ &= \frac{n(E_1)}{n(S)} + \frac{n(E_2)}{n(S)} \\ &= P(E_1) + P(E_2) \end{aligned}$$

EXAMPLE 1. A box contains 6 red marbles, 4 green, and 5 blue. One marble is drawn at random from the box. What is the probability of drawing a red or a blue marble?

SOLUTION. The probability of the event, E_1, "drawing a red marble," is

$$P(E_1) = \tfrac{6}{15}$$

The probability of the event, E_2, "drawing a blue marble," is

$$P(E_2) = \tfrac{5}{15}$$

Since E_1 and E_2 are mutually exclusive events, the probability of E_1 or E_2 is

$$\begin{aligned} P(E_1 \text{ or } E_2) = P(E_1 \cup E_2) &= P(E_1) + P(E_2) \\ &= \tfrac{6}{15} + \tfrac{5}{15} \\ &= \tfrac{11}{15} \end{aligned}$$

THEOREM 15.4. If E_1 and E_2 are any two events, then

$$P(E_1 \cup E_2) = P(E_1) + P(E_2) - P(E_1 \cap E_2).$$

PROOF. If E_1 and E_2 are two events that are not mutually exclusive, they have some elements in common and $E_1 \cap E_2 \neq \phi$. Now

$$P(E_1 \cup E_2) = \frac{n(E_1 \cup E_2)}{n(S)}$$

But $n(E_1 \cup E_2) = n(E_1) + n(E_2) - n(E_1 \cap E_2)$, since $n(E_1 \cap E_2)$ is included twice in $n(E_1) + n(E_2)$. Hence,

$$
\begin{aligned}
P(E_1 \cup E_2) &= \frac{n(E_1 \cup E_2)}{n(S)} \\
&= \frac{n(E_1) + n(E_2) - n(E_1 \cap E_2)}{n(S)} \\
&= \frac{n(E_1)}{n(S)} + \frac{n(E_2)}{n(S)} - \frac{n(E_1 \cap E_2)}{n(S)} \\
&= P(E_1) + P(E_2) - P(E_1 \cap E_2)
\end{aligned}
$$

We see that Theorem 15.3 is a special case of Theorem 15.4, since if E_1 and E_2 are mutually exclusive events then $E_1 \cap E_2 = \varnothing$ and the number of elements in the empty set is 0.

EXAMPLE 2. If two cards are drawn from an ordinary deck of 52 playing cards, what is the probability that both are black or both are aces?

SOLUTION. A deck of cards consists of 52 cards and an outcome of this experiment consists of two cards. The number of elements in the sample space is the number of combinations of 52 things taken 2 at a time. Then

$$n(S) = {}_{52}C_2 = \frac{52!}{2!50!}$$

Let E_1 be the event, "both cards are black." There are 26 black cards in a deck of playing cards. The number of elements in E_1 is the number of combinations of 26 things taken 2 at a time. Hence

$$n(E_1) = {}_{26}C_2 = \frac{26!}{2!24!}$$

Let E_2 be the event "both cards are aces." There are 4 aces in a deck of cards. The number of elements in E_2 is the number of combinations of 4 things taken 2 at a time. Then

$$n(E_2) = {}_4C_2 = \frac{4!}{2!2!}$$

The events E_1 and E_2 have one element in common, the pair of black aces. Hence, $n(E_1 \cap E_2) = 1$. Then

$$P(E_1 \cup E_2) = P(E_1) + P(E_2) - P(E_1 \cap E_2)$$

$$= \frac{\frac{26!}{2!24!}}{\frac{52!}{2!50!}} + \frac{\frac{4!}{2!2!}}{\frac{52!}{2!50!}} - \frac{1}{\frac{52!}{2!50!}}$$

$$= \frac{\frac{26!}{2!24!} + \frac{4!}{2!2!} - 1}{\frac{52!}{2!50!}}$$

$$= \frac{325 + 6 - 1}{1326}$$

$$= \frac{330}{1326}$$

$$= \frac{55}{221}$$

EXERCISES 15.6

1. Which of the following pairs of events are mutually exclusive?
   (a) In tossing a coin twice: throwing tails; throwing heads.
   (b) In rolling a die twice: rolling an even number; rolling a 6.
   (c) In rolling a die twice: rolling a 1; rolling a 6.
   (d) In drawing a card from a deck of 52 playing cards and then drawing a second card: drawing a king; drawing a heart.
2. In a bag are 5 red beads, 6 white, and 9 blue. One bead is picked at random. What is the probability of each of the following events?
   (a) picking a red bead
   (b) picking a blue bead
   (c) picking either a red or a blue bead
3. Jackets come in three sizes: small, medium, and large. In a lot of 500 medium jackets, 50 have been made too small and 25 have been made too large. The remainder are acceptable according to company standards. An inspector selects a medium sized jacket at random. What is the probability that it will be:
   (a) too small?
   (b) too large?
   (c) either too large or too small?
4. Two coins are tossed. What is the probability of each of the following events?
   (a) both heads

(b) one head and one tail

(c) both heads or one head and one tail

5. Two cards are drawn at random from an ordinary deck of 52 playing cards. What is the probability of each of the following events?

(a) Both cards are red.

(b) Both cards are kings.

(c) Both cards are red or both cards are kings.

(d) Both cards are jacks or both cards are kings.

(e) Both cards are red or both cards are black.

6. Three coins are tossed. What is the probability of each of the following events?

(a) two heads or heads on only one coin

(b) two tails or tails on only one coin

7. Two dice are tossed. What is the probability of each of the following events?

(a) a 5 on one die or a sum of 7

(b) a sum of 8 or a 6 on one die

(c) a sum of 7 or a sum of 11

(d) a 3 on one die or a sum of 2

8. Twenty-five slips of paper are numbered 1 through 25 and placed in a box. One slip is drawn at random from the box. What is the probability of each of the following events?

(a) drawing a number that is prime or a number that is divisible by 3

(b) drawing a number that is divisible by 2 or that is divisible by 3

(c) drawing a number less than 10 or a number that is even

15.8 ◆ Conditional Probability

In some experiments we are interested in events knowing that one event has already occurred.

EXAMPLE 1. Three coins are tossed. What is the probability that all three coins show tails if we know that the first coin shows tails?

SOLUTION. A sample space for this experiment is

$$S = \{HHH, HHT, HTH, HTT, THH, THT, TTH, TTT\}$$

where HHT means heads on the first two coins and tails on the third. To find the probability in question we do not need all of the eight outcomes in the sample space S, but only the subset E_2 containing the outcomes in which the first coin shows tails. Then

$$E_2 = \{TTT, TTH, THT, THH\}$$

We see that $n(E_2) = 4$. Of these elements, those in which all coins show tails are the elements of the event

$$E_1 = \{TTT\}$$

The probability that we seek is

$$P(E_1) = \tfrac{1}{4}$$

We can derive a general formula for the probability of event E_1 occurring given that event E_2 has occurred. We call this the **conditional probability** of E_1 given E_2 has occurred and denote it by $P(E_1 \mid E_2)$. We read this symbol, "the probability of E_1 given E_2 has occurred."

Let S be a sample space of an experiment and E_1 and E_2 two events of S, $E_2 \neq \varnothing$. Since we know that E_2 has occurred, all the elements in S cannot occur; only those elements in S which are also in E_2 can occur. Hence the relevant space in this case is that defined by the event E_2 of S. We also see that not all elements in the event E_1 can occur. Only those outcomes of E_1 which are also elements of E_2 can possibly occur. The set of these elements is $E_1 \cap E_2$. The probability of E_1 occurring given that E_2 has already occurred is

(1) $$P(E_1 \mid E_2) = \frac{n(E_1 \cap E_2)}{n(E_2)}$$

If we divide numerator and denominator of the right member of Equation 1 by $n(S)$, we obtain

$$P(E_1 \mid E_2) = \frac{\dfrac{n(E_1 \cap E_2)}{n(S)}}{\dfrac{n(E_2)}{n(S)}}$$

but, by definition, $\dfrac{n(E_1 \cap E_2)}{n(S)} = P(E_1 \cap E_2)$ and $\dfrac{n(E_2)}{n(S)} = P(E_2)$; hence

$$P(E_1 \mid E_2) = \frac{P(E_1 \cap E_2)}{P(E_2)}$$

EXAMPLE 2. It is known that 5% of the students enrolled in a certain mathematics class are engineering majors and 1% of the students in the class are women majoring in engineering. What is the probability that a student selected at random is a woman if we know that the student is an engineering major?

SOLUTION. Let E_2 be the event, "the student is an engineering student," and E_1 the event, "the student is a woman." Then

$$P(E_1 \mid E_2) = \frac{P(E_1 \cap E_2)}{P(E_2)}$$

$$= \frac{\frac{1}{100}}{\frac{5}{100}}$$

$$= \tfrac{1}{5}$$

EXERCISES 15.7

1. Four coins are tossed. What is the probability that all four coins show heads if we know that the first coin shows heads?

2. Two dice are tossed. What is the probability that the sum is 11 if one of the dice shows a 5?

3. A die is tossed. Find the probability that the number showing is divisible by 3 if it is known that the outcome is an even number.

4. Three coins are tossed and we see that not all three fall heads and not all three fall tails. What is the probability that there is an odd number of tails showing?

5. A box contains 5 slips which are numbered 1, 2, 3, 4, and 5, respectively. Three slips are drawn and one of them is numbered 3. Find the probability that the slip numbered 5 is also among them.

6. A committee of three is selected from six persons designated by A, B, C, D, E, and F, respectively. Find the probability of A being selected given that B has been selected.

7. A student must choose 2 elective courses from the following list: mathematics, physical science, art, music, or geography. What is the probability that:
   (a) mathematics is chosen?
   (b) mathematics is chosen if neither of the fine arts courses is chosen?

8. A letter is selected at random from the word ABOUT. What is the probability of selecting:
   (a) a "b" given that a vowel has been selected?
   (b) an "o" given that a vowel has been selected?

9. A green and a white die are tossed.
   (a) What is the probability that the sum showing is under 5 given that the white die shows 2?
   (b) What is the probability that the sum is under 8 given that the green die shows 6?
   (c) What is the probability that the sum is under 6 given that the white die shows 2?
   (d) What is the probability that the sum is under 6 given that the green die shows 6?

10. There are 4 balls in a box, 2 are black and 2 are white. One ball is drawn from the box and not returned. Any ball in the box has the same chance of being drawn as the other.
   (a) What is the probability of drawing a black ball on the second draw knowing that a white ball has been drawn?
   (b) What is the probability of drawing a black ball on the second draw knowing that a black ball has been drawn?

11. Given that 8% of the light bulbs produced in a factory are red and that 2% of all bulbs are red and defective, what is the probability that a bulb selected at random is defective if we know that it is red?

12. Two cards are drawn from an ordinary deck of 52 playing cards.
   (a) If the first card is replaced before the second is drawn, what is the probability of drawing a queen followed by the jack of hearts?
   (b) If the first card is not replaced before the second is drawn, what is the probability of drawing a queen followed by the jack of hearts?

13. A supermarket usually stocks special items to encourage sales. A given market advertises a special on whipping cream and strawberries if purchased together. The probability of a customer buying both whipping cream and strawberries is 0.05, while the probability of buying strawberries alone is 0.30. What is the probability that a customer who has purchased strawberries will buy whipping cream also?

15.9 ◆ Independent and Dependent Events

Let us perform the experiment of tossing three coins. A sample space is

$$S = \{HHH, HHT, HTH, HTT, THH, THT, TTH, TTT\}$$

Let E_1 be the event "heads on the first coin" and E_2 be the event "tails on the third coin." Then

$$E_1 = \{HHH, HHT, HTH, HTT\}$$
$$E_2 = \{HHT, HTT, THT, TTT\}$$

and $\qquad E_1 \cap E_2 = \{HTT, HHT\}$

We see that

$$n(E_1) = 4 \qquad n(E_2) = 4 \qquad n(S) = 8$$

and

$$n(E_1 \cap E_2) = 2$$

Then the probability of the event heads on the first coin and tails on the third coin, which we denote by $E_1 \cap E_2$, is

$$P(E_1 \text{ and } E_2) = P(E_1 \cap E_2) = \tfrac{2}{8} = \tfrac{1}{4}$$

Observe that $P(E_1 \cap E_2)$ is also equal to

$$P(E_1) \cdot P(E_2) = \tfrac{4}{8} \cdot \tfrac{4}{8} = \tfrac{1}{4}$$

This result is not a coincidence. It often happens in situations in which we are interested in the probability that two events will occur together, that is, we are interested in $P(E_1 \text{ and } E_2) = P(E_1 \cap E_2)$. If E_1 and E_2 are two events of a sample space S they may have no relation to each other or they may be so related that the occurrence of one affects the occurrence of the other. If the events are not related, then the probability of the first will be independent of whether the second event has occurred.

DEFINITION 15.7. If E_1 and E_2 are events in a sample space S, then E_1 and E_2 are called **independent events** if and only if

$$P(E_1 \text{ and } E_2) = P(E_1 \cap E_2) = P(E_1) \cdot P(E_2).$$

If $P(E_1 \cap E_2) \neq P(E_1) \cdot P(E_2)$, the events E_1 and E_2 are called **dependent events.**

EXAMPLE 1. There are two decks of ordinary playing cards. From each deck one card is drawn at random. What is the probability of drawing two red cards?

SOLUTION. Since the occurrence of either of these events is not affected by the occurrence of the other, the events are independent. The probability of the event E_1, "drawing a red card from the first deck," is $\tfrac{25}{52} = \tfrac{1}{2}$. The probability of the event E_2, "drawing a red card from the second deck," is $\tfrac{26}{52} = \tfrac{1}{2}$. Hence

$$P(E_1 \cap E_2) = \tfrac{1}{2} \cdot \tfrac{1}{2} = \tfrac{1}{4}$$

EXAMPLE 2. Two dice are tossed. Let E_1 be the event, "a 6 on the first die," and E_2 be the event, "a 1 on the second die." Are E_1 and E_2 independent events?

SOLUTION. The sample space will be the set of 36 possible outcomes. For example, $(2, 5)$ which denotes 2 on the first die and 5 on the second die is one outcome. Then

$$E_1 = \{(6, 1), (6, 2), (6, 3), (6, 4), (6, 5), (6, 6)\}$$
and
$$E_2 = \{(1, 1), (2, 1), (3, 1), (4, 1), (5, 1), (6, 1)\}$$
$$E_1 \cap E_2 = \{(6, 1)\}$$

We have $n(E_1) = 6$, $n(E_2) = 6$, and $n(S) = 36$, and $n(E_1 \cap E_2) = 1$. Now

$$P(E_1) = \tfrac{6}{36} = \tfrac{1}{6}$$
$$P(E_2) = \tfrac{6}{36} = \tfrac{1}{6}$$
$$P(E_1 \cap E_2) = \tfrac{1}{36}$$

Since

$$P(E_1 \cap E_2) = P(E_1) \cdot P(E_2)$$

E_1 and E_2 are independent events.

EXERCISES 15.8

1. A coin and a die are tossed. Find the probability of each of the following events.
   (a) The coin shows tails and the die shows 1.
   (b) The coin shows heads and the die shows 5.
   (c) The coin shows heads and the die shows a number greater than 3.
2. Two dice are tossed. Find the probability that:
   (a) 6 shows on the first die and 1 shows on the second.
   (b) a number greater than 5 shows on the first die and a number less than 3 shows on the second.
   (c) a prime number shows on the first die and a composite number shows on the second.
3. There are two standard decks of 52 playing cards. From each deck one card is drawn. What is the probability of drawing each of the following?
   (a) two aces
   (b) two red cards
   (c) a red card and a black card
   (d) one spade and one heart
   (e) two queens
4. The first of two bags contains 4 red and 3 white beads; the second, 3 green and 5 blue beads. From each bag one bead is drawn. What is the probability of obtaining:
   (a) a red and a green bead?
   (b) a white and a blue bead?
   (c) a white and a green bead?
5. One box contains 2 red, 1 white, and 3 blue marbles. A second box contains 1 green, 2 black, and 3 yellow marbles. From each box one marble is drawn. What is the probability of each of the following events?
   (a) a red and a black marble
   (b) a white and a yellow marble
   (c) a blue and a green marble

6. In each of two laundry bags there are some socks, not sorted into pairs. In one bag there are 5 black socks and 6 white socks. In the other bag there are 9 black socks and 8 white socks. One sock is picked at random from each bag. What is the probability of each of the following events?
   (a) Both are black.
   (b) Both are white.
   (c) One is black and one is white.

7. There are 20 shoes thrown in the bottom of a closet. Of these 8 are black and 12 are white. If it is dark and you reach in the closet and take out two shoes, what is the probability that:
   (a) both are black?
   (b) both are white?
   (c) one is black and one is white?

8. A certain mathematics problem is to be solved by two students, A and B. The probability that A will solve the problem is $\frac{3}{4}$, and the probability that B will solve it is $\frac{1}{8}$.
   (a) What is the probability that the problem will not be solved?
   (b) What is the probability that it will be solved by B and not by A?
   (c) What is the probability that it will be solved by A and not by B?
   (d) What is the probability that it will be solved by both A and B?
   (e) What is the probability that the problem will be solved?

CHAPTER REVIEW

1. How many three-digit numerals can be formed from the digits 2, 3, 5, and 7?

2. How many three-digit numerals can be formed from the digits 2, 3, 5, and 7 if repetition of digits is not allowed?

3. Evaluate:
   (a) $P(9; 7)$ (b) $P(12; 4)$ (c) $P(15; 3)$

4. An internal telephone dial system allows four-digit extension numbers. It is discovered that dialing 0 or dialing the same digit twice in the same number blows all the fuses. How many extensions can this system handle without blowing the fuses?

5. Evaluate:
   (a) $\binom{13}{3}$ (b) $\binom{13}{10}$ (c) $\binom{29}{27}$

6. A committee of 15 persons adopts a resolution by a vote of 9 in favor and 6 opposed. In how many ways could such a vote have occurred?

7. Describe a sample space for the experiment of tossing four coins.

8. Four coins are tossed. What is the subset defining each of the following events?
   (a) three heads and one tail   (b) all heads

9. Three coins are tossed. What is the probability that two show heads and one shows tails?

10. A marble is drawn from a box containing 4 clear marbles, 10 red ones, 23 blue ones, and 2 purple ones. What is the probability that the marble drawn is not blue?

11. A manufacturer receives a lot of 20 bolts of material. Unknown to him, 5 of the bolts contain flaws. He picks 2 bolts at random. What is the probability that the first will be perfect and the second will contain flaws?

12. Two dice are tossed. What is the probability that one shows 3 and the other shows 4?

13. In Problem 12, what is the probability that one die shows 1 and the other 5?

14. In Problem 10, two marbles are drawn. What is the probability that the first drawn is red and the second is purple?

15. Two cards are drawn from a standard deck of 52 playing cards. What is the probability that the first is an ace and the second is a king?

16. There are three boys an two girls at a beach party. Two of them are chosen to go get some hamburgers at a drive-in.
    (a) In how many ways can the choice be made?
    (b) How many pairs consist of two boys?
    (c) How many pairs consist of two girls?
    (d) How many pairs consist of one boy and one girl?
    (e) What is the probability that two boys will be selected?
    (f) What is the probability that a boy and a girl will be selected?
    (g) What is the probability that at least one boy is picked?

17. A student must choose two courses from mathematics, physics, chemistry, biology, and basket weaving. What is the probability that mathematics is chosen but basket weaving is not?

18. Ten percent of the pistols produced at a small arms factory are 25 calibre. Two percent of all 25 calibre pistols are defective. What is the probability that a pistol selected at random is 25 calibre and defective?

19. A boxcar contains a shipment of 35 urns. Six are red, 12 are blue, 8 are white, and 9 are puce. The car is derailed and 6 urns are smashed. What is the probability that 3 white and 3 red urns are smashed?

20. The probability that it will rain today is $\frac{2}{3}$. The probability that it will not rain tomorrow is $\frac{1}{2}$. What is the probability that it will not rain both days?

Answers to Odd-Numbered Problems

EXERCISES 1.1

1. {Sunday, Monday, Tuesday, Wednesday, Thursday, Friday, Saturday}
3. {3, 6, 9, 12, 15, 18, 21, 24, 27}
5. {January, February, March, April, May, June, July, August, September, October, November, December}
7. {1, 2, 3, 4, 5, 6, 7, 8, 9, 10, 11}
9. {California, Oregon, Washington, Alaska, Hawaii}
11. $\{x \mid x$ is one of the first four letters of the English alphabet$\}$
13. $\{x \mid x$ is a natural number and x is less than 6$\}$
15. $\{x \mid x$ is an even natural number$\}$
17. $\{x \mid x$ is a natural number and x is greater than 10 and less than 17$\}$
19. $\{x \mid x$ is a natural number greater than 14$\}$ **21.** b; e
23. $\{1, 2, 3\}$; $\{1, 2, 4\}$; $\{1, 2, 5\}$; $\{1, 3, 4\}$; $\{1, 3, 5\}$; $\{1, 4, 5\}$; $\{2, 3, 4\}$; $\{2, 3, 5\}$; $\{2, 4, 5\}$; $\{3, 4, 5\}$ **25.** a; b; d

EXERCISES 1.2

1. (a) $\{1, 2, 3, 4, 5, 6, 7, 8, 9\}$;   (b) $\{1, 2, 3, 4, 5, 6, 7, 8, 10, 12\}$;
   (c) $\{1, 3, 5, 7\}$;   (d) $\{2, 4, 6, 8\}$;   (e) $\{1, 2, 3, 4, 5, 6, 7, 8, 9, 10, 12\}$;   (f) ϕ
3. $\{2, 3, 4, 6, 8, 9, 10\}$ **5.** $\{3\}$ **7.** $\{1, 2, 4, 5, 7, 8, 10\}$ **9.** $\{6, 9\}$
11. $\{1, 3, 5, 6, 7, 8, 9, 10\}$ **13.** $A = B = \phi$
15. A is a subset of B or $A \subseteq B$ **17.** A **19.** U **21.** A **23.** ϕ
27. a; b; c; e

EXERCISES 1.3

1. $\{7, 8, 9, \ldots, 17\}$ **3.** $\{7, 8, 9, \ldots\}$ **5.** $\{\ldots, -4, -3, -2, -1\}$
7. $\{3, 6, 9, 12, \ldots\}$ **9.** $\{2, 4, 6, 8, \ldots\}$
11. $\{x \mid x$ is a natural number and $x > 5\}$ **13.** $\{x \mid x$ is a positive integer$\}$
15. $\{x \mid x$ is a real number and $-2 < x < 2\}$
17. $\{x \mid x$ is an integer and x is even$\}$ **19.** b

EXERCISES 1.4

1. addition   **3.** substitution   **5.** multiplication   **7.** transitive
9. symmetric   **11.** closure property for addition
13. associative property of multiplication   **15.** additive identity axiom
17. commutative property of addition   **19.** distributive property
21. distributive property   **23.** additive inverse axiom
25. distributive property   **27.** a; b; c; d   **29.** yes

EXERCISES 1.5

1. (a) commutative property of multiplication;   (b) Theorem 1.5;
(c) multiplicative identity axiom
3. (a) hypothesis;   (b) hypothesis;   (c) multiplicative inverse axiom;
(d) multiplication property of equality;   (e) associative property of
multiplication;   (f) multiplicative inverse axiom;   (g) multiplicative
identity axiom
5. (a) multiplicative inverse axiom;   (b) multiplicative inverse axiom and
symmetric property of equality;   (c) transitive property of equality;
(d) commutative property of multiplication;   (e) Theorem 1.3 or
cancellation property of multiplication
7. (a) associative property of addition;   (b) commutative property of addition;
(c) commutative property of addition
9. (a) hypothesis;   (b) multiplication property of equality;   (c) associative
property of multiplication;   (d) multiplication inverse axiom;
(e) multiplicative identity axiom
11. (a) commutative property of multiplication;   (b) associative property of
multiplication;   (c) associative property of multiplication;
(d) multiplicative inverse axiom;   (e) multiplicative identity axiom;
(f) multiplicative inverse axiom
13. $a + (-a) = 0$ additive inverse axiom
$0 = b + (-b)$ additive inverse axiom and symmetric property of
equality
$a + (-a) = b + (-b)$ transitive property of equality
$a + (-a) = a + (-b)$ substitution property, and hypothesis
$-a = -b$ Problem 2, Exercise 1.5 or cancellation property of
addition
15. $(a - b) + c = [a + (-b)] + c$ definition of subtraction
$= a + [(-b) + c]$ associative property of addition
$= a + \{(-b) + [-(-c)]\}$ Theorem 1.1
$= a + \{-[b + (-c)]\}$ Theorem 1.10
$= a + [-(b - c)]$ definition of subtraction
$= a - (b - c)$ definition of subtraction
17. Let $a \neq 0$ have two multiplicative inverses $\dfrac{1}{a_1}$ and $\dfrac{1}{a_2}$. Then

$$a \cdot \frac{1}{a_1} = 1 \qquad \text{multiplicative inverse axiom}$$

$$1 = a \cdot \frac{1}{a_2} \quad \text{multiplicative inverse axiom and symmetric property of equality}$$

$$a \cdot \frac{1}{a_1} = a \cdot \frac{1}{a_2} \quad \text{transitive property of equality}$$

$$\frac{1}{a_1} = \frac{1}{a_2} \qquad \text{Theorem 1.3}$$

EXERCISES 1.6

1. (a) Theorem 1.11 and definition of greater than;   (b) Theorem 1.11 and definition of greater than;   (c) Theorem 1.13;   (d) definition of subtraction;   (e) commutative property of addition;   (f) associative property of addition;   (g) associative property of addition;   (h) commutative property of addition;   (i) associative property of addition;   (j) Problem 1, Exercise 1.5;   (k) distributive property;   (l) Problem 1, Exercise 1.5;   (m) definition of subtraction;   (n) Theorem 1.12 and step (c).

3. Let $a > 0$ and $b < 0$. Since $b < 0$, $-b > 0$ by Theorem 1.15. Then $(a)(-b)$ is positive by Theorem 1.14. But $a(-b) = -ab$ by Theorem 1.5. Since $-ab$ is positive, ab is negative by Theorem 1.15.

5. Since $a < b$, $b - a$ is positive by Theorem 1.11. Then

$b - a > 0$	definition of positive
$(-1)(b - a) < (-1)0$	multiplication property of order for negative numbers
$(-1)(b - a) < 0$	Theorem 1.4 (multiplication property of 0)
$(-1)[b + (-a)] < 0$	definition of subtraction
$(-1)(b) + (-1)(-a) < 0$	distributive property
$-b + [-(-a)] < 0$	Problem 1, Exercise 1.5
$\{-b + [-(-a)]\} + (-a) < 0 + (-a)$	addition property of order
$-b + [-(-a) + (-a)] < 0 + (-a)$	associative property of addition
$-b + 0 < 0 + (-a)$	additive inverse axiom
$-b < -a$	additive identity axiom
$-a > -b$	definition of greater than

7.

$a + c < b + c$	hypothesis
$[a + c] + (-c) < [b + c] + (-c)$	addition property of order
$a + [c + (-c)] < b + [c + (-c)]$	associative property of addition
$a + 0 < b + 0$	additive inverse axiom
$a < b$	additive identity axiom

9.

$ac > bc$	hypothesis
$c < 0$	hypothesis

$\dfrac{1}{c}$ exists and is less than 0 because the reciprocal of a negative number is negative

$$(ac)\left(\frac{1}{c}\right) < (bc)\left(\frac{1}{c}\right) \qquad$$ multiplication property of order with negative numbers

$$a\left(c \cdot \frac{1}{c}\right) < b\left(c \cdot \frac{1}{c}\right) \qquad$$ associative property of multiplication

$$a \cdot 1 < b \cdot 1 \qquad$$ multiplicative inverse axiom

$$a < b \qquad$$ multiplicative identity axiom

11. $\quad a < b \qquad$ hypothesis

$$2a < a + b \qquad$$ addition property of order

$$a < \frac{a + b}{2} \qquad$$ multiplication property of order with positive numbers

$$a + b < 2b \qquad$$ hypothesis and addition property of order

$$\frac{a + b}{2} < b \qquad$$ multiplication property of order with positive numbers

$$a < \frac{a + b}{2} < b \qquad$$ steps 3 and 5

EXERCISES 1.7

1. (a) 6;　(b) 5;　(c) 3;　(d) 7;　(e) 6;　(f) 17　　**3.** 12 and -12
7. (a) 11;　(b) 9;　(c) -2;　(d) 12;　(e) 27;　(f) 20;　(g) 64;　(h) 72
11. $x = -y$ or $y = -x$.　　**13.** $x = 0$ or $y = 0$ or both.

CHAPTER 1　REVIEW

1. (a) multiplicative inverse axiom;　(b) associative property of addition;
(c) commutative property of multiplication;　(d) distributive property;
(e) additive identity axiom;　(f) additive inverse axiom;　(g) multiplicative
identity axiom;　(h) associative property of multiplication;　(i) commutative
property of addition;　(j) closure property of multiplication
3. (a) hypothesis;　(b) multiplicative inverse axiom;　(c) multiplication
property of equality;　(d) associative property of multiplication;
(e) multiplicative inverse axiom;　(f) multiplicative identity axiom
5. $\{x \mid x \in R$ and $7.8 < x < 12.7\}$　　**7.** (a) addition;　(b) symmetric;
(c) transitive;　(d) substitution;　(e) addition　　**9.** (a) 15;　(b) 15;　(c) 18;
(d) -8;　(e) 2　　**11.** zero; integers　・　**13.** opposite; additive inverse
15. $-a$ or the opposite of a　　**17.** ab　　**19.** rational numbers

EXERCISES 2.1

1. 27　　**3.** $\frac{1}{8}$　　**5.** a^7　　**7.** $10^7 = 10,000,000$　・　**9.** $2^6 = 64$

11. $15a^3b^7$　　**13.** $-2592a^{13}b^{37}$　　**15.** a^2　　**17.** $\frac{b}{a^2}$　　**19.** $\frac{1}{x^2y^2}$

21. $\frac{p^2r}{q^6}$　　**23.** $\frac{1}{3m^{3k}}$　　**25.** $\frac{27a^{k-2}b^{8k-2}}{25}$　　**27.** $\frac{72}{625x^{13}y^5}$　　**29.** $\frac{6561x^{16}y^{25}}{16}$

EXERCISES 2.2

1. $\dfrac{1}{a^7}$ **3.** $\dfrac{2}{a^2b^5}$ **5.** 1 **7.** x^4y^4 **9.** $64a^6b^9$ **11.** $\dfrac{y^6}{x^{17}}$

13. $\dfrac{a^2 + ab + b^2}{a^2b^2}$ **15.** $\dfrac{3x^2 + 1}{x}$ **17.** $\dfrac{x + y}{xy}$ **19.** $\dfrac{1}{y - x}$ **21.** 0

23. $9x^{2n}$

EXERCISES 2.3

1. $5\sqrt{2}$ **3.** $-5\sqrt[3]{3}$ **5.** 10 **7.** $5p^2r^4\sqrt{2pr}$ **9.** $3\sqrt[3]{3}$

11. $-\dfrac{3a^2\sqrt[3]{a}}{b^3}$ **13.** $\dfrac{|y|}{x^2}$ **15.** $2x^4y^5\sqrt[3]{2y}$ **17.** $2x^2y\sqrt[4]{2y^3}$ **19.** $|a - b|$

21. $\dfrac{x - 2\sqrt{xy} + y}{x - y}$ **23.** $\dfrac{\sqrt[4]{x}}{x}$ **25.** $-8\sqrt[3]{4}$ **27.** $\dfrac{1 - \sqrt{x}}{1 - x}$

29. $\dfrac{x(1 - \sqrt{x})}{1 - x} + \dfrac{2\sqrt{1 + x}}{1 + x}$ **31.** $\dfrac{12 - 3\sqrt{x + 1}}{15 - x}$

33. $\dfrac{\sqrt{x + y} - \sqrt{x - y}}{2y}$ **35.** $\frac{1}{8}(\sqrt{x + 2} + \sqrt{x - 2})(\sqrt{x + 3} + \sqrt{x - 5})$

EXERCISES 2.4

1. 9 **3.** 125 **5.** 4 **7.** $\frac{1}{25}$ **9.** $\dfrac{1}{25x^{8/3}}$ **11.** $\dfrac{9a^2}{b^4}$ **13.** a^3

15. $b^{1/2}$ $x^{9/2}$ **17.** $x^{2/3}$ **19.** $\dfrac{y^{16/3}}{3^{2/3}x^4}$ **21.** $x - y$ **23.** $\dfrac{x^2 + y^2}{x^2y^2(x - y)^2}$

25. $x^{3/4} + y^{3/4}$ **27.** $a^2 + b^2$ **29.** $u^{4/3} - 2u^{2/3}w^{2/3} + w^{4/3}$

EXERCISES 2.5

1. $a^2 - b^2$ **3.** $a^2 - 2ab + b^2$ **5.** $y^2 + y - 12$ **7.** $y^2 - 10y + 16$
9. $4x^2 + 8xy + 3y^2$ **11.** $15x^2 + 4xy - 4y^2$ **13.** $9x^2 + 12xy + 4y^2$
15. $x^4 + 2x^2y^2 + y^4$ **17.** $12a^2 + 10ab - 28b^2$ **19.** $a^3 + b^3$
21. $x^3 + 8y^3$ **23.** $\frac{1}{4}x^4 + 2x^2y^3 + 4y^6$ **25.** $x - 2\sqrt{xy} + y$
27. $(\sqrt[4]{a})^3 - (\sqrt[4]{b})^3$ or $a^{3/4} - b^{3/4}$ **29.** $9a^2 - 24a\sqrt{b} + 16b$
31. $a^{1/2} - b^{1/2}$ or $\sqrt{a} - \sqrt{b}$ **33.** $4x^2 + y^2 + 9z^2 + 4xy - 12zx - 6yz$
35. $a^4 + b^4 + c^4 - 2a^2b^2 + 2a^2c^2 - 2b^2c^2$
37. $4a^4b^2 + 4c^4 + 12a^2b - 8a^2bc^2 - 12c^2 + 9$
39. $16 - x^4 - 4x^2y^3 - 4y^6$

EXERCISES 2.6

1. $(3x + 5)(3x - 5)$ **3.** $16(z^3 - 2t^2)(z^3 + 2t^2)$ **5.** $(y - 1)^2$
7. $(y + 5)(y - 2)$ **9.** $(a + 5)^2$ **11.** $(x - 9)(x + 3)$

13. $(2a + b)(a - b)$ **15.** $(5a - 1)^2$ **17.** $(4a + 3b)^2$

19. $5(a + b)(a + 3b)$ **21.** $(1 + 3a)^2$ **23.** $(x - y)(x^2 + xy + y^2)$

25. $(z + 5b)(z^2 - 5zb + 25b^2)$ **27.** $(3x - 2y)(2x + 5y)$

29. $(x^2 + y^2)(x + y)(x - y)$

31. $(2x + y)(2x - y)(4x^2 - 2xy + y^2)(4x^2 + 2xy + y^2)$

33. $(2a - c - d)(4a^2 + 2ac + 2ad + c^2 + 2cd + d^2)$

35. $(3x - 14)(2x - 3)$

37. $(3x - 2y - 11)(9x^2 + 6xy + 4y^2 - 3x + 20y + 37)$

39. $2(4x - 5)(-6x + 11)$

EXERCISES 2.7

1. $\frac{25}{72}$ **3.** $\frac{a + b}{a^2 + 3b^2}$ **5.** $\frac{y + 1}{y^2 + y + 1}$ **7.** $\frac{x + 3}{x - 3}$ **9.** $\frac{a - 2}{a + 2}$

11. $\frac{3a + 2b}{a + 5b}$ **13.** $\frac{(x^2 - xy + y^2)(x^2 + xy + y^2)}{x^2 + y^2}$ **15.** $\frac{3x - 4y}{2(4x + 5y)}$

17. $\frac{5}{9x^4 y^3}$ **19.** $\frac{1}{2y^3}$ **21.** $\frac{45x^2}{8y^2}$ **23.** $\frac{1}{x + y}$ **25.** $\frac{(x - 3)(x + 1)}{(x - 2)^2}$

27. $\frac{(x + 4)(2x - 5)}{x - 4}$ **29.** 1 **31.** $\frac{10x + 1}{6}$ **33.** $\frac{1 - x}{x(x + 1)}$

35. $\frac{6x^3 + 5y^3 - 4xz^2}{18x^2 y^2 z}$ **37.** $\frac{6(3a^2 - 4)}{a(9a^2 - 4)}$

39. $\frac{a + 3b + 1}{3(4 - a)}$ or $-\frac{a + 3b + 1}{3(a - 4)}$ **41.** $\frac{x(y - x)}{x^3 + y^3}$

43. $\frac{4xy}{y^2 - x^2}$ or $-\frac{4xy}{x^2 - y^2}$ **45.** $\frac{4(x + 1)}{x^2 + x + 1}$ **47.** 1 **49.** $\frac{41}{23}$

51. $\frac{x^2 - 1}{x^2 + 1}$ **53.** $\frac{x + 1}{x}$ **55.** $x - 2$ **57.** $x + y$ **59.** $\frac{2xy}{x^2 + y^2}$

CHAPTER 2   REVIEW

1. (a) x^{11}; (b) x^3 **3.** (a) x^3; (b) x^2 **5.** (a) $5\sqrt{5}$; (b) $x^2|y|$

7. (a) $a^{11/15}$; (b) $\frac{1}{a^{1/12}}$ **9.** (a) $a^{5/4} c^{3/4}$; (b) $\frac{x^{1/12} y^{1/8}}{2^{1/4}}$

11. $2(4 - x)(4 + x)$ **13.** $(6x^2 + 5)(x^2 - 12)$ **15.** $\frac{2a}{3}$

17. $\frac{(x - 4)(x + 3)}{(2x + 1)(x - 1)}$ **19.** $\frac{2x^2 - x - 1}{20(x - 1)^3}$

EXERCISES 3.1

1. $\{3\}$ **3.** $\{-1\}$ **5.** $\{-1\}$ **7.** $\{-2\}$ **9.** $\{0\}$ **11.** $\{7\}$

13. $\{11\}$ **15.** $\{10\}$ **17.** $\{-\frac{5}{2}\}$ **19.** $\{\frac{5}{4}\}$ **21.** 39 years old

23. 4, 5, 6   **25.** 11 nickels; 18 dimes; 33 quarters   **27.** 7 P.M.
29. 19, 21, 23

EXERCISES 3.2

1. $\{-3, 3\}$ **3.** $\{2, -8\}$ **5.** $\{8, -14\}$ **7.** $\left\{-\frac{7}{3}, 5\right\}$
9. $\{-4, -3\}$ **11.** $\left\{\frac{4}{5}, 2\right\}$ **13.** $\left\{0, \frac{2}{5}\right\}$ **15.** $\{-3, 6\}$
17. $\left\{-\frac{7}{3}, -\frac{5}{3}\right\}$ **19.** $\left\{\frac{18}{5}, 6\right\}$

EXERCISES 3.3

1. $\{x \mid x < -3\}$

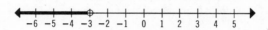

3. $\{x \mid x > 3\}$

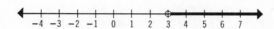

5. $\{x \mid x < -1\}$

7. $\{x \mid x \leq 7\}$

9. $\{x \mid x < 5\}$

11. $\left\{x \mid x < \frac{17}{2}\right\}$

13. $\{x \mid x \mid < -5\}$

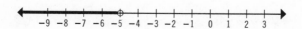

15. $\{x \mid x < \frac{9}{2}\}$

17. $\{x \mid x \geq \frac{5}{4}\}$

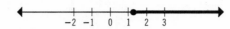

19. $\{x \mid x > -2\}$

21. 9 years old   **23.** between \$28,800 and \$38,400   **25.** 18

EXERCISES 3.4

1. $\{x \mid x > 5\} \cup \{x \mid x < 1\}$ **3.** $\{x \mid -6 < x < 4\}$
5. $\{x \mid -8 < x < 1\}$ **7.** $\{x \mid x \geq 8\} \cup \{x \mid x \leq -4\}$
9. $\{x \mid -2 < x < -\frac{4}{5}\}$ **11.** $\{x \mid -2 \leq x \leq 1\}$ **13.** $\{x \mid -\frac{1}{2} < x < \frac{7}{2}\}$
15. $\{x \mid x > \frac{7}{3}\} \cup \{x \mid x < \frac{1}{3}\}$ **17.** $\{x \mid x > -\frac{1}{8}\} \cup \{x \mid x < -\frac{13}{8}\}$
19. $\{x \mid x > \frac{8}{5}\} \cup \{x \mid x < \frac{6}{5}\}$

EXERCISES 3.5

1. $\{1, 5\}$ **3.** $\{-2, 1\}$ **5.** $\{-4 - 2\sqrt{3}, -4 + 2\sqrt{3}\}$
7. $\{1 - \sqrt{2}, 1 + \sqrt{2}\}$ **9.** $\{\frac{3}{4} - \frac{1}{4}\sqrt{13}, \frac{3}{4} + \frac{1}{4}\sqrt{13}\}$
11. $\{-\frac{1}{6} - \frac{1}{6}\sqrt{61}, -\frac{1}{6} + \frac{1}{6}\sqrt{61}\}$ **13.** $\{\frac{5}{2} - \frac{1}{2}\sqrt{53}, \frac{5}{2} + \frac{1}{2}\sqrt{53}\}$
15. $\{-\frac{1}{2} - \frac{1}{2}\sqrt{3}, -\frac{1}{2} + \frac{1}{2}\sqrt{3}\}$ **17.** $\{-\frac{3}{16} - \frac{1}{16}\sqrt{105}, -\frac{3}{16} + \frac{1}{16}\sqrt{105}\}$
19. $\{\frac{1}{5} - \frac{1}{5}\sqrt{21}, \frac{1}{5} + \frac{1}{5}\sqrt{21}\}$ **21.** -10 or 10 **23.** -1 **25.** 2 or -2
27. $\frac{5}{2}$ or $-\frac{5}{2}$ **29.** 2 or -2 **31.** 20 quarters, 40 nickels   **33.** 94
35. 200 by 900

EXERCISES 3.6

1. ϕ **3.** $\{x \mid x < -2 \text{ or } x > 5\}$ **5.** $\{x \mid -5 \leq x \leq -3\}$
7. $\{x \mid 2 \leq x \leq 3\}$ **9.** $\{x \mid x < -3 \text{ or } x > 7\}$ **11.** $\{x \mid -4 \leq x \leq \frac{3}{2}\}$
13. $\{x \mid 3 - \sqrt{14} \leq x \leq 3 + \sqrt{14}\}$ **15.** $\{x \mid -3 \leq x \leq 2\}$
17. $\{x \mid x < -1 - \sqrt{6} \text{ or } x > -1 + \sqrt{6}\}$
19. $\{x \mid x < \frac{5}{6} - \frac{1}{6}\sqrt{37} \text{ or } x > \frac{5}{6} + \frac{1}{6}\sqrt{37}$

EXERCISES 3.7

1. $\{9\}$ **3.** $\{-2\sqrt{5}, 2\sqrt{5}\}$ **5.** $\{3\}$ **7.** $\{-4\sqrt{5}, 4\sqrt{5}\}$

9. $\{-\frac{2}{3}\sqrt{5}, \frac{2}{3}\sqrt{5}\}$ **11.** $\{2, 5\}$ **13.** ϕ **15.** $\{2\}$ **17.** $\{-1, 3\}$
19. $\{0\}$

EXERCISES 3.8

1. $\{5\}$ **3.** $\{17\}$ **5.** $\{5\}$ **7.** $\{\frac{16}{9}\}$ **9.** $\{1\}$ **11.** $\{2\}$
13. $\{27\}$ **15.** ϕ **17.** $\{\frac{7}{3}\}$ **19.** $\{3\sqrt{2}\}$

EXERCISES 3.9

1. $\{-1, 1\}$ **3.** $\{-2\sqrt{3}, 2\sqrt{3}\}$ **5.** $\{-3, 3\}$
7. $\{-\frac{1}{3}, -\frac{1}{5}\sqrt{5}, \frac{1}{5}\sqrt{5}, \frac{1}{3}\}$ **9.** $\{16\}$ **11.** $\{81\}$ **13.** $\{\frac{1}{2}, 1\}$ **15.** $\{4\}$
17. $\{-2, 2\}$ **19.** $\{-3, -1, 2, 4\}$

CHAPTER 3 REVIEW

1. $\{-3\}$ **3.** $\frac{1}{6}$ **5.** $\{1\}$

7. $\{x \mid x \geq 9\}$

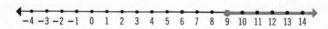

9. $\{x \mid 1 < x < 5\}$

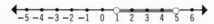

11. $\{-1, 4\}$ **13.** $\{-\frac{1}{6} + \frac{1}{6}\sqrt{13}, -\frac{1}{6} - \frac{1}{6}\sqrt{13}\}$

15. $\{x \mid x < -2 \text{ or } x > 3\}$

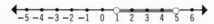

17. $\{2 - \sqrt{3}, 2 + \sqrt{3}\}$ **19.** $\{-1, 2\}$

EXERCISES 4.1

1. $\{(1, 3), (1, 6), (1, 7), (2, 3), (2, 6), (2, 7)\}$
3. $\{(a, e), (a, f), (b, e), (b, f), (c, e), (c, f)\}$
5. $\{(c, d), (c, o), (c, g), (a, d), (a, o), (a, g), (t, d), (t, o), (t, g)\}$ **7.** 12   **9.** 0

11.

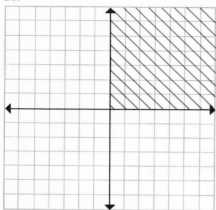

13.

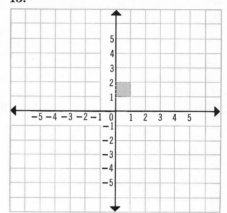

15.

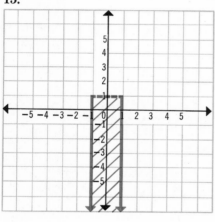

17.

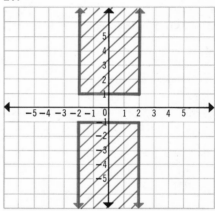

19.

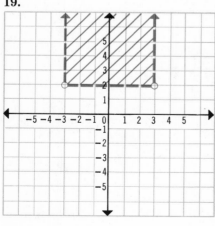

21.

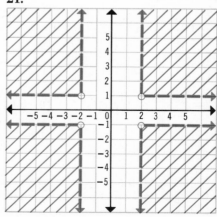

23. $\{(6, 1), (7, 2), (8, 3), (9, 4), (10, 5)\}$
25. $\{(1, 1), (2, 1), (3, 1), (4, 1), (5, 1), (6, 1), (7, 1), (8, 1), (9, 1), (10, 1)\}$
27. $\{(1, 2), (2, 1)\}$ **29.** ϕ

EXERCISES 4.2

1. domain: $\{-3, -2, -1, 0, 1, 2\}$; range: $\{0, 1, 4, 9\}$
3. domain and range: set of real numbers R
5. domain: $\{x \mid x \neq 5\}$; range: $\{y \mid y \neq 3\}$
7. domain: set of real numbers R; range: $\{y \mid 0 < y \leq 1\}$
9. domain and range: $\{t \mid t \geq 0\}$ **11.** domain and range: $\{t \mid -3 \leq t \leq 3\}$
13. domain: $\{x \mid -3 \leq x \leq 3\}$; range: $\{y \mid -2 \leq y \leq 2\}$
15. domain and range: $\{t \mid t \geq 0\}$
17. domain: $\{x \mid x \geq 2\}$; range: $\{y \mid y \geq 1\}$
19. domain: $\{x \mid x \neq 2 \text{ and } x \neq -1\}$; range: $\{y \mid y \neq -1\}$

21.

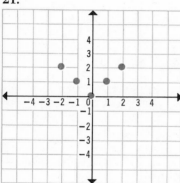

23.

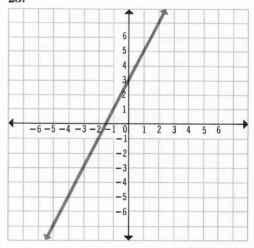

25.

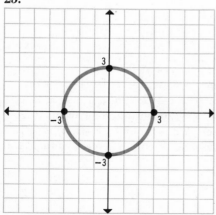

EXERCISES 4.3

1. function   **3.** not a function   **5.** function   **7.** function
9. not a function   **11.** $\{2, 5, 8\}$ **13.** $\{9, 16, 25\}$ **15.** $\{10, 13, 16\}$
17. function   **19.** function   **21.** not a function   **23.** not a function
25. function

27.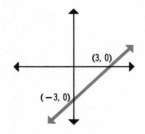

29.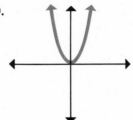

EXERCISES 4.4

1. (a) 4;   (b) -2;   (c) -6;   (d) $2a + 2b$;   (e) $2x + 2h$; (f) $2h$

3. (a) 4;   (b) $\frac{4}{3}$;   (c) $16 + 8h + h^2$;   (d) $\dfrac{1}{h - 3}$;   (e) $(3 + h)^2 + \dfrac{1}{3 + h}$;

   (f) -4;   (g) $\frac{1}{9}$;   (h) $\dfrac{1}{x^2}$

5. (a) $2xh + h^2 - 2h$; (b) $2x + h - 2$

7. (a) 7;   (b) $3\sqrt{y} + 1$;   (c) $3z^2 + 1$;   (d) $3|a| + 1$;   (e) $\sqrt{3x}$;   (f) $\sqrt{3x^2 + 1}$;
   (g) $3x^2 + 1 - \sqrt{3x + 1}$;   (h) $-3x + 1$ **9.** a; c

11. (a) 7;   (b) -4 and 4;   (c) 4;   (d) -4 and 3

EXERCISES 4.5

1. $d = \sqrt{34}$; $m = \frac{5}{3}$ **3.** $d = 17$; $m = -\frac{15}{8}$ **5.** $d = 3\sqrt{10}$; $m = \frac{1}{3}$
7. $d = 10$; $m = -\frac{3}{4}$ **9.** $d = 13$; $m = \frac{12}{5}$ **11.** $x - y + 1 = 0$
13. $x + y = 0$ **15.** $y + 2 = 0$ **17.** $x + y + 4 = 0$
19. $y = \frac{2}{3}x - \frac{4}{3}$; $b = -\frac{4}{3}$; $m = \frac{2}{3}$ **21.** $y = x - 7$; $m = 1$; $b = -7$
23. $y = \frac{3}{2}x - 9$; $m = \frac{3}{2}$; $b = -9$ **25.** $y = x + 2$; $m = 1$; $b = 2$
27. Distance between $A(2, -1)$ and $B(3, 4)$ is $\sqrt{26}$; distance between $B\,(3, 4)$
   and $C(-7, 6)$ is $\sqrt{104}$; distance between $A(2, -1)$ and $C(-7, 6)$ is $\sqrt{130}$;
   since $(AB)^2 + (BC)^2 = (AC)^2$, the Pythagorean theorem assures us that ABC
   is a right triangle.

29.

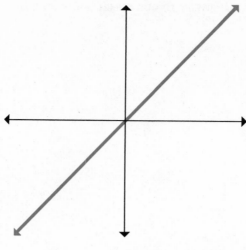

31.

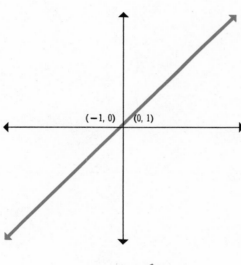

33.

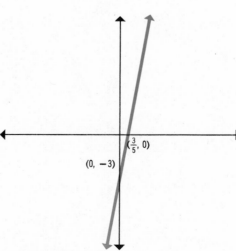

EXERCISES 4.6

1.

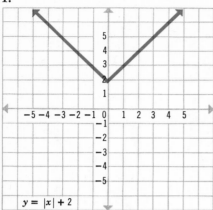

$y = |x| + 2$

3.

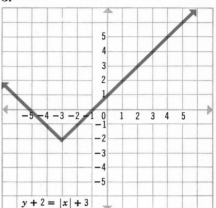

$y + 2 = |x| + 3$

5.

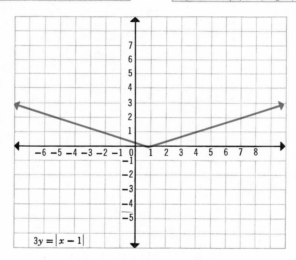

$3y = |x - 1|$

7.

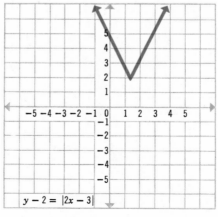

$y - 2 = |2x - 3|$

9.

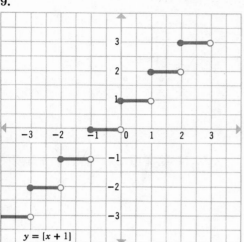

$y = [x + 1]$

11.

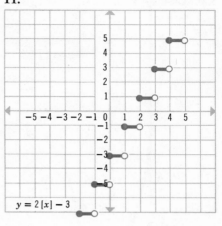

$y = 2[x] - 3$

13.

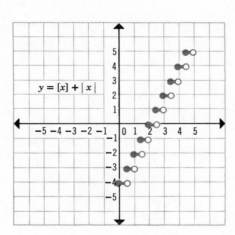

$y = [x] + |x|$

15.

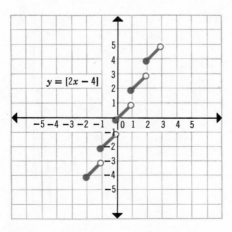

$y = [2x - 4]$

EXERCISES 4.7

1.

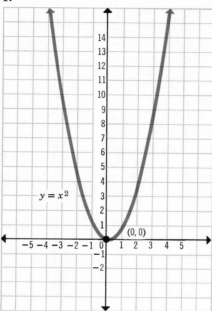

$y = x^2$

$(0, 0)$

3.

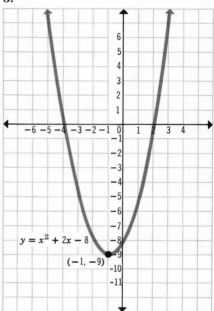

$y = x^2 + 2x - 8$

$(-1, -9)$

5.

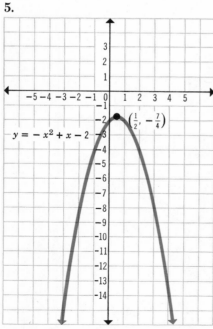

$y = -x^2 + x - 2$

$\left(\frac{1}{2}, -\frac{7}{4}\right)$

7.

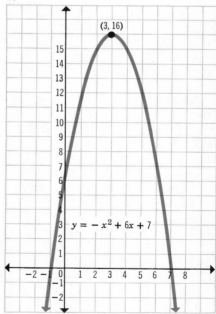

$(3, 16)$

$y = -x^2 + 6x + 7$

9.

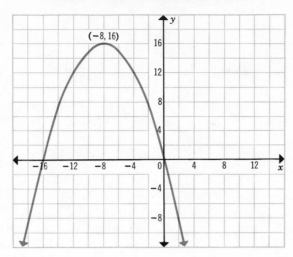

11. 8 and 8   **13.** 8 by 16   **15.** 3 and 5   **17.** $-\frac{7}{5}$ and $\frac{3}{2}$
19. -2 and $\frac{1}{2}$

EXERCISES 4.8

1.

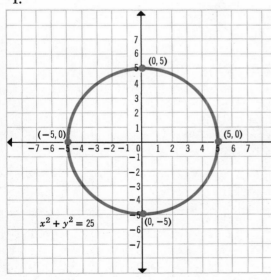

3.

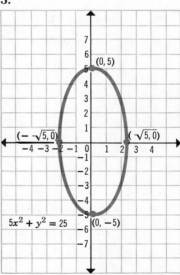

5.

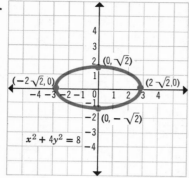

$(0, \sqrt{2})$
$(-2\sqrt{2}, 0)$
$(2\sqrt{2}, 0)$
$(0, -\sqrt{2})$
$x^2 + 4y^2 = 8$

7.

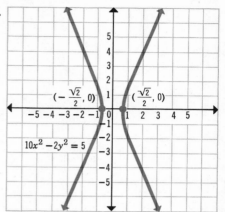

$(-\frac{\sqrt{2}}{2}, 0)$
$(\frac{\sqrt{2}}{2}, 0)$
$10x^2 - 2y^2 = 5$

9.

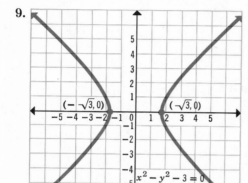

$(-\sqrt{3}, 0)$
$(\sqrt{3}, 0)$
$x^2 - y^2 - 3 = 0$

CHAPTER 4   REVIEW

1. $A \times B = \{(1, 0), (1, 2), (1, 6), (2, 0), (2, 2), (2, 6), (5, 0), (5, 2), (5, 6)\}$
  $B \times A = \{(0, 1), (0, 2), (0, 5), (2, 1), (2, 2), (2, 5), (6, 1), (6, 2), (6, 5)\}$

3(a)

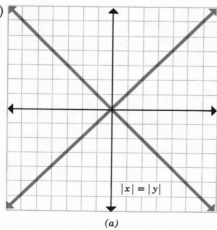

$|x| = |y|$

(a)

3(b)

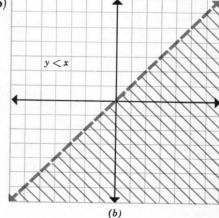

$y < x$

(b)

5. (a) not a function; (b) function; (c) function; (d) function; (e) not a function

7.

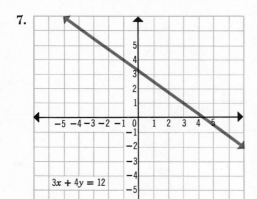

$3x + 4y = 12$

9. (a) 5; (b) $\sqrt{26}$; (c) 6; (d) 5

11. (a) $-\frac{1}{2}$; (b) 0; (c) none

13. $3x - y + 10 = 0$

15(a)

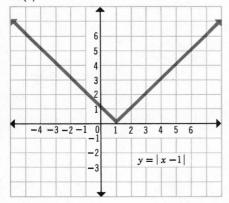

$y = |x - 1|$

15(b)

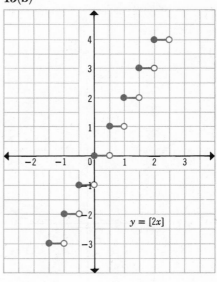

$y = [2x]$

17.

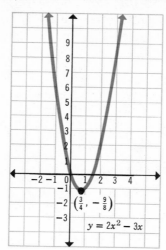

$$y = 2x^2 - 3x$$

$\left(\frac{3}{4}, -\frac{9}{8}\right)$

19. 40,000 square feet

EXERCISES 5.1

1. increasing   **3.** increasing   **5.** decreasing   **7.** decreasing
9. increasing

11.

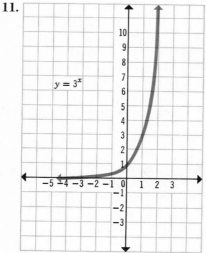

$y = 3^x$

13.

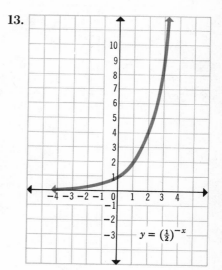

$y = \left(\frac{1}{2}\right)^{-x}$

15.

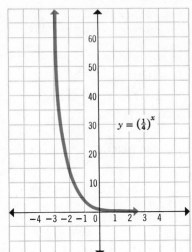

$y = \left(\frac{1}{4}\right)^x$

17.

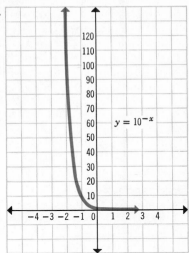

$y = 10^{-x}$

19.

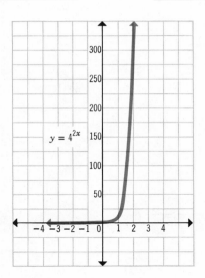

$y = 4^{2x}$

21. $\left(\frac{1}{4}\right)^x = (4^{-1})^x = 4^{(-1)x} = 4^{-x}$

23. $3^{x-2} = 3^x \cdot 3^{-2} = 3^x \left(\frac{1}{3^2}\right) = 3^x \cdot \frac{1}{9} = \frac{1}{9} \cdot 3_x$ **25.** $\{-4\}$ **27.** $\{\frac{1}{2}\}$

29. $\{3\}$

EXERCISES 5.2

  1. $\{(2, 1), (4, 2), (5, 3), (8, 4)\}$; function has an inverse function
  3. $\{(-2, 0), (-1, 1), (2, 2), (3, 3)\}$; function has an inverse function
  5. $\{(-2, -2), (-1, -1), (0, 0), (0, 1)\}$; function has no inverse function

7. $\{(1, 1), (2, 2), (3, 3), (15, 15)\}$; function has an inverse function
9. $\{(0, -1), (1, -\frac{1}{2}), (2, -\frac{1}{3}), (4, -\frac{1}{4}), (11, -\frac{1}{10})\}$; function has an inverse function
11. $\{(x, y) \mid y = \pm\sqrt{x}\}$; function has no inverse function
13. $\{(x, y) \mid y = \sqrt[3]{x}\}$; function has an inverse function
15. $\{(x, y) \mid y = 1 \pm \sqrt{x - 1}\}$; function has no inverse
17. $\left\{(x, y) \mid y = \dfrac{1}{x}\right\}$; function has an inverse function

19. $\left\{(x, y) \mid y + 2 = \dfrac{1}{x + 1}\right\}$; function has an inverse function

EXERCISES 5.3

1. $\log_2 8 = 3$ **3.** $\log_5 25 = 2$ **5.** $\log_{1/4} \frac{1}{16} = 2$ **7.** $\log_5 5 = 1$
9. $\log_{10} \frac{1}{100} = -2$ **11.** $2^5 = 32$ **13.** $5^0 = 1$ **15.** $(\frac{1}{3})^{-3} = 27$
17. $8^{4/3} = 16$ **19.** $y = 0$ **21.** $b = 10$ **23.** $x = \frac{1}{1000}$ or $x = 0.001$
25. $x = 125$

27. $\log_b \left(\dfrac{x}{z}\right)$ **29.** $\log_b \dfrac{\sqrt{x}}{\sqrt[3]{y^2}}$ **31.** $\log_b \sqrt[4]{\dfrac{x^3}{y^3}}$ or $\log_b \sqrt[4]{\left(\dfrac{x}{y}\right)^3}$

33. $\log_b x + \log_b y - \log_b z$ **35.** $\frac{1}{2} \log_b x$ **37.** $\frac{1}{2} (\log_b x - \log_b y)$
39. $\log_b 2 + \frac{3}{2} \log_b x + \frac{3}{2} \log_b y$ **41.** $\frac{2}{5} [\log_b (x - 1) - \log_b x]$

43.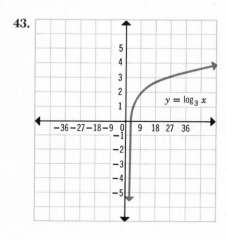

45.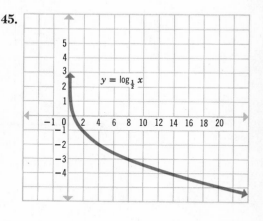

EXERCISES 5.4

1. 0.5866 **3.** 1.2504 **5.** $8.9694 - 10$ **7.** $8.9704 - 10$
9. $6.9259 - 10$ **11.** 5.8956 **13.** $7.7619 - 10$ **15.** 5.0913
17. 3.7111 **19.** $5.4343 - 10$ **21.** 4 **23.** 9500 **25.** 0.00694
27. 0.0548 **29.** 6.663 **31.** 16.19 **33.** $2.659 \cdot 10^{-13}$ **35.** 84.06
37. 4704 **39.** 20.08

EXERCISES 5.5

1. 1171   **3.** 3.784   **5.** 1.349   **7.** 0.02681   **9.** 40.33   **11.** 0.586
13. -1.167×10^6 **15.** 0.005291   **17.** -4.202 **19.** 1.778

EXERCISES 5.6

1. $\{5\}$ **3.** $\{4.755\}$ **5.** $\{37.5588\}$ **7.** $\{\frac{5}{14}\}$ **9.** $\{0.6825\}$
11. $\{500\}$ **13.** $\{100,000\}$ **15.** $\{\sqrt{10}\}$ or $\{0.3162\}$
17. $\{-\frac{4999}{2998}\}$ or $\{-1.6677\}$ **19.** $\{-\frac{7}{4} + \frac{5}{4}\sqrt{33}, -\frac{7}{4} - \frac{5}{4}\sqrt{33}\}$

EXERCISES 5.7

1. 6.5901   **3.** 6.0472   **5.** 4.7361   **7.** 4.59498   **9.** 4.4773
11. 8.3272   **13.** 1.78097   **15.** 2.042967   **17.** $\frac{1}{2}$ **19.** $\frac{1}{4}$

CHAPTER 5   REVIEW

1. $\{(0,0), (3,2), (4,4), (5,1)\}$

3.

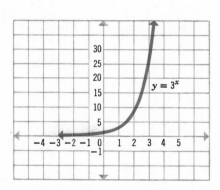

5. (a) $\log_2 16 = 4$;   (b) $\log_2 x = y$ **7.** (a) 0.3160;   (b) 0.9499;   (c) 0.7267
9. (a) 0.5027;   (b) 0.8472;   (c) 0.6992   **11.** (a) 8.73;   (b) 2.76;   (c) 5.38
13. (a) 4.153;   (b) 6.814;   (c) 1.385   **15.** 1.859   **17.** $\{-10.0496\}$
19. 1.4651

EXERCISES 6.1

1. $\frac{\pi}{4}$ **3.** C **5.** M **7.** T **9.** B **11.** T **13.** P **15.** L

17. T **19.** B **21.** Y **23.** $\frac{\pi}{3}$ **25.** $\frac{7\pi}{3}$ **27.** $-\frac{7\pi}{6}$ **29.** $\frac{4\pi}{3}$

31. $-\frac{14\pi}{3}$ **33.** $-\frac{\pi}{2}$ **35.** $\frac{5\pi}{3}$ **37.** $-\frac{37\pi}{3}$ **39.** $(\frac{1}{2}\sqrt{3}, -\frac{1}{2})$

EXERCISES 6.2

1. $(-\frac{1}{2}\sqrt{2}, \frac{1}{2}\sqrt{2})$ **3.** $(\frac{1}{2}\sqrt{2}, -\frac{1}{2}\sqrt{2})$ **5.** $(-\frac{1}{2}, -\frac{1}{2}\sqrt{3})$ **7.** $(-\frac{1}{2}\sqrt{3}, \frac{1}{2})$
9. $(\frac{1}{2}\sqrt{3}, -\frac{1}{2})$ **11.** $(\frac{1}{2}, \frac{1}{2}\sqrt{3})$

13.

θ	0	$\dfrac{\pi}{6}$	$\dfrac{\pi}{4}$	$\dfrac{\pi}{3}$	$\dfrac{\pi}{2}$	$\dfrac{2\pi}{3}$
$C(\theta)$	$(1,0)$	$(\frac{1}{2}\sqrt{3}, \frac{1}{2})$	$(\frac{1}{2}\sqrt{2}, \frac{1}{2}\sqrt{2})$	$(\frac{1}{2}, \frac{1}{2}\sqrt{3})$	$(0,1)$	$(-\frac{1}{2}, \frac{1}{2}\sqrt{3})$

θ	$\dfrac{3\pi}{4}$	$\dfrac{5\pi}{6}$	π	$\dfrac{7\pi}{6}$	$\dfrac{5\pi}{4}$
$C(\theta)$	$(-\frac{1}{2}\sqrt{2}, \frac{1}{2}\sqrt{2})$	$(-\frac{1}{2}\sqrt{3}, \frac{1}{2})$	$(-1,0)$	$(-\frac{1}{2}\sqrt{3}, -\frac{1}{2})$	$(-\frac{1}{2}\sqrt{2}, -\frac{1}{2}\sqrt{2})$

θ	$\dfrac{4\pi}{3}$	$\dfrac{3\pi}{2}$	$\dfrac{5\pi}{3}$	$\dfrac{7\pi}{4}$	$\dfrac{11\pi}{6}$	2π
$C(\theta)$	$(-\frac{1}{2}, -\frac{1}{2}\sqrt{3})$	$(0,-1)$	$(\frac{1}{2}, -\frac{1}{2}\sqrt{3})$	$(\frac{1}{2}\sqrt{2}, -\frac{1}{2}\sqrt{2})$	$(\frac{1}{2}\sqrt{3}, -\frac{1}{2})$	$(1,0)$

15. $(-1,0)$ **17.** $(1,0)$ **19.** $(-\frac{1}{2}\sqrt{3}, -\frac{1}{2})$ **21.** $(\frac{1}{2}, -\frac{1}{2}\sqrt{3})$
23. $(\frac{1}{2}\sqrt{2}, -\frac{1}{2}\sqrt{2})$

EXERCISES 6.3

1. $\frac{1}{2}\sqrt{3}$ **3.** -1 **5.** -1 **7.** $-\frac{1}{2}\sqrt{3}$ **9.** $-\frac{1}{2}$ **11.** $\dfrac{\sqrt{3}}{2}$ or $\dfrac{1}{2}\sqrt{3}$
13. $-\frac{1}{2}\sqrt{3}$ or $-\frac{1}{2}\sqrt{3}$ **15.** $-\frac{1}{2}\sqrt{2}$ **17.** 1 **19.** $-\frac{1}{2}\sqrt{2}$ **21.** $\frac{1}{2}\sqrt{2}$
23. $\frac{1}{2}\sqrt{2}$ **25.** $-\frac{1}{2}\sqrt{2}$ **27.** $\frac{1}{2}\sqrt{3}$ **29.** $\frac{1}{2}$ **31.** $\frac{1}{2}\sqrt{2}$ **33.** $\frac{1}{2}\sqrt{3}$
35. $\frac{1}{2}\sqrt{2}$ **37.** -1 **39.** $\frac{1}{2}\sqrt{3}$

EXERCISES 6.4

1. $\frac{1}{4}(\sqrt{2} + \sqrt{6})$ **3.** $\frac{1}{4}(\sqrt{6} - \sqrt{2})$ **5.** $\frac{1}{4}(\sqrt{6} + \sqrt{2})$
7. $\frac{1}{4}(\sqrt{2} + \sqrt{6})$ **9.** $\frac{1}{4}(\sqrt{2} + \sqrt{6})$ **11.** $\frac{1}{4}(\sqrt{2} - \sqrt{6})$ **13.** $-\dfrac{\sqrt{2}}{2}$
15. $\frac{1}{2}$ **17.** -1 **19.** $\frac{1}{4}(\sqrt{2} - \sqrt{6})$ **21.** $-\frac{12}{13}$ **23.** $\frac{4}{5}$ **25.** $\frac{3}{5}$
27. $-\frac{5}{13}$ **29.** $-\frac{4}{5}$

EXERCISES 6.5

1. $\frac{1}{2}\sqrt{2 - \sqrt{2}}$ **3.** $\frac{1}{2}\sqrt{2 - \sqrt{3}}$ **5.** $\frac{1}{2}\sqrt{2 + \sqrt{2}}$ **7.** $\sin\dfrac{\pi}{3}$ or $\dfrac{\sqrt{3}}{2}$

9. $\sin 5\theta$ **11.** $\sin \dfrac{\pi}{2}$ or 1 **13.** $\cos \dfrac{6\pi}{7}$ **15.** $\sin \dfrac{8\pi}{9}$ **17.** $a^2 - b^2$

19. $-a$ **21.** $-b$ **23.** $2ab$

25. $\sin(\pi - \theta) = \sin \pi \cos \theta - \cos \pi \sin \theta$
$\qquad\qquad = 0 \cdot \cos \theta - (-1) \sin \theta$
$\qquad\qquad = \sin \theta$

27. $\sin 2\theta = \sin(\theta + \theta)$
$\qquad\quad = \sin \theta \cos \theta + \cos \theta \sin \theta$
$\qquad\quad = 2 \sin \theta \cos \theta$

29. $\cos 2\theta = \cos(\theta + \theta)$
$\qquad\quad = \cos \theta \cos \theta - \sin \theta \sin \theta$
$\qquad\quad = \cos^2 \theta - \sin^2 \theta$
$\qquad\quad = (1 - \sin^2 \theta) - \sin^2 \theta$
$\qquad\quad = 1 - 2 \sin^2 \theta$

EXERCISES 6.6

1.

θ	0	$\dfrac{\pi}{6}$	$\dfrac{\pi}{4}$	$\dfrac{\pi}{3}$	$\dfrac{\pi}{2}$	$\dfrac{2\pi}{3}$	$\dfrac{3\pi}{4}$	$\dfrac{5\pi}{6}$	π
$\tan \theta$	0	$\tfrac{1}{3}\sqrt{3}$	1	$\sqrt{3}$	not defined	$-\sqrt{3}$	-1	$-\tfrac{1}{3}\sqrt{3}$	0

θ	$\dfrac{7\pi}{6}$	$\dfrac{5\pi}{4}$	$\dfrac{4\pi}{3}$	$\dfrac{3\pi}{2}$	$\dfrac{5\pi}{3}$	$\dfrac{7\pi}{4}$	$\dfrac{11\pi}{6}$	2π
$\tan \theta$	$\tfrac{1}{3}\sqrt{3}$	1	$\sqrt{3}$	not defined	$-\sqrt{3}$	-1	$-\tfrac{1}{3}\sqrt{3}$	0

3.

θ	0	$\dfrac{\pi}{6}$	$\dfrac{\pi}{4}$	$\dfrac{\pi}{3}$	$\dfrac{\pi}{2}$	$\dfrac{2\pi}{3}$	$\dfrac{3\pi}{4}$	$\dfrac{5\pi}{6}$	π
$\sec \theta$	1	$\tfrac{2}{3}\sqrt{3}$	$\sqrt{2}$	2	not defined	-2	$-\sqrt{2}$	$-\tfrac{2}{3}\sqrt{3}$	-1

θ	$\dfrac{7\pi}{6}$	$\dfrac{5\pi}{4}$	$\dfrac{4\pi}{3}$	$\dfrac{3\pi}{2}$	$\dfrac{5\pi}{3}$	$\dfrac{7\pi}{4}$	$\dfrac{11\pi}{6}$	2π
$\sec \theta$	$-\tfrac{2}{3}\sqrt{3}$	$-\sqrt{2}$	-2	not defined	2	$\sqrt{2}$	$\tfrac{2}{3}\sqrt{3}$	1

5. $\sin \theta = \tfrac{5}{13}$; $\cos \theta = \tfrac{12}{13}$; $\csc \theta = \tfrac{13}{5}$; $\sec \theta = \tfrac{13}{12}$; $\cot \theta = \tfrac{12}{5}$

7. $\sin \theta = -\tfrac{4}{5}$; $\cos \theta = -\tfrac{3}{5}$; $\tan \theta = \tfrac{4}{3}$; $\csc \theta = -\tfrac{5}{4}$; $\cot \theta = \tfrac{3}{4}$

9. $\sin \theta = -\tfrac{5}{13}$; $\cos \theta = -\tfrac{12}{13}$; $\tan \theta = \tfrac{5}{12}$; $\csc \theta = -\tfrac{13}{5}$; $\sec \theta = -\tfrac{13}{12}$

11. $\sin\theta = -\frac{12}{13}$; $\tan\theta = -\frac{12}{5}$; $\csc\theta = -\frac{13}{12}$; $\sec\theta = \frac{13}{5}$; $\cot\theta = -\frac{5}{12}$

13. $\sin\theta = -\frac{4}{5}$; $\cos\theta = -\frac{3}{5}$; $\tan\theta = \frac{4}{3}$; $\sec\theta = -\frac{5}{3}$; $\cot\theta = \frac{3}{4}$

15. $\sin\theta = -\frac{2}{5}\sqrt{5}$; $\cos\theta = -\frac{1}{5}\sqrt{5}$; $\tan\theta = 2$; $\cos\theta = -\frac{1}{2}\sqrt{5}$; $\sec\theta = -\sqrt{5}$

17. $\sin\theta = -\frac{3}{5}$; $\cos\theta = -\frac{4}{5}$; $\tan\theta = \frac{3}{4}$; $\csc\theta = -\frac{5}{3}$; $\sec\theta = -\frac{5}{4}$

19. $\sin\theta = \frac{1}{6}\sqrt{35}$; $\cos\theta = -\frac{1}{6}$; $\tan\theta = -\sqrt{35}$; $\csc\theta = \frac{6}{35}\sqrt{35}$; $\cot\theta = -\frac{1}{35}\sqrt{35}$

21.
$$\tan(\theta_1 - \theta_2) = \frac{\sin(\theta_1 - \theta_2)}{\cos(\theta_1 - \theta_2)}$$

$$= \frac{\sin\theta_1 \cos\theta_2 - \cos\theta_1 \sin\theta_2}{\cos\theta_1 \cos\theta_2 + \sin\theta_1 \sin\theta_2}$$

$$= \frac{\dfrac{\sin\theta_1 \cos\theta_2}{\cos\theta_1 \cos\theta_2} - \dfrac{\cos\theta_1 \sin\theta_2}{\cos\theta_1 \cos\theta_2}}{\dfrac{\cos\theta_1 \cos\theta_2}{\cos\theta_1 \cos\theta_2} + \dfrac{\sin\theta_1 \sin\theta_2}{\cos\theta_1 \cos\theta_2}}$$

$$= \frac{\dfrac{\sin\theta_1}{\cos\theta_1} - \dfrac{\sin\theta_2}{\cos\theta_2}}{1 + \dfrac{\sin\theta_1}{\cos\theta_1} \cdot \dfrac{\sin\theta_2}{\cos\theta_2}}$$

$$= \frac{\tan\theta_1 - \tan\theta_2}{1 + \tan\theta_1 \tan\theta_2}$$

EXERCISES 6.8

1. $\left\{\dfrac{\pi}{6}, \dfrac{11\pi}{6}\right\}$ **3.** $\left\{\dfrac{\pi}{4}, \dfrac{3\pi}{4}, \dfrac{5\pi}{4}, \dfrac{7\pi}{4}\right\}$ **5.** $\left\{0, \dfrac{7\pi}{6}, \dfrac{11\pi}{6}\right\}$ **7.** $\{0, \pi\}$

9. $\left\{0, \dfrac{\pi}{6}, \dfrac{5\pi}{6}, \pi\right\}$ **11.** $\left\{0, \dfrac{\pi}{6}, \dfrac{5\pi}{6}, \pi, \dfrac{3\pi}{2}\right\}$ **13.** $\left\{0, \dfrac{3\pi}{4}, \pi, \dfrac{7\pi}{4}\right\}$

15. $\{0, \pi\}$ **17.** $\left\{0, \dfrac{4\pi}{3}\right\}$ **19.** $\left\{\dfrac{3\pi}{2}\right\}$ **21.** $\left\{0, \dfrac{\pi}{6}, \dfrac{5\pi}{6}, \pi, \dfrac{7\pi}{6}, \dfrac{11\pi}{6}\right\}$

23. $\left\{0, \dfrac{\pi}{6}, \dfrac{\pi}{2}, \dfrac{5\pi}{6}, \pi, \dfrac{7\pi}{6}, \dfrac{3\pi}{2}, \dfrac{11\pi}{6}\right\}$ **25.** $\left\{\dfrac{\pi}{4}, \dfrac{\pi}{2}, \dfrac{5\pi}{4}, \dfrac{3\pi}{2}\right\}$

27. $\{0\}$ **29.** $\left\{\dfrac{\pi}{3}, \dfrac{2\pi}{3}, \dfrac{4\pi}{3}, \dfrac{5\pi}{3}\right\}$ **31.** ϕ. **33.** $\left\{\dfrac{\pi}{6}, \dfrac{5\pi}{6}\right\}$ **35.** $\left\{\dfrac{\pi}{3}, \dfrac{5\pi}{3}\right\}$

CHAPTER 6　REVIEW

1. (a) $\left(-\frac{1}{2}\sqrt{2}, \frac{1}{2}\sqrt{2}\right)$; (b) $(-1, 0)$; (c) $\left(-\frac{1}{2}\sqrt{3}, -\frac{1}{2}\right)$; (d) $\left(-\frac{1}{2}\sqrt{2}, \frac{1}{2}\sqrt{2}\right)$

3. $\frac{1}{2}\sqrt{3}$ **5.** $\frac{1}{2}$ **7.** (a) 2π; (b) π; (c) 2π **9.** $\frac{1}{4}(\sqrt{2} + \sqrt{6})$

11. $\cos 2x \cos 2y + \sin 2x \sin 2y$

13. $\sin\theta = -\frac{1}{2}\sqrt{3}$; $\cos\theta = \frac{1}{2}$; $\tan\theta = -\sqrt{3}$; $\csc\theta = -\frac{2}{3}\sqrt{3}$; $\cot\theta = -\frac{1}{3}\sqrt{3}$

19. $\left\{\dfrac{3\pi}{4}, \dfrac{7\pi}{4}\right\}$

EXERCISES 7.1

1. $36°$ **3.** $300°$ **5.** $0°$ **7.** $-540°$ **9.** $-150°$ **11.** $157.6°$
13. $528.8°$ **15.** $-172.5°$ **17.** 0.21 **19.** 0.84 **21.** 1.48
23. -2.27

25.

Degrees	0°	30°	45°	60°	90°	120°	135°	150°
Radians	0	$\dfrac{\pi}{6}$	$\dfrac{\pi}{4}$	$\dfrac{\pi}{3}$	$\dfrac{\pi}{2}$	$\dfrac{2\pi}{3}$	$\dfrac{3\pi}{4}$	$\dfrac{5\pi}{6}$

Degrees	180°	210°	225°	240°	270°	300°	315°	330°
Radians	π	$\dfrac{7\pi}{6}$	$\dfrac{5\pi}{4}$	$\dfrac{4\pi}{3}$	$\dfrac{3\pi}{2}$	$\dfrac{5\pi}{3}$	$\dfrac{7\pi}{4}$	$\dfrac{11\pi}{6}$

EXERCISES 7.2

1. $\sin \theta = \frac{3}{5}$; $\cos \theta = \frac{4}{5}$; $\tan \theta = \frac{3}{4}$; $\csc \theta = \frac{5}{3}$; $\sec \theta = \frac{5}{4}$; $\cot \theta = \frac{4}{3}$

3. $\sin \theta = -\frac{5}{13}$; $\cos \theta = \frac{12}{13}$; $\tan \theta = -\frac{5}{12}$; $\csc \theta = -\frac{13}{5}$; $\sec \theta = \frac{13}{12}$; $\cot \theta = -\frac{12}{5}$

5. $\sin \theta = -\frac{5}{13}$; $\cos \theta = -\frac{12}{13}$; $\tan \theta = \frac{5}{12}$; $\csc \theta = -\frac{13}{5}$; $\sec \theta = -\frac{13}{12}$; $\cot \theta = \frac{12}{5}$

7. $\sin \theta = \frac{2}{5}\sqrt{5}$; $\cos \theta = -\frac{1}{5}\sqrt{5}$; $\tan \theta = -2$; $\csc \theta = \frac{1}{2}\sqrt{5}$; $\sec \theta = -\sqrt{5}$; $\cot \theta = -\frac{1}{2}$

9. $\sin \theta = -\frac{1}{5}\sqrt{5}$; $\cos \theta = \frac{2}{5}\sqrt{5}$; $\tan \theta = -\frac{1}{2}$; $\csc \theta = -\sqrt{5}$; $\sin \theta = \frac{1}{2}\sqrt{5}$; $\cot \theta = -2$

11. $\cos \theta = \frac{3}{5}$; $\tan \theta = \frac{4}{3}$; $\csc \theta = \frac{5}{4}$; $\sec \theta = \frac{5}{3}$; $\cot \theta = \frac{3}{4}$

13. $\sin \theta = -\dfrac{\sqrt{95}}{12}$; $\tan \theta = \dfrac{\sqrt{95}}{7}$; $\csc \theta = -\dfrac{12\sqrt{95}}{95}$; $\sec \theta = -\frac{12}{7}$;

$\cot \theta = \dfrac{7\sqrt{95}}{95}$

15. $\sin \theta = \dfrac{2\sqrt{229}}{229}$; $\cos \theta = -\dfrac{15\sqrt{229}}{229}$; $\tan \theta = -\frac{2}{15}$; $\csc \theta = \frac{1}{2}\sqrt{229}$;

$\sec \theta = -\dfrac{\sqrt{229}}{15}$

17. $\cos \theta = -\frac{12}{13}$; $\tan \theta = -\frac{5}{12}$; $\csc \theta = \frac{13}{5}$; $\sec \theta = -\frac{13}{12}$; $\cot \theta = -\frac{12}{5}$

19. $\sin \theta = -\dfrac{7\sqrt{130}}{130}$; $\cos \theta = -\dfrac{9\sqrt{130}}{130}$; $\csc \theta = -\dfrac{\sqrt{130}}{7}$;

$\sec \theta = -\dfrac{\sqrt{130}}{9}$; $\cot \theta = \frac{9}{7}$

21. $30°$ **23.** $45°$ **25.** $45°$ **27.** $30°$ **29.** $60°$ **31.** $10°$

33. $33°$ **35.** $47° \, 30'$ **37.** $87° \, 50'$ **39.** $63° \, 25'$ **41.** $\dfrac{\pi}{4}$ **43.** $\dfrac{\pi}{3}$

45. $\dfrac{\pi}{4}$ **47.** $\dfrac{\pi}{4}$ **49.** $\dfrac{\pi}{12}$

EXERCISES 7.3

1. $\sin 60° = \frac{1}{2}\sqrt{3};\quad \cos 60° = \frac{1}{2};\quad \tan 60° = \sqrt{3}$
    $\sec 60° = 2;\quad \csc 60° = \frac{2}{3}\sqrt{3};\quad \cot 60° = \frac{1}{3}\sqrt{3}$

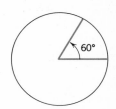

3. $\sin \dfrac{\pi}{6} = \frac{1}{2};\quad \cos \dfrac{\pi}{6} = \frac{1}{2}\sqrt{3};\quad \tan \dfrac{\pi}{6} = \frac{1}{3}\sqrt{3};$

   $\csc \dfrac{\pi}{6} = 2;\quad \sec \dfrac{\pi}{6} = \frac{2}{3}\sqrt{3};\quad \cot \dfrac{\pi}{6} = \sqrt{3}$

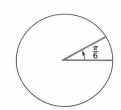

5. $\sin 315° = -\frac{1}{2}\sqrt{2};\quad \csc 315° = -\sqrt{2}$
   $\cos 315° = \frac{1}{2}\sqrt{2};\quad \sec 315° = \sqrt{2}$
   $\tan 315° = -1;\quad \cot 315° = -1$

7. $\sin(-315°) = \frac{1}{2}\sqrt{2};\quad \csc(-315°) = \sqrt{2}$
   $\cos(-315°) = \frac{1}{2}\sqrt{2};\quad \sec(-315°) = \sqrt{2}$
   $\tan(-315°) = 1;\quad \cot(-315°) = 1$

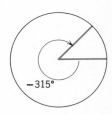

9. $\sin 330° = -\frac{1}{2};\quad \csc 330° = -2;$
   $\cos 330° = \frac{1}{2}\sqrt{3};\quad \sec 330° = \frac{2}{3}\sqrt{3};$
   $\tan 330° = -\frac{1}{3}\sqrt{3};\quad \cot 330° = -\sqrt{3}$

11. $\sin\dfrac{\pi}{3} = \tfrac{1}{2}\sqrt{3}; \quad \csc\dfrac{\pi}{3} = \tfrac{2}{3}\sqrt{3};$

$\cos\dfrac{\pi}{3} = \tfrac{1}{2}; \quad \sec\dfrac{\pi}{3} = 2;$

$\tan\dfrac{\pi}{3} = \sqrt{3}; \quad \cot\dfrac{\pi}{3} = \tfrac{1}{3}\sqrt{3}$

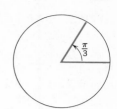

13. $\sin\dfrac{3\pi}{4} = \tfrac{1}{2}\sqrt{2}; \quad \csc\dfrac{3\pi}{4} = \sqrt{2}$

$\cos\dfrac{3\pi}{4} = -\tfrac{1}{2}\sqrt{2}; \quad \sec\dfrac{3\pi}{4} = -\sqrt{2}$

$\tan\dfrac{3\pi}{4} = -1; \quad \cot\dfrac{3\pi}{4} = -1$

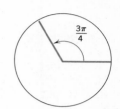

15. $\sin(-300°) = \tfrac{1}{2}\sqrt{3}; \quad \csc(-300°) = \tfrac{2}{3}\sqrt{3}$
$\cos(-300°) = \tfrac{1}{2}; \quad \sec(-300°) = 2$
$\tan(-300°) = \sqrt{3}; \quad \cot(-300°) = \tfrac{1}{3}\sqrt{3}$

17. $\sin(-120°) = -\tfrac{1}{2}\sqrt{3}; \quad \csc(-120°) = -\tfrac{2}{3}\sqrt{3}$
$\cos(-120°) = -\tfrac{1}{2}; \quad \sec(-120°) = -2$
$\tan(-120°) = \sqrt{3}; \quad \cot(-120°) = \tfrac{1}{3}\sqrt{3}$

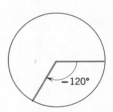

19. $\sin\left(-\dfrac{4\pi}{3}\right) = \tfrac{1}{2}\sqrt{3}; \quad \csc\left(-\dfrac{4\pi}{3}\right) = \tfrac{2}{3}\sqrt{3}$

$\cos\left(-\dfrac{4\pi}{3}\right) = -\tfrac{1}{2}; \quad \sec\left(-\dfrac{4\pi}{3}\right) = -2$

$\tan\left(-\dfrac{4\pi}{3}\right) = -\sqrt{3}; \quad \cot\left(-\dfrac{4\pi}{3}\right) = -\tfrac{1}{3}\sqrt{3}$

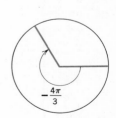

21. $(\tfrac{1}{2})^2 + (\tfrac{1}{2}\sqrt{3})^2 = \tfrac{1}{4} + \tfrac{3}{4} = 1$ **23.** $1^2 + 1 = 1 + 1 = 2 = (\sqrt{2})^2$

25. $\dfrac{-\tfrac{1}{2}\sqrt{3}}{\tfrac{1}{2}} = -\sqrt{3}$ **27.** $1 - (-\tfrac{1}{2}\sqrt{3})^2 = 1 - \tfrac{3}{4} = \tfrac{1}{4} = (-\tfrac{1}{2})^2$

29. $(\tfrac{1}{3}\sqrt{3})(\sqrt{3}) = 1$ **31.** $<$ **33.** $=$ **35.** $>$ **37.** $>$ **39.** $<$

EXERCISES 7.4

1. 0.5348   **3.** 0.4841   **5.** 1.031   **7.** 0.1334   **9.** 1.668
11. 0.3121   **13.** 0.9787   **15.** 0.1287   **17.** 0.3961   **19.** 4.682
21. 0.9853   **23.** 0.6330   **25.** 0.9601   **27.** 1.047   **29.** 1.281
31. 7° 30′   **33.** 15° 50′   **35.** 23°   **37.** 66° 20′   **39.** 45°   **41.** 55°
43. 39° 30′   **45.** 18°   **47.** 22°   **49.** 63° 30′   **51.** 18° 10′
53. 27° 20′   **55.** 55° 30′   **57.** 30′   **59.** 63° 30′

EXERCISES 7.5

1. 0.8274   **3.** −0.5658   **5.** −1.003   **7.** −0.0958   **9.** 2.161
11. −0.1883   **13.** −3.021   **15.** −0.1736   **17.** −0.1334
19. −6.765   **21.** 159°   **23.** 335°   **25.** 243° 30′   **27.** 249° 10′
29. 312° 40′

EXERCISES 7.6

1. 0.3462   **3.** 0.7432   **5.** 1.810   **7.** 0.6421   **9.** 0.9985   **11.** 1.184
13. 0.9839   **15.** 4.625   **17.** 0.9343   **19.** 0.6400   **21.** 3.130
23. 0.5087   **25.** 0.7155   **27.** 0.1608   **29.** 8° 24′   **31.** 22° 21′
33. 25° 45′   **35.** 54° 27′   **37.** 7° 30′   **39.** 44° 15′   **41.** 27° 33′
43. 44° 41′   **45.** 8° 5′   **47.** 25° 58′

CHAPTER 7   REVIEW

1. (a) $\dfrac{\pi}{12}$;   (b) $\dfrac{31\pi}{18}$;   (c) $\dfrac{107\pi}{30}$;   (d) $-\dfrac{83\pi}{180}$

3. $\sin \theta = \dfrac{\sqrt{3}}{2}$; $\cos \theta = -\frac{1}{2}$; $\tan \theta = -\sqrt{3}$; $\cot \theta = -\frac{1}{3}\sqrt{3}$;
$\sec \theta = -2$; $\csc \theta = \frac{2}{3}\sqrt{3}$

5. $\sin \theta = -\frac{1}{2}$; $\cos \theta = \frac{1}{2}\sqrt{3}$; $\tan \theta = -\frac{1}{3}\sqrt{3}$; $\cot \theta = -\sqrt{3}$;
$\sec \theta = \frac{2}{3}\sqrt{3}$; $\csc \theta = -2$

7. $\sin \theta = -\frac{1}{2}\sqrt{2}$; $\cos \theta = \frac{1}{2}\sqrt{2}$; $\tan \theta = -1$; $\cot \theta = -1$; $\sec \theta = \sqrt{2}$;
$\csc \theta = -\sqrt{2}$

9. $-\frac{1}{2}\sqrt{2}$ **11.** 0.7969   **13.** 59°   **15.** −1.098   **17.** 1.002
19. 0.9878

EXERCISES 8.1

1. $B = 60°$, $c = 8\sqrt{3}$, $a = 4\sqrt{3}$
3. $B = 51° 10′$, $a = 250.355$, $c = 399.324$
5. $A = 38° 10′$, $B = 51° 50′$, $c = 635.970$ **7.** 12.19 feet
9. 78.8 feet or 79 feet   **11.** 445.36 miles per hour or 445 m.p.h.
13. 50° 11′   **15.** 555.75 feet   **17.** 87.2 feet   **19.** 22° 10′

EXERCISES 8.2

1. $9.8255 - 10$ **3.** 0.1710 **5.** $9.7072 - 10$ **7.** $9.4482 - 10$
9. 0.0235 **11.** $33°\ 40'$ **13.** $52°\ 20'$ **15.** $41°\ 40'$ **17.** $40°\ 37'$
19. $7°\ 24'$

EXERCISES 8.3

1. $b \doteq 29.5;\quad c \doteq 28.2;\quad C \doteq 70°$ **3.** $b = 3.16;\quad c = 3.92;\quad C = 62°\ 10'$
5. $C = 126°;\quad a = 277.7;\quad c = 382.2$
7. $C = 55°\ 19';\quad B = 49°\ 41'\quad c = 80.87$
9. $B = 22°\ 13';\quad C = 128°\ 23';\quad C = 51.36$ **11.** 615.2 feet
13. 7423 feet; 3345 feet   **15.** 18 miles; 113 miles

EXERCISES 8.4

1. $C = 24°\ 10';\quad A = 125°\ 10';\quad B = 30°\ 40'$
3. $a = 160;\quad B = 10°;\quad C = 20°$ **5.** $a = 23;\quad B = 42°;\quad C = 80°$
7. $c = 44;\quad B = 40°;\quad C = 35°$ **9.** $a = 230;\quad B = 11°\ 50';\quad C = 101°$
11. 1342 feet   **13.** 215 miles per hour; $206°$ **15.** $65°\ 25';\ 57°\ 5'$

EXERCISES 8.5

1. 7.62 pounds   **3.** 63 miles per hour   **5.** 180.1 miles per hour; $8°\ 50'$
7. 16.2 miles per hour; $20°\ 30'$ from direction of boat
9. $77°\ 30';$ 1 minute 10 seconds   **11.** 794.85 pounds

CHAPTER 8   REVIEW

1. $B = 34°;\ b = 44.74;\ a = 66.32$ **3.** $120°$ **5.** no triangle
7. $18°\ 12';\quad 51°\ 19'$ **9.** 563.5 miles

EXERCISES 9.1

1. $\frac{1}{2}\sqrt{2}$ **3.** $-\sqrt{3}$ **5.** 2 **7.** $\frac{1}{2}$ **9.** 1 **11.** $\sqrt{2}$ **13.** 2π
15. 2π **17.** $\frac{\pi}{3}$ **19.** $\frac{2\pi}{5}$

EXERCISES 9.2

1. $\pi < \theta < 2\pi$ **3.** $\frac{3\pi}{2} < \theta < 2\pi$ **5.** $\theta = \frac{\pi}{2}$ and $\pi < \theta < 2\pi$

7. $0 \le \theta < \frac{\pi}{2}$ and $\frac{3\pi}{2} < \theta \le 2\pi$ **9.** $0 \le \theta < \frac{\pi}{2}$ and $\pi \le \theta < \frac{3\pi}{2}$

11. $\dfrac{\pi}{2} < \theta < \dfrac{3\pi}{2}$ **13.** $0 < \theta < \pi$ and $\pi < \theta < 2\pi$

15. $0 < \theta < \dfrac{\pi}{2}$ and $\dfrac{3\pi}{2} < \theta < 2\pi$

EXERCISES 9.3

1.

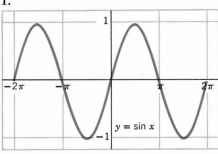

3.

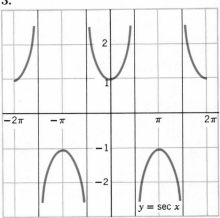

5.

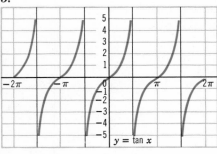

7.

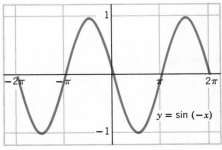

9.

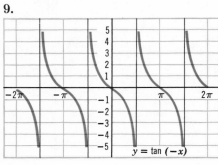

11.

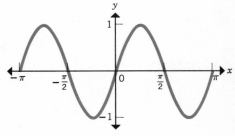

13.

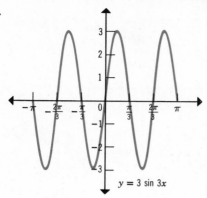

$y = 3 \sin 3x$

15.

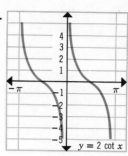

$y = 2 \cot x$

17.

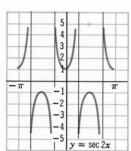

$y = \sec 2x$

19.

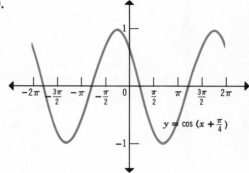

$y = \cos \left(x + \frac{\pi}{4} \right)$

CHAPTER 9   REVIEW

1. (a) $\frac{1}{2}\sqrt{2}$;   (b) $\frac{1}{2}\sqrt{3}$;   (c) $-\frac{1}{3}\sqrt{3}$;   (d) $\frac{1}{2}\sqrt{2}$;   (e) $-\frac{1}{2}$;   (f) $\frac{2}{3}\sqrt{3}$;   (g) -1;
(h) $\frac{2}{3}\sqrt{3}$

3.

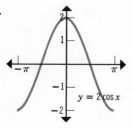

$y = 2 \cos x$

5.

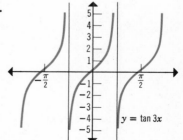

$y = \tan 3x$

7.

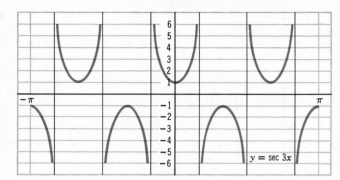

$y = \sec 3x$

9.

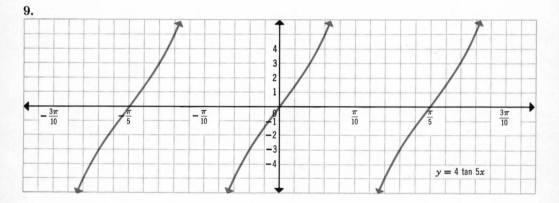

$y = 4 \tan 5x$

11.

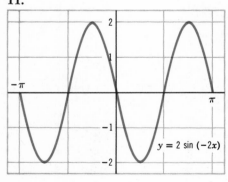

$y = 2 \sin (-2x)$

13.

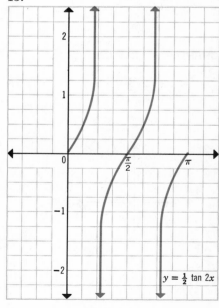

$y = \frac{1}{2} \tan 2x$

15.

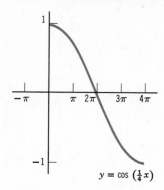

$$y = \cos\left(\tfrac{1}{4}x\right)$$

EXERCISES 10.1

1. has an inverse function   **3.** does not have an inverse function
5. does not have an inverse function   **7.** has an inverse function
9. does not have an inverse function

EXERCISES 10.2

1. $\dfrac{\pi}{6}$ **3.** $\dfrac{2\pi}{3}$ **5.** 0 **7.** $\dfrac{3\pi}{4}$ **9.** $\dfrac{5\pi}{4}$ **11.** $\dfrac{\pi}{3}$ **13.** $\dfrac{3\pi}{4}$ **15.** $\dfrac{\pi}{2}$

17. $\dfrac{7\pi}{6}$ **19.** π **21.** $x = \arctan y$ **23.** $x = \tfrac{1}{3}\operatorname{arc\,cot} 2y$

25. $x = \sin\dfrac{y}{2}$

27.

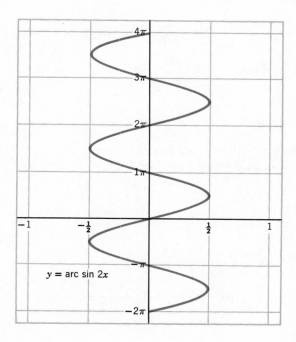

$$y = \operatorname{arc\,sin} 2x$$

29.

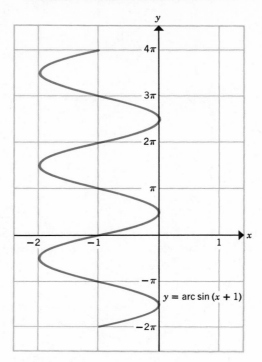

$y = \arcsin(x + 1)$

EXERCISES 10.3

1. $\dfrac{\pi}{6}$ 3. π 5. $-\dfrac{\pi}{3}$ 7. $-\dfrac{\pi}{6}$ 9. -2.0421 11. x 13. x

15. $\sqrt{3}$ 17. $\dfrac{1}{\sqrt{1-x^2}}$ 19. $\dfrac{1}{x},\ x \neq 0$

CHAPTER 10   REVIEW

1. (a) has an inverse;   (b) does not have an inverse;
   (c) does not have an inverse;   (d) does not have an inverse

3. (a) $\dfrac{2\pi}{3}$;   (b) 0;   (c) $\dfrac{\pi}{6}$;   (d) $\dfrac{\pi}{2}$;   (e) $\dfrac{\pi}{6}$;   (f) $\dfrac{3\pi}{2}$;

   (g) $\dfrac{\pi}{4}$;   (h) $\dfrac{\pi}{4}$;   (i) $\dfrac{\pi}{4}$;   (j) $\dfrac{3\pi}{2}$

5. (a) 0;   (b) $\dfrac{\pi}{4}$;   (c) $\dfrac{\pi}{6}$;   (d) $\dfrac{\pi}{3}$;   (e) $-\dfrac{\pi}{4}$

7. (a) $\frac{1}{3}\sqrt{5}$;   (b) $-\frac{3}{10}\sqrt{10}$;   (c) $2x$;   (d) $-\dfrac{\sqrt{11}}{6}$

9. $\text{Arc cos}\left(-\frac{1}{2}\right) = \frac{2}{3}\pi = 2\left(\dfrac{\pi}{3}\right) = 2\,\text{Arc cos}\,\frac{1}{2}$

11. False; the domain of the Arc sine is $\{x \mid -1 \le x \le 1\}$ while the domain of the Arc cosecant is $\{x \mid x \ge 1 \text{ or } x \le -1\}$.

13. True. If $y = \text{Arc sin } x$, then $x = \sin y$; $-\dfrac{\pi}{2} \le y \le \dfrac{\pi}{2}$. But $\sin(-y) = -\sin y$

so $-x = \sin(-y)$ or $-y = \text{Arc sin}(-x)$.

15. True only if $0 \le x \le 1$.

EXERCISES 11.1

1. -2 **3.** -6 **5.** 10 **7.** -2 **9.** 11 **11.** -14 **13.** $-5x$
15. $5x$ **17.** $2x^2 - x$ **19.** $3x^3 - 2$ **21.** $-17x^2$ **23.** 1 **25.** $\{6\}$

EXERCISES 11.2

1. 57 **3.** -44 **5.** -77 **7.** 62 **9.** -33 **11.** 89
13. $a_1b_2c_3 + a_3b_1c_2 + a_2b_3c_1 - a_3b_2c_1 - a_1b_3c_2 - a_2b_1c_3$
15. $\left\{\dfrac{-2 + \sqrt{61}}{3}, \dfrac{-2 - \sqrt{61}}{3}\right\}$ **17.** 0

EXERCISES 11.3

1. -319 **3.** 0 **5.** $2\sin x \sin 2x \sin 3x\,(\sin 2x - 2\sin x)$ **7.** $\sin x$
9. $(d - c)(d - b)(d - a)(c - b)(b - a)(c - a)$ **11.** 0
13. $2\,(a^3 + b^3)$ **15.** 0 **17.** 0

EXERCISES 11.4

1. D-6 applied to the second column and first row.

3. $\begin{vmatrix} 3a & 3b & 3c \\ 4a & 4b & 4c \\ d & e & f \end{vmatrix} = 12\begin{vmatrix} a & b & c \\ a & b & c \\ d & e & f \end{vmatrix}$ by D-6

$$= 0 \quad \text{by D-5}$$

5. $\begin{vmatrix} 5k & p & 6k \\ 5m & t & 6m \\ 5n & 5 & 6n \end{vmatrix} = 30\begin{vmatrix} k & p & k \\ m & t & m \\ n & 5 & n \end{vmatrix}$ by D-6

$$= 0 \quad \text{by D-5}$$

7. Second and third rows alike, therefore the determinant is equal to zero by D-5.

9.
$$
\begin{vmatrix} b+c & c+a & a+b \\ b_1+c_1 & c_1+a_1 & a_1+b_1 \\ b_2+c_2 & c_2+a_2 & b_2+b_2 \end{vmatrix} = \begin{vmatrix} b+c & a-b & a-c \\ b_1+c_1 & a_1-b_1 & a_1-c_1 \\ b_2+c_2 & a_2-b_2 & a_2-c_2 \end{vmatrix} \quad \text{by D-7}
$$

$$
= \begin{vmatrix} b+c & a-b & a+b \\ b_1+c_1 & a_1-b_1 & a_1+b_1 \\ b_2+c_2 & a_2-b_2 & a_2+b_2 \end{vmatrix} \quad \text{by D-7}
$$

$$
= \begin{vmatrix} b+c & 2a & a+b \\ b_1+c_1 & 2a_1 & a_1+b_1 \\ b_2+c_2 & 2a_2 & a_2+b_2 \end{vmatrix} \quad \text{by D-7}
$$

$$
= 2 \begin{vmatrix} b+c & a & a+b \\ b_1+c_1 & a_1 & a_1+b_1 \\ b_2+c_2 & a_2 & a_2+b_2 \end{vmatrix} \quad \text{by D-6}
$$

$$
= 2 \begin{vmatrix} b+c & a & b \\ b_1+c_1 & a_1 & b_1 \\ b_2+c_2 & a_2 & b_2 \end{vmatrix} \quad \text{by D-7}
$$

$$
= 2 \begin{vmatrix} c & a & b \\ c_1 & a_1 & b_1 \\ c_2 & a_2 & b_2 \end{vmatrix} \quad \text{by D-7}
$$

$$
= -2 \begin{vmatrix} a & c & b \\ a_1 & c_1 & b_1 \\ a_2 & c_2 & b_2 \end{vmatrix} \quad \text{by D-3}
$$

$$
= 2 \begin{vmatrix} a & b & c \\ a_1 & b_1 & c_1 \\ a_2 & b_2 & c_2 \end{vmatrix} \quad \text{by D-3}
$$

11. $abcd(c-a)(c-b)(b-a)(d-a)(d-b)(d-c)$

13. $(x_1-1)(x_2-1)(x_3-1)(x_4-1)$

EXERCISES 11.5

1. $\{(1, -2)\}$ **3.** $\{(5, 3)\}$ **5.** $\{(\sin \theta, \cos \theta)\}$ **7.** $\{(2, 3, 5)\}$
9. $\{(5, -4, -10)\}$ **11.** $\{(3, -3, 0)\}$ **13.** $\{(a, 1, -1)\}$
15. $\{(x, y, z, w) \mid x = 4, y = 3, z = -2, \text{ and } w = -4\}$
17. $\{(x, y, z, w) \mid x = 0, y = 0, z = -4, \text{ and } w = 1\}$
19. $\{(x, y, z, w) \mid x = 7, y = -1, z = -1, \text{ and } w = -1\}$

CHAPTER 11   REVIEW

1. 3   **3.** -44 **5.** $(c - a)(c - b)(b - a)(ab + ac + bc)$

7. $\cos x \cos y - \sin x \sin y = \cos (x + y)$
9. $\tan x \cot x - \tan y \cot y = 1 - 1 = 0$ **11.** $\{(2, 1)\}$ **13.** $\{(0, 0, 0)\}$

15. $\{-7\}$ **17.** $\begin{vmatrix} 3 & 1 & -1 \\ 2 & 1 & 0 \\ 6 & -1 & 0 \end{vmatrix}$

19. (a) $\begin{vmatrix} a_{12} & a_{13} & a_{14} \\ a_{32} & a_{33} & a_{34} \\ a_{42} & a_{43} & a_{44} \end{vmatrix} ; \begin{vmatrix} a_{11} & a_{13} & a_{14} \\ a_{31} & a_{33} & a_{34} \\ a_{41} & a_{43} & a_{44} \end{vmatrix} ; \begin{vmatrix} a_{11} & a_{12} & a_{14} \\ a_{31} & a_{32} & a_{34} \\ a_{41} & a_{42} & a_{44} \end{vmatrix} ; \begin{vmatrix} a_{11} & a_{12} & a_{13} \\ a_{31} & a_{32} & a_{33} \\ a_{41} & a_{42} & a_{44} \end{vmatrix} ;$

(b) $\begin{vmatrix} a_{22} & a_{23} & a_{24} \\ a_{32} & a_{33} & a_{34} \\ a_{42} & a_{43} & a_{44} \end{vmatrix} ; (-1) \begin{vmatrix} a_{21} & a_{23} & a_{24} \\ a_{31} & a_{33} & a_{34} \\ a_{41} & a_{43} & a_{44} \end{vmatrix} ; \begin{vmatrix} a_{21} & a_{22} & a_{24} \\ a_{31} & a_{32} & a_{34} \\ a_{41} & a_{42} & a_{44} \end{vmatrix} ;$

$(-1) \begin{vmatrix} a_{21} & a_{22} & a_{23} \\ a_{31} & a_{32} & a_{33} \\ a_{41} & a_{42} & a_{43} \end{vmatrix}$

EXERCISES 12.1

1. $(7, 9)$ **3.** $(10, -8)$ **5.** $(2, 1)$ **7.** $(5, -1)$ **9.** $(3, 29)$
11. $(10, 0)$ **13.** $\left(\frac{74}{73}, \frac{27}{73}\right)$ **15.** $\left(-\frac{1}{3}, -2\right)$ **17.** $(3, -1)$ **19.** $(-1, -1)$
21. $(1, -5)$ **23.** $(-5, 1)$ **25.** $(3, -5)$ **27.** $(2, 7)$ **29.** $\left(0, \frac{1}{3}\right)$
31. $\left(-\frac{1}{5}, 0\right)$ **33.** $\left(-\frac{8}{89}, \frac{5}{89}\right)$

EXERCISES 12.2

1. $3 - 4i$ **3.** $5 + 7i$ **5.** $2\sqrt{2} - 3\sqrt{3}\,i$ **7.** $-4 + \sqrt{5}\,i$ **9.** $-4i$

11. $8 + 8i$ **13.** $-9 + i$ **15.** $-1 + 12i$ **17.** $-9 + 8i$ **19.** 13

21. $2 - 26i$ **23.** 5 **25.** $\frac{3}{5} + \frac{4}{5}i$ **27.** $\frac{8}{11} - \frac{1}{11}\sqrt{2}\, i$

29. $\dfrac{-8}{25} - \dfrac{31}{25}\sqrt{2}\, i$ **31.** $\{-2i, 2i\}$ **33.** $\{-2\sqrt{5}\, i, 2\sqrt{5}\, i\}$

35. $\{\sqrt{56}\, i, -\sqrt{56}\, i\}$ or $\{2\sqrt{14}\, i, -2\sqrt{14}\, i\}$ **37.** $\{-4\sqrt{3}\, i, 4\sqrt{3}\, i\}$

EXERCISES 12.3

1. $3i$ **3.** $-4i$ **5.** $-3\sqrt{3}\, i$ **7.** -6 **9.** 7 **11.** $\frac{1}{3} - \frac{1}{6}\sqrt{2}\, i$

13. $-\dfrac{5}{6} + \dfrac{\sqrt{5}}{6}\, i$ **15.** $\dfrac{9}{14} + \dfrac{\sqrt{3}}{14}\, i$ **17.** $-10 + 24i$ **19.** $41 + 8\sqrt{5}\, i$

21. $40 + 29\sqrt{10}\, i$ **23.** $\dfrac{6 - \sqrt{6}}{11} + \dfrac{3\sqrt{3} + 2\sqrt{2}}{11}i$ **25.** $\dfrac{11}{7} + \dfrac{9\sqrt{3}}{7}i$

27. $\dfrac{11}{7} - \dfrac{2\sqrt{10}}{7}i$ **29.** $\dfrac{\sqrt{7}}{46} + \dfrac{3\sqrt{56}}{46}$

EXERCISES 12.4

1. $\{1 + \sqrt{39}\, i, 1 - \sqrt{39}\, i\}$ **3.** $\{-3 + i, -3 - i\}$

5. $\{-4 + 3i, -4 - 3i\}$

7. $\left\{\dfrac{1 + 2i}{3}, \dfrac{1 - 2i}{3}\right\}$ **9.** $\left\{\dfrac{-3 + \sqrt{151}\, i}{8}, \dfrac{-3 - \sqrt{151}\, i}{8}\right\}$

11. $\left\{\dfrac{1 + \sqrt{3}\, i}{2}, \dfrac{1 - \sqrt{3}\, i}{2}\right\}$ **13.** $\left\{\dfrac{3 + \sqrt{39}\, i}{4}, \dfrac{3 - \sqrt{39}\, i}{4}\right\}$

15. $\left\{\dfrac{-5 + \sqrt{59}\, i}{6}, \dfrac{-5 - \sqrt{59}\, i}{6}\right\}$ **17.** $\left\{\dfrac{-1 + \sqrt{15}\, i}{8}, \dfrac{-1 - \sqrt{51}\, i}{8}\right\}$

19. $\left\{\dfrac{-5 + \sqrt{71}\, i}{4}, \dfrac{-5 - \sqrt{71}\, i}{4}\right\}$ **21.** $\{1 + \frac{7}{8}\sqrt{2}\, i, 1 - \frac{7}{8}\sqrt{2}\, i\}$

23. $\left\{\dfrac{2 + i}{3}, \dfrac{2 - i}{3}\right\}$ **25.** $\{-1 + 2\sqrt{2}\, i, -1 - 2\sqrt{2}\, i\}$

27. $\left\{\dfrac{3 + \sqrt{131}\, i}{10}, \dfrac{3 - \sqrt{131}\, i}{10}\right\}$ **29.** $\{1 + 7i, 1 - 7i\}$

31. $\{1 + 3i, 1 - 3i\}$ **33.** $\{-\sqrt{2} - \sqrt{7}\, i, -\sqrt{2} + \sqrt{7}\, i\}$

35. $\left\{\dfrac{17 + \sqrt{791}\, i}{12}, \dfrac{17 - \sqrt{791}\, i}{12}\right\}$ **37.** $\left\{\dfrac{-3 + 3\sqrt{3}\, i}{2}, \dfrac{-3 - 3\sqrt{3}\, i}{2}\right\}$

39. $\left\{\dfrac{-3\sqrt{3} + \sqrt{69}\, i}{12}, \dfrac{-3\sqrt{3} - \sqrt{69}\, i}{12}\right\}$

EXERCISES 12.5

1. $\sqrt{13}$ **3.** $\sqrt{5}$ **5.** 5 **7.** $\sqrt{22}$ **9.** $\sqrt{19}$ **11.** $225°$ **13.** $0°$

15. $240°$ **17.** $300°$ **19.** $270°$

21.

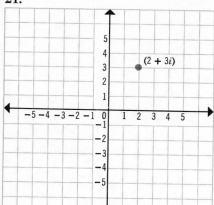

23.

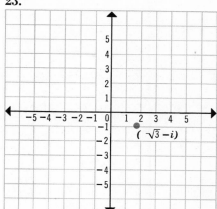

25.

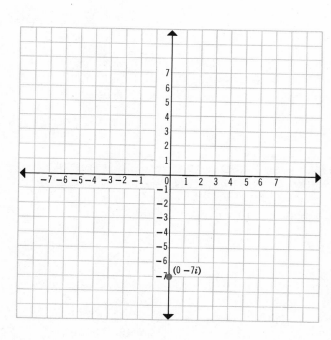

27.

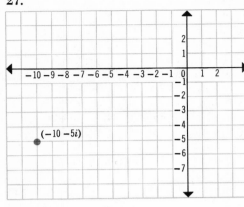

29.

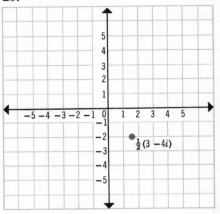

31. less than

33. $|z| = \sqrt{a^2 + b^2}$
$-z = -a - bi$
$|-z| = \sqrt{(-a)^2 + (-b)^2} = \sqrt{a^2 + b^2} = |z|$

35. $|z|^2 = (\sqrt{a^2 + b^2})^2 = a^2 + b^2$
$|z\,\overline{z}| = |(a + bi)(a - bi)| = |a^2 + b^2| = a^2 + b^2 = |z|^2$

EXERCISES 12.6

1. $\sqrt{2}\,(\cos 45° + i \sin 45°)$ **3.** $3\,(\cos 0° + i \sin 0°)$
5. $\sqrt{2}\,(\cos 315° + i \sin 315°)$ **7.** $2\,(\cos 30° + i \sin 30°)$
9. $7\sqrt{2}\,(\cos 315° + i \sin 315°)$ **11.** $2\,(\cos 270° + i \sin 270°)$
13. $5\,(\cos 180° + i \sin 180°)$ **15.** $2\,(\cos 60° + i \sin 60°)$
17. $\cos 0° + i \sin 0°$ **19.** $\cos 300° + i \sin 300°$

EXERCISES 12.7

1. $27\,(\cos 90° + i \sin 90°)$
3. $2^{10}\,(\cos 270° + i \sin 270°)$ or $1024\,(\cos 1350° + i \sin 1350°)$
5. $\cos 200° + i \sin 200°$ **7.** $\cos 300° + i \sin 300°$
9. $64\,(\cos 90° + i \sin 90°)$ or $64\,(\cos 810° + i \sin 810°)$
11. $32\,(\cos 135° + i \sin 135°)$ **13.** $16\,(\cos 120° + i \sin 120°)$
15. $\cos 0° + i \sin 0°$ **17.** $\frac{1}{16}\,(\cos 180° + i \sin 180°)$
19. $4\sqrt{2}\,(\cos 45° + i \sin 45°)$

EXERCISES 12.8

1. $2, -1 + \sqrt{3}\,i, -1 - \sqrt{3}\,i$
3. $\cos 48° + i \sin 48°, \cos 120° + i \sin 120°, \cos 192° + i \sin 192°,$
$\cos 264° + i \sin 264°, \cos 336° + i \sin 336°$

5. $-1, \frac{1}{2} + \frac{1}{2}\sqrt{3}\,i, \frac{1}{2} - \frac{1}{2}\sqrt{3}\,i$

7. $\sqrt{2} + \sqrt{2}\,i, \sqrt{2} - \sqrt{2}\,i, -\sqrt{2} - \sqrt{2}\,i, -\sqrt{2} + \sqrt{2}\,i$

9. $-3, \frac{3}{2} + \frac{3}{2}\sqrt{3}\,i, \frac{3}{2} - \frac{3}{2}\sqrt{3}\,i$

11. $\sqrt[4]{12}\,(\cos 165° + i\sin 165°), \sqrt[4]{12}\,(\cos 345° + i\sin 345°)$

13. $2^{1/5}\,(\cos 48° + i\sin 48°); 2^{1/5}\,(\cos 120° + i\sin 120°);$
$2^{1/5}\,(\cos 192° + i\sin 192°); 2^{1/5}\,(\cos 264° + i\sin 264°);$
$2^{1/5}\,(\cos 336° + i\sin 336°)$
$\cos 192° + i\sin 192°; \cos 264° + i\sin 264°; \cos 336° + i\sin 336°$

15. $2\,(\cos 15° + i\sin 15°); 2\,(\cos 135° + i\sin 135°); 2\,(\cos 225° + i\sin 225°)$

CHAPTER 12   REVIEW

1. (a) $(-1, -7);$ (b) $(-5, -3);$ (c) $(20, 60);$ (d) $\left(-\frac{2}{25}, \frac{11}{25}\right)$

3. (a) $3i;$ (b) $2\sqrt{3}\,i;$ (c) $-5i;$ (d) $3\sqrt{3}\,i;$ (e) $-9i;$ (f) $-8i$

5. (a) $r = 2, \theta = 135°;$ (c) $r = 2, \theta = 90°;$

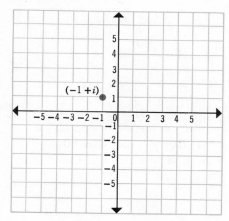

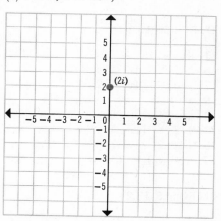

(b) $r = 3\sqrt{2}, \theta = 225°;$ (d) $r = 2\sqrt{2}, \theta = 45°$

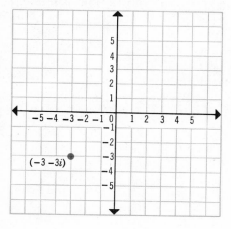

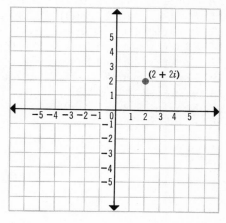

7. (a) $6 (\cos 70° + i \sin 70°)$; (b) $\frac{1}{8}(\cos 105° + i \sin 105°)$;
(c) $4 (\cos 120° + i \sin 120°)$; (d) $5 (\cos 280° + i \sin 280°)$

9. (a) $8 (\cos 90° + i \sin 90°) = 8i$;
(b) $\frac{1}{2}[\cos (-25°) + i \sin (-25°)]$ or $\frac{1}{2}[\cos 335° + i \sin 335°]$

11. $-2, 1 + \sqrt{3}\, i, 1 - \sqrt{3}$

13. $\sqrt[4]{2} (\cos 157° \, 30' + i \sin 157° \, 30')$;
$\sqrt[4]{2} (\cos 337° \, 30' + i \sin 337° \, 30')$.

15. (a) $3 + 4i$; (b) $-1 + 5i$; (c) $1 + 2i$; (d) $\frac{2}{5} + \frac{3}{5}i$; (e) $-2 - 3i$;
(f) $\frac{1}{2} - \frac{1}{2}i$; (g) $\sqrt{2} (\cos 45° + i \sin 45°)$; (h) $1 - i$;
(i) $4 (\cos 180° + i \sin 180°) = -4$;
(j) $\sqrt{2} (\cos 45° + i \sin 45°)$; $\sqrt{2} (\cos 225 + i \sin 225°)$;

(k) $\sqrt[3]{5}$; $-\dfrac{\sqrt[3]{5}}{2} + \dfrac{\sqrt[3]{5} \cdot \sqrt{3}}{2}\, i$; $-\dfrac{\sqrt[3]{5}}{2} - \dfrac{\sqrt[3]{5}\sqrt{3}}{2}\, i$

17. (a) no; (b) yes   **19.** (a) false; (b) true; (c) true; (d) true

EXERCISES 13.1

1. $4x^2 + 13x + 3$; $R = 12$ **3.** $x^3 + 7x - 43$; $R = 224$

5. $-7x^3 - 4x^2 - 4x - 3$; $R = -1$

7. $-4x^4 + 4x^3 + 22x^2 - 22x + 37$; $R = -49$

9. $x^8 + x^7 + x^6 + x^5 + x^4 + x^3 + x^2 + x + 1$; $R = 2$

11. $x^5 + 5x^4 + 20x^3 + 80x^2 + 320x + 1279$; $R = 5118$

13. $11x^7 - 11x^6 + 11x^5 - 13x^4 + 13x^3 - 13x^2 + 16x - 16$; $R = -7$

15. $14x^6 + 25x^5 + 67x^4 + 123x^3 + 255x^2 + 510x + 1020$; $R = 2041$

17. $6x^5 - 9x^2$; $R = -6$ **19.** $6x^3 + 8x + 4$; $R = 10$

EXERCISES 13.2

1. $P(1) = -3$; $P(2) = 8$; $P(3) = 59$

3. $P(-2) = 206$; $P(-3) = 2150$; $P(1) = -10$

5. $P(-1) = -67$; $P(-2) = -183$; $P(2) = 29$

7. $P(0) = 0$; $P(-1) = 0$; $P(1) = 0$ **9.** $P(1) = 13$; $P(2) = 125$; $P(3) = 761$

11. $P(-1) = 28$; $P(2) = 691$; $P(-3) = 7146$

13. $P(-\frac{3}{2}) = 350$; $P(\frac{1}{2}) = \frac{195}{4}$; $P(-1) = 138$

15. Factor: $x - 1$; not factors: $x + 1$; $x - 2$

17. Factor: $x - 1$; not factors: $x + 1$; $x - 3$

19. None are factors   **21.** None are factors

23. Factors: $x + 1$; $x - 1$; not a factor: x **25.** None are factors

27. None are factors   **29.** Yes

EXERCISES 13.3

1.

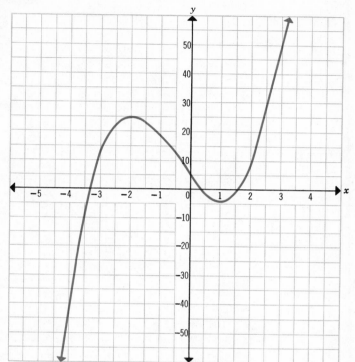

3.

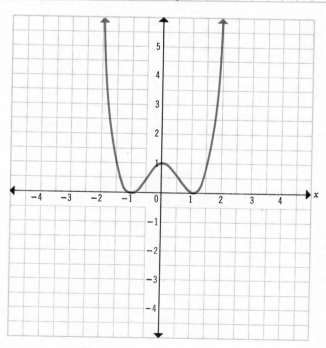

5.

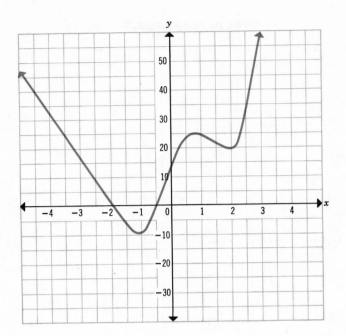

7.

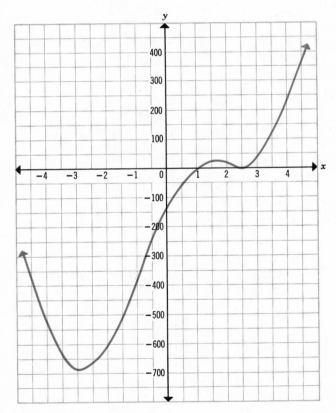

9.

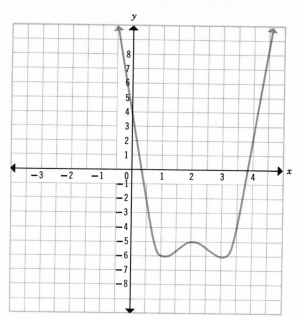

11.

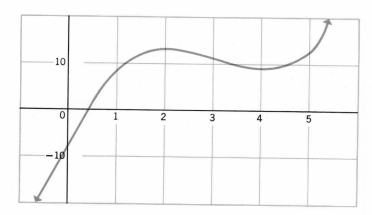

13.

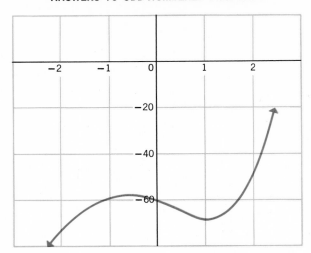

15.

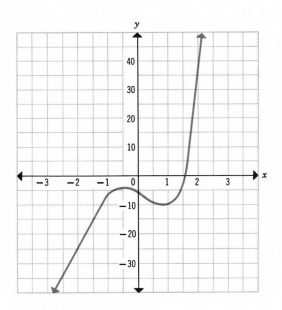

EXERCISES 13.4

1. $-3i$ **3.** $-i$; 1; -1 **5.** $-1 - \sqrt{3}\,i$; 5 **7.** $2 - 3i$; -1; -1

9. $\dfrac{1}{2} - \dfrac{\sqrt{3}}{2}\,i$; $-\dfrac{1}{2} + \dfrac{\sqrt{3}}{2}\,i$; 1; -1 **11.** $2 - i$ **13.** $-2i$; $1 - 3i$

EXERCISES 13.5

1. -6; -1; -1 **3.** 1; 3; 6 **5.** $\tfrac{1}{3}$; $\tfrac{1}{2}$; 1 **7.** $-\tfrac{1}{2}$; $\tfrac{1}{4}$; $\tfrac{1}{3}$

9. $1; \frac{1}{2}; -\frac{2}{3}; \quad -\frac{1}{2} + \frac{\sqrt{5}}{2}i; \quad -\frac{1}{2} - \frac{\sqrt{5}}{2}i$ **11.** $-\frac{3}{2}; -\sqrt{7}; \sqrt{7}$

13. $-1; 6; \frac{1}{2}$ **15.** $\frac{1}{3}; -3; 2\sqrt{3}; -2\sqrt{3}$ **17.** $-1; 1; i; -i$ **19.** $0; 1$

21. If $\frac{p}{q}$ is a rational root of $P(x) = a_0 x^n + a_1 x^{n-1} + \cdots + a_n$, then q must

divide a_0 by Theorem 13.6. If $a_0 = 1$, then $q = 1$ or $q = -1$. Any

rational number of the form $\frac{+p}{\pm 1}$ is an integer. But p must divide a_n by

Theorem 13.6. Hence, any root $\pm p$ must divide a_n.

EXERCISES 13.6

1. 1.246 **3.** 2.095 **5.** 3.069 **7.** -3.050 **9.** 1.540 **11.** 0.347
13. 4.064 **15.** 0.315

CHAPTER 13 REVIEW

1. $2x^4 - 4x^3 + 5x^2 - 10x + 22$; remainder: -43
3. (a) factor; (b) not a factor; (c) factor; (d) not a factor
5. $-3; -2; 1; 2$ **7.** $1 - i$ **9.** -3 **11.** -2.46

13.

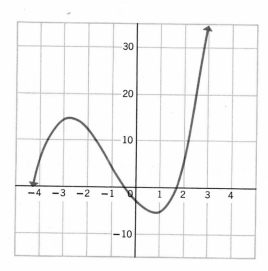

15. $\{-6, -5, 7\}$
17. $\pm 1; \pm 2; \pm 4; \pm 8; \pm 16; \pm\frac{1}{2}; \pm\frac{1}{3}; \pm\frac{1}{4}; \pm\frac{1}{6}; \pm\frac{1}{9}; \pm\frac{1}{12}; \pm\frac{1}{18}; \pm\frac{1}{27}; \pm\frac{1}{36};$
$\pm\frac{1}{54}; \pm\frac{1}{108}; \pm\frac{2}{3}; \pm\frac{2}{9}; \pm\frac{2}{27}; \pm\frac{4}{3}; \pm\frac{4}{9}; \pm\frac{8}{3}; \pm\frac{8}{9}; \pm\frac{8}{27}; \pm\frac{16}{3}; \pm\frac{16}{9}, \pm\frac{16}{27}; \pm\frac{4}{27}$
19. No; there must be one real root, since complex roots appear in conjugate pairs.

EXERCISES 14.1

1. (1) $1^3 = \dfrac{1^2 \cdot 2^2}{4}$

(2) $(1^3 + 2^3 + 3^3 + \cdots + k^3) + (k + 1)^3 = \dfrac{k^2 (k + 1)^2}{4} + (k + 1)^3$

$$= (k + 1)^2 \left[\dfrac{k^2}{4} + (k + 1) \right]$$

$$= \dfrac{(k + 1)^2 (k^2 + 4k + 4)}{4}$$

$$= \dfrac{(k + 1)^2 (k + 2)^2}{4}$$

3. (1) $1 \cdot 2 = \dfrac{1 \cdot 2 \cdot 3}{3}$

(2) $[(1)(2) + (2)(3) + (3)(4) + \cdots + k(k + 1)] + (k + 1)(k + 2)$

$$= \dfrac{k (k + 1)(k + 2)}{3} + (k + 1)(k + 2)$$

$$= (k + 1)(k + 2)\left(\dfrac{k}{3} + 1 \right)$$

$$= \dfrac{(k + 1)(k + 2)(k + 3)}{3}$$

5. (1) $1^2 = \dfrac{1 \cdot (4 - 1)}{3}$

(2) $[1^2 + 3^2 + 5^2 + \cdots + (2k - 1)^2] + (2k + 1)^2$

$$= \dfrac{k (4k^2 - 1)}{3} + (2k + 1)^2$$

$$= \dfrac{4k^3 - k + 12k^2 + 12k + 3}{3}$$

$$= \dfrac{4k^3 + 12k^2 + 11k + 3}{3}$$

$$\dfrac{(k + 1)[4 (k + 1)^2 - 1]}{3} = \dfrac{(k + 1)(4k^2 + 8k + 4 - 1)}{3}$$

$$= \dfrac{(k + 1)(4k^2 + 8k + 3)}{3}$$

$$= \dfrac{4k^3 + 12k^2 + 11k + 3}{3}$$

7. (1) $1 \cdot 3 = \frac{1}{6} \cdot 1 \cdot 2 \cdot 9$

(2) $[1 \cdot 3 + 2 \cdot 4 + 3 \cdot 5 + \cdots + (k)(k + 2)] + (k + 1)(k + 3)$

$\qquad = \frac{1}{6}k (k + 1)(2k + 7) + (k + 1)(k + 3)$

$\qquad = (k + 1)[\frac{1}{6}k (2k + 7) + (k + 3)]$

$$= \frac{(k+1)}{6} [2k^2 + 7k + 6k + 18]$$

$$= \frac{(k+1)}{6} (2k^2 + 13k + 18)$$

$$= \tfrac{1}{6} (k+1)(k+2)(2k+9)$$
$$= \tfrac{1}{6} (k+1)(k+2)[2(k+1)+7]$$

9. (1) $\dfrac{1}{1 \cdot 2} = \dfrac{1}{1+1}$

(2) $\left[\dfrac{1}{(1)(2)} + \dfrac{1}{(2)(3)} + \cdots + \dfrac{1}{k(k+1)} \right] + \dfrac{1}{(k+1)(k+2)}$

$$= \frac{k}{k+1} + \frac{1}{(k+1)(k+2)}$$

$$= \frac{k(k+2)+1}{(k+1)(k+2)}$$

$$= \frac{k^2 + 2k + 1}{(k+1)(k+2)}$$

$$= \frac{(k+1)^2}{(k+1)(k+2)}$$

$$= \frac{k+1}{k+2}$$

11. (1) $1 \cdot 6 = (3-2)(3+3) = 3(1+2-1)$

(2) $[(1)(6) + (4)(9) + \cdots + (3k-2)(3k+3)] + (3k+1)(3k+6)$

$$= 3k(k^2 + 2k - 1) + (3k+1)(3k+6)$$
$$= 3[k(k^2 + 2k - 1) + (3k+1)(k+2)]$$
$$= 3(k^3 + 2k^2 - k + 3k^2 + 7k + 2)$$
$$= 3(k^3 + 5k^2 + 6k + 2)$$

But

$$3(k+1)[(k+1)^2 + 2(k+1) - 1]$$
$$= 3(k+1)(k^2 + 2k + 1 + 2k + 2 - 1)$$
$$= 3(k+1)(k^2 + 4k + 2)$$
$$= 3(k^3 + 5k^2 + 6k + 2)$$

13. (1) $\dfrac{1}{(1)(3)} = \dfrac{1}{2+1}$

(2) $\left[\dfrac{1}{(1)(3)} + \dfrac{1}{(3)(5)} + \cdots + \dfrac{1}{(2k-1)(2k+1)} \right] + \dfrac{1}{(2k+1)(2k+3)}$

$$= \frac{k}{2k+1} + \frac{1}{(2k+1)(2k+3)}$$

$$= \frac{k(2k+3)+1}{(2k+1)(2k+3)}$$

$$= \frac{2k^2 + 3k + 1}{(2k+1)(2k+3)}$$

$$= \frac{(2k + 1)(k + 1)}{(2k + 1)(2k + 3)}$$

$$= \frac{k + 1}{2k + 3} = \frac{k + 1}{2(k + 1) + 1}$$

15. (1) $3^1 = \frac{3}{2}(3 - 1)$

(2) $[3 + 3^2 + 3^3 + \cdots + 3^k] + 3^{k+1} = \frac{3}{2}(3^k - 1) + 3^{k+1}$

$$= \frac{1}{2} \cdot 3^{k+1} - \frac{3}{2} + 3^{k+1}$$

$$= \frac{3}{2} \cdot 3^{k+1} - \frac{3}{2}$$

$$= \frac{3}{2} \cdot (3^{k+1} - 1)$$

17. (1) $\frac{1}{2} = 1 - \frac{1}{2}$

(2) $\left[\frac{1}{2} + \frac{1}{2^2} + \cdots + \frac{1}{2^k}\right] + \frac{1}{2^{k+1}} = 1 - \frac{1}{2^k} + \frac{1}{2^{k+1}}$

$$= 1 - \frac{2}{2^{k+1}} + \frac{1}{2^{k+1}}$$

$$= 1 - \frac{1}{2^{k+1}}$$

19. (1) $1 = 1 + (1 - 1)(2^1)$

(2) $[1 + (2)(2) + (3)(2^2) + \cdots + (k)(2^{k-1})] + (k + 1)(2^k)$

$$= 1 + (k - 1)(2^k) + (k + 1)(2^k)$$

$$= 1 + k \cdot 2^k - 2^k + k \cdot 2^k + 2^k$$

$$= 1 + 2k \cdot 2^k$$

$$= 1 + k \cdot 2^{k+1}$$

EXERCISES 14.2

1. $1; 3; 6; 10$ **3.** $\frac{1}{3}; \frac{1}{2}; \frac{3}{5}; \frac{2}{3}$ **5.** $-2; 4; -8; 16$ **7.** $\frac{1}{2}; \frac{1}{4}; \frac{1}{8}; \frac{1}{16}$

9. $\frac{\sqrt[3]{2}}{2}; \frac{\sqrt[3]{4}}{6}; \frac{1}{12}; \frac{\sqrt[3]{2}}{60}$ **11.** 100 **13.** 46 **15.** 4.5 **17.** -39

19. 16 **21.** $17 - 5n$ **23.** $1 + 0.001n$ **25.** $z - 3n + 3$

27. $\frac{2}{5^7}$ **29.** $-\frac{2}{3^9}$ **31.** $\sin^7 \theta$ **33.** $(0.4)^4$ **35.** $\sec^{30} \theta$ **37.** 42

39. $\frac{9}{4}$ **41.** semiannually $(\$100)(1 + \frac{1}{2})^2 = \$225$

quarterly $(\$100)(1 + \frac{1}{4})^4 = \$244.33$

monthly $(\$100)(1 + \frac{1}{12})^{12} = \$260.30$

weekly $(\$100)(1 + \frac{1}{52})^{52} = \$273.00$

daily $(\$100)(1 + \frac{1}{365})^{365} = \$273.00$

hourly $(\$100)(1 + \frac{1}{8760})^{8760} = \$273.00$

every minute $(\$100)(1 + \frac{1}{525600})^{525600} = \$273.00$

every second $(\$100)(1 + \frac{1}{31536000})^{31536000} = \$273.00$

43. $14{,}912$

EXERCISES 14.3

1. $[3(1) + 1] + [3(2) + 1] + [3(3) + 1] + [3(4) + 1] + [3(5) + 1]$
$$= 4 + 7 + 10 + 13 + 16$$

3. $3 + 9 + 27 + 81 + 243$ 5. $\dfrac{1}{2} + \dfrac{2}{3} + \dfrac{3}{4} + \dfrac{4}{5} + \dfrac{5}{6} + \cdots + \dfrac{n}{n + 1}$

7. 280 9. $80\left(1 - \dfrac{1}{2^{15}}\right)$ 11. $\dfrac{6}{5}\left(1 - \dfrac{1}{5^{12}}\right)$ 13. $\sin^2\theta \cdot \dfrac{1 - \csc^{16}\theta}{1 - \csc^2\theta}$

15. 78 17. 246 19. 15 21. 45

23. $a = 1; d = 1; S_n = \dfrac{n}{2}[2 + (n - 1)(1)] = \dfrac{n}{2}(n + 1)$

25. $a = 1; r = \frac{2}{3}; \ S_n = \dfrac{1 - (\frac{2}{3})^n}{1 - \frac{2}{3}} = 3\left[1 - \left(\dfrac{2}{3}\right)^n\right]$

27. $17\frac{80}{81}$ feet.   29. 324 31. $\$15,400$ 33. $\$4.70$

EXERCISES 14.4

1. $\frac{18}{5}$ 3. $\frac{32}{9}$ 5. $\frac{5}{6}$ 7. 1 9. $\frac{1}{9}$ 11. $\frac{4}{3}$ 13. $\frac{293}{45}$ 15. $\frac{2779}{333}$
17. $\frac{53}{99}$ 19. $\frac{410}{333}$ 21. 0 23. 0 25. 1

EXERCISES 14.5

1. $x^8 + 8x^7 + 28x^6 + 56x^5 + 70x^4 + 56x^3 + 28x^2 + 8x + 1$
3. $x^5 + 10x^4 + 40x^3 + 80x^2 + 80x + 32$
5. $8a^6 - 36a^4b^3 + 54a^2b^6 - 27b^9$

7. $\dfrac{1}{x^7} + \dfrac{7}{x^5} + \dfrac{21}{x^3} + \dfrac{35}{x} + 35x + 21x^3 + 7x^5 + x^7$

9. $78{,}125x^7 + 109{,}375x^6y + 65{,}625x^5y^2 + 21{,}875x^4y^3 + 4375x^3y^4$
$\quad + 525x^2y^5 + 35xy^6 + y^7$
11. $a^8 - 24a^7b + 252a^6b^2 - 1512a^5b^3 + 5670a^6b^4$
13. $3^{15}x^{15} - 15 \cdot 3^{14}x^{14}y + 105 \cdot 3^{13}x^{13}y^2 - 455 \cdot 3^{13}x^{12}y^3 + 1365 \cdot 3^{11}x^{11}y^4$
15. $x^{18} - 36x^{16} + 612x^{14} - 6528x^{12} + 48{,}960x^{10}$
17. $x^{14} - 14x^{11} + 91x^8 - 364x^5 + 1001x^2$ 19. $480{,}700x^{18}y^7$
21. $320{,}320x^3$ 23. $792 \cdot 2^{17} \cdot a^7 \cdot b^{10}$ 25. $14{,}307{,}150 \cdot 2^{21} \cdot 3^9 \cdot x^{21} \cdot y^9$
27. $6048x^5y^2$

EXERCISES 14.6

1. $1 + \frac{1}{2}x - \frac{1}{8}x^2 + \frac{1}{16}x^3 - \frac{5}{128}x^4$ 3. $1 + \frac{1}{3}x - \frac{1}{9}x^2 + \frac{5}{81}x^3 - \frac{10}{243}x^4$
5. $1 - \frac{1}{2}x + \frac{3}{8}x^2 - \frac{5}{16}x^3 + \frac{35}{128}x^4$ 7. $1 + 2x + 3x^2 + 4x^3 + 5x^4$
9. 7.0711 11. 1.005 13. 2.924 15. 0.8375

17. $\quad \dfrac{1}{1 - x^2} = 1 + x^2 + x^4 + x^6 + \cdots$

$$(1 - x^2)^{-1} = 1 + (-1)(-x^2) + \frac{(-1)(-2)}{2 \cdot 1}(-x^2)^2$$

$$+ \frac{(-1)(-2)(-3)}{3 \cdot 2 \cdot 1}(-x^2)^3 + \cdots$$

$$= 1 + x^2 + x^4 + x^6 + \cdots$$

CHAPTER 14   REVIEW

1. (1) $\dfrac{1}{(1)(2)} = \dfrac{1}{1 + 1}$

(2) $\left[\dfrac{1}{(1)(2)} + \dfrac{1}{(2)(3)} + \cdots + \dfrac{1}{k(k + 1)} \right] + \dfrac{1}{(k + 1)(k + 2)}$

$$= \frac{k}{k + 1} + \frac{1}{(k + 1)(k + 2)}$$

$$= \frac{1}{k + 1}\left(k + \frac{1}{k + 2} \right)$$

$$= \frac{k(k + 2) + 1}{(k + 1)(k + 2)}$$

$$= \frac{k^2 + 2k + 1}{(k + 1)(k + 2)}$$

$$= \frac{(k + 1)^2}{(k + 1)(k + 2)}$$

$$= \frac{k + 1}{k + 2}$$

3. (a) $-3, 9, -27, 81$; (b) $3, 6, 10, 15$; (c) $\frac{1}{4}, \frac{1}{16}, \frac{1}{64}, \frac{1}{256}$; (d) $\frac{1}{4}, -\frac{1}{8}, \frac{1}{16}, -\frac{1}{32}$

5. (a) $\frac{3}{128}$; (b) $4 \cdot 3^{14}$; (c) $\dfrac{1}{2^{20}}$; (d) $\csc^{10} \theta$ **7.** $\frac{1}{256}$

9. (a) 315; (b) 255; (c) $\frac{61}{27}$; (d) $-\frac{51}{128}$ **11.** (a) $\frac{5}{9}$; (b) $\frac{8}{9}$; (c) $\frac{13}{99}$; (d) $\frac{706}{495}$

13. (a) $x^{18} - 27x^{16} + 324x^{14}$; (b) $\cos^{12} x - 12 \cos^{11} x + 66 \cos^{10} x$

15. (a) $1 - \frac{1}{2}x - \frac{1}{8}x^2 - \frac{1}{16}x^3$; (b) $1 + 3x + 6x^2 + 10x^3$;

(c) $1 - 3x + 6x^2 - 10x^3$; (d) $1 - \frac{1}{3}x - \frac{1}{9}x^2 - \frac{5}{81}x^3$

17. 14 days or 2 weeks   **19.** 8190

EXERCISES 15.1

1. 24   **3.** 360   **5.** 24   **7.** 20   **9.** 240   **11.** 720   **13.** 60

15. 36

EXERCISES 15.2

1. (a) 6,720; (b) 332,640; (c) 1,816,214,400; (d) 720;

(e) $19 \cdot 18 \cdot 17 \cdot 16 \cdot 15 \cdot 14 \cdot 13$;

(f) $25 \cdot 24 \cdot 23 \cdot 22 \cdot 21 \cdot 20 \cdot 19 \cdot 18 \cdot 17 \cdot 16 \cdot 15$ **3.** 120   **5.** 3024

7. $10! = 3,628,800$ **9.** $2^3 = 8$ **11.** 10,080   **13.** 50,400   **15.** 34

17. 6720   **19.** 60,480

EXERCISES 15.3

1. (a) 35;   (b) 84;   (c) 220;   (d) 6435;   (e) 167,960;   (f) 31,824   **3.** 5
5. 8   **7.** 4,845   **9.** 3003   **11.** 45   **13.** (a) 21;   (b) 39;   (c) 126
15. 24,310   **17.** 2416

EXERCISES 15.4

1. $\{(H, H), (H, T), (T, H), (T, T)\}$
3. $\{AS, 2S, 3S, \ldots, KS, AH, 2H, 3H, \ldots, KH, AD, 2D, 3D, \ldots, KD, AC, 2C,$
$3C, \ldots, KC\}$
5. $\{ab, ac, ad, ae, af, ag, bc, bd, be, bf, bg, cd, ce, cf, cg, de, df, dg, ef, eg, fg\}$
7. 2300   **9.** (a) $\{(2, 6), (3, 5), (4, 4), (5, 3), (6, 2)\}$;
   (b) $\{(1, 6), (2, 5), (3, 4), (4, 3), (5, 2), (1, 6)\}$;
   (c) $\{(1, 5), (1, 6), (2, 5), (2, 6), (3, 5), (3, 6), (4, 5), (4, 6), (5, 1), (5, 2), (5, 3),$
     $(5, 4), (5, 5), (5, 6), (6, 1), (6, 2), (6, 3), (6, 4), (6, 5), (6,6)\}$;
   (d) $\{(1, 6), (2, 6), (3, 6), (4, 6), (5, 6), (6, 6), (6, 1), (6, 2), (6, 3), (6, 4), (6, 5)\}$
11. (a) $\{(T, T, T)\}$;   (b) $\{(T, H, H), (H, T, H), (H, H, T)\}$;   (c) $\{(T, T, H),$
     $(T, H, T), (H, T, T), (H, H, T), (H, T, H), (T, H, H), (H, H, H)\}$;
   (d) $\{(T, H, H), (H, T, H), (H, H, T), (T, T, H), (T, H, T), (H, T, T), (T, T, T)\}$
13. (a) $\{(H, H, H, H)\}$;   (b) $\{(H, H, T, T), (H, T, H, T), (T, H, H, T),$
     $(H, T, T, H), (T, H, T, H), (T, T, H, H)\}$;   (c) $\{(H, H, T, T), (H, T, H, T),$
     $(T, H, H, T), (H, T, T, H), (T, H, T, H), (T, T, H, H), (T, T, T, T),$
     $(H, H, H, H)\}$;   (d) $\{(T, H, H, H), (T, T, H, H), (T, H, T, H), (T, H, H, T),$
     $(H, T, H, T), (H, H, T, T), (H, T, T, H), (T, T, T, H), (T, T, H, T),$
     $(T, H, T, T), (H, T, T, T), (T, T, T, T), (H, H, H, T), (H, H, T, H),$
     $(H, H, T, T), (H, T, H, H)\}$

EXERCISES 15.5

1. (a) $\frac{1}{4}$;   (b) $\frac{1}{2}$;   (c) $\frac{1}{13}$;   (d) $\frac{1}{52}$;   (e) $\frac{3}{13}$;   (f) $\frac{1}{26}$ **3.** (a) $\frac{1}{10}$;   (b) $\frac{7}{10}$;   (c) $\frac{2}{5}$
5. (a) $\frac{1}{4}$;   (b) $\frac{1}{2}$;   (c) $\frac{3}{4}$ **7.** (a) $\frac{3}{5}$;   (b) $\frac{9}{10}$;   (c) $\frac{3}{5}$;   (d) $\frac{2}{5}$
9. (a) $\frac{1}{2}$;   (b) $\frac{1}{2}$;   (c) $\frac{1}{5}$;   (d) $\frac{4}{5}$;   (e) $\frac{1}{2}$ **11.** (a) $\frac{1}{3}$;   (b) $\frac{1}{35}$;   (c) $\frac{2}{21}$;   (d) $\frac{8}{15}$
13. (a) $\frac{95}{138}$;   (b) $\frac{1}{46}$;   (c) $\frac{20}{69}$ **15.** (a) $\frac{49}{198}$;   (b) $\frac{2}{33}$;   (c) $\frac{7}{330}$;   (d) $\frac{1}{110}$;   (d) $\frac{10}{99}$

EXERCISES 15.6

1. a; c   **3.** (a) $\frac{1}{10}$;   (b) $\frac{1}{20}$;   (c) $\frac{3}{20}$
5. (a) $\frac{25}{102}$;   (b) $\frac{1}{221}$;   (c) $\frac{55}{221}$;   (d) $\frac{2}{221}$;   (e) $\frac{25}{51}$
7. (a) $\frac{5}{12}$;   (b) $\frac{7}{18}$;   (c) $\frac{2}{9}$;   (d) $\frac{1}{3}$

EXERCISES 15.7

1. $\frac{1}{8}$ **3.** $\frac{1}{3}$ **5.** $\frac{1}{2}$ **7.** (a) $\frac{2}{5}$;   (b) $\frac{2}{3}$ **9.** (a) $\frac{1}{3}$;   (b) $\frac{1}{6}$;   (c) $\frac{1}{2}$;   (d) 0
11. $\frac{1}{4}$ **13.** $\frac{1}{6}$

EXERCISES 15.8

1. (a) $\frac{1}{12}$; (b) $\frac{1}{12}$; (c) $\frac{1}{4}$ **3.** (a) $\frac{1}{169}$; (b) $\frac{1}{4}$; (c) $\frac{1}{4}$; (d) $\frac{1}{16}$; (e) $\frac{1}{169}$
5. (a) $\frac{1}{9}$; (b) $\frac{1}{12}$; (c) $\frac{1}{12}$ **7.** (a) $\frac{14}{95}$; (b) $\frac{33}{95}$; (c) $\frac{48}{95}$

CHAPTER 15   REVIEW

1. 64   **3.** (a) 181,440;   (b) 11,880;   (c) 2,730   **5.** (a) 286;   (b) 286;   (c) 406
7. {HHHH, HHHT, HHTH, HTHH, THHH, TTHH, THTH, THHT, HTTH,
HTHT, HHTT, HTTT, THTT, TTHT, TTTH, TTTT}
9. $\frac{3}{8}$ **11.** $\frac{15}{38}$ **13.** $\frac{1}{18}$ **15.** $\frac{4}{663}$ **17.** $\frac{3}{10}$ **19.** $\frac{4}{5797}$

APPENDIX:
Tables

Table 1   Four-Place Values of Trigonometric Functions
Angle θ in Degrees and Radians

Angle θ									
Degrees	**Radians**	**sin θ**	**csc θ**	**tan θ**	**cot θ**	**sec θ**	**cos θ**		
0° 00′	.0000	.0000	No value	.0000	No value	1.000	1.0000	1.5708	90° 00′
10	029	029	343.8	029	343.8	000	000	679	50
20	058	058	171.9	058	171.9	000	000	650	40
30	.0087	.0087	114.6	.0087	114.6	1.000	1.0000	1.5621	30
40	116	116	85.95	116	85.94	000	.9999	592	20
50	145	145	68.76	145	68.75	000	999	563	10
1° 00′	.0175	.0175	57.30	.0175	57.29	1.000	.9998	1.5533	89° 00′
10	204	204	49.11	204	49.10	000	998	504	50
20	233	233	42.98	233	42.96	000	997	475	40
30	.0262	.0262	38.20	.0262	38.19	1.000	.9997	1.5446	30
40	291	291	34.38	291	34.37	000	996	417	20
50	320	320	31.26	320	31.24	001	995	388	10
2° 00′	.0349	.0349	28.65	.0349	28.64	1.001	.9994	1.5359	88° 00′
10	378	378	26.45	378	26.43	001	993	330	50
20	407	407	24.56	407	24.54	001	992	301	40
30	.0436	.0436	22.93	.0437	22.90	1.001	.9990	1.5272	30
40	465	465	21.49	466	21.47	001	989	243	20
50	495	494	20.23	495	20.21	001	988	213	10
3° 00′	.0524	.0523	19.11	.0524	19.08	1.001	.9986	1.5184	87° 00′
10	553	552	18.10	553	18.07	002	985	155	50
20	582	581	17.20	582	17.17	002	983	126	40
30	.0611	.0610	16.38	.0612	16.35	1.002	.9981	1.5097	30
40	640	640	15.64	641	15.60	002	980	068	20
50	669	669	14.96	670	14.92	002	978	039	10
4° 00′	.0698	.0698	14.34	.0699	14.30	1.002	.9976	1.5010	86° 00′
10	727	727	13.76	729	13.73	003	974	981	50
20	756	756	13.23	758	13.20	003	971	952	40
30	.0785	.0785	12.75	.0787	12.71	1.003	.9969	1.4923	30
40	814	814	12.29	816	12.25	003	967	893	20
50	844	843	11.87	846	11.83	004	964	864	10
5° 00′	.0873	.0872	11.47	.0875	11.43	1.004	.9962	1.4835	85° 00′
10	902	901	11.10	904	11.06	004	959	806	50
20	931	929	10.76	934	10.71	004	957	777	40
30	.0960	.0958	10.43	.0963	10.39	1.005	.9954	1.4748	30
40	989	987	10.13	992	10.08	005	951	719	20
50	.1018	.1016	9.839	.1022	9.788	005	948	690	10
6° 00′	.1047	.1045	9.567	.1051	9.514	1.006	.9945	1.4661	84° 00′
		cos θ	**sec θ**	**cot θ**	**tan θ**	**csc θ**	**sin θ**	**Radians**	**Degrees**
								Angle θ	

Table I—*continued*

Degrees	Radians	sin θ	csc θ	tan θ	cot θ	sec θ	cos θ		
6° 00′	.1047	.1045	9.567	.1051	9.514	1.006	.9945	1.4661	**84° 00′**
10	076	074	9.309	080	9.255	006	942	632	50
20	105	103	9.065	110	9.010	006	939	603	40
30	.1134	.1132	8.834	.1139	8.777	1.006	.9936	1.4573	30
40	164	161	8.614	169	8.556	007	932	544	20
50	193	190	8.405	198	8.345	007	929	515	10
7° 00′	.1222	.1219	8.206	.1228	8.144	1.008	.9925	1.4486	**83° 00′**
10	251	248	8.016	257	7.953	008	922	457	50
20	280	276	7.834	287	7.770	008	918	428	40
30	.1309	.1305	7.661	.1317	7.596	1.009	.9914	1.4399	30
40	338	334	7.496	346	7.429	009	911	370	20
50	367	363	7.337	376	7.269	009	907	341	10
8° 00′	.1396	.1392	7.185	.1405	7.115	1.010	.9903	1.4312	**82° 00′**
10	425	421	7.040	435	6.968	010	899	283	50
20	454	449	6.900	465	6.827	011	894	254	40
30	.1484	.1478	6.765	.1495	6.691	1.011	.9890	1.4224	30
40	513	507	6.636	524	6.561	012	886	195	20
50	542	536	6.512	554	6.435	012	881	166	10
9° 00′	.1571	.1564	6.392	.1584	6.314	1.012	.9877	1.4137	**81° 00′**
10	600	593	277	614	197	013	872	108	50
20	629	622	166	644	084	013	868	079	40
30	.1658	.1650	6.059	.1673	5.976	1.014	.9863	1.4050	30
40	687	679	5.955	703	871	014	858	1.4021	20
50	716	708	855	733	769	015	853	992	10
10° 00′	.1745	.1736	5.759	.1763	5.671	1.015	.9848	1.3963	**80° 00′**
10	774	765	665	793	576	016	843	934	50
20	804	794	575	823	485	016	838	904	40
30	.1833	.1822	5.487	.1853	5.396	1.017	.9833	1.3875	30
40	862	851	403	883	309	018	827	846	20
50	891	880	320	914	226	018	822	817	10
11° 00′	.1920	.1908	5.241	.1944	5.145	1.019	.9816	1.3788	**79° 00′**
10	949	937	164	974	066	019	811	759	50
20	978	965	089	.2004	4.989	020	805	730	40
30	.2007	.1994	5.016	.2035	4.915	1.020	.9799	1.3701	30
40	036	.2022	4.945	065	843	021	793	672	20
50	065	051	876	095	773	022	787	643	10
12° 00′	.2094	.2079	4.810	.2126	4.705	1.022	.9781	1.3614	**78° 00′**
10	123	108	745	156	638	023	775	584	50
20	153	136	682	186	574	024	769	555	40
30	.2182	.2164	4.620	.2217	4.511	1.024	.9763	1.3526	30
40	211	193	560	247	449	025	757	497	20
50	240	221	502	278	390	026	750	468	10
13° 00′	.2269	.2250	4.445	.2309	4.331	1.026	.9744	1.3439	**77° 00′**
		cos θ	sec θ	cot θ	tan θ	csc θ	sin θ	Radians	Degrees
									Angle θ

Table I—*continued*

Angle θ									
Degrees	**Radians**	**sin θ**	**csc θ**	**tan θ**	**cot θ**	**sec θ**	**cos θ**		
13° 00′	.2269	.2250	4.445	.2309	4.331	1.026	.9744	1.3439	**77° 00′**
10	298	278	390	339	275	027	737	410	50
20	327	306	336	370	219	028	730	381	40
30	.2356	.2334	4.284	.2401	4.165	1.028	.9724	1.3352	30
40	385	363	232	432	113	029	717	323	20
50	414	391	182	462	061	030	710	294	10
14° 00′	.2443	.2419	4.134	.2493	4.011	1.031	.9703	1.3265	**76° 00′**
10	473	447	086	524	3.962	031	696	235	50
20	502	476	039	555	914	032	689	206	40
30	.2531	.2504	3.994	.2586	3.867	1.033	.9681	1.3177	30
40	560	532	950	617	821	034	674	148	20
50	589	560	906	648	776	034	667	119	10
15° 00′	.2618	.2588	3.864	.2679	3.732	1.035	.9659	1.3090	**75° 00′**
10	647	616	822	711	689	036	652	061	50
20	676	644	782	742	647	037	644	032	40
30	.2705	.2672	3.742	.2773	3.606	1.038	.9636	1.3003	30
40	734	700	703	805	566	039	628	974	20
50	763	728	665	836	526	039	621	945	10
16° 00′	.2793	.2756	3.628	.2867	3.487	1.040	.9613	1.2915	**74° 00′**
10	822	784	592	899	450	041	605	886	50
20	851	812	556	931	412	042	596	857	40
30	.2880	.2840	3.521	.2962	3.376	1.043	.9588	1.2828	30
40	909	868	487	994	340	044	580	799	20
50	938	896	453	.3026	305	045	572	770	10
17° 00′	.2967	.2924	3.420	.3057	3.271	1.046	.9563	1.2741	**73° 00′**
10	996	952	388	089	237	047	555	712	50
20	.3025	979	357	121	204	048	546	683	40
30	.3054	.3007	3.326	.3153	3.172	1.048	.9537	1.2654	30
40	083	035	295	185	140	049	528	625	20
50	113	062	265	217	108	050	520	595	10
18° 00′	.3142	.3090	3.236	.3249	3.078	1.051	.9511	1.2566	**72° 00′**
10	171	118	207	281	047	052	502	537	50
20	200	145	179	314	018	053	492	508	40
30	.3229	.3173	3.152	.3346	2.989	1.054	.9483	1.2479	30
40	258	201	124	378	960	056	474	450	20
50	287	228	098	411	932	057	465	421	10
19° 00′	.3316	.3256	3.072	.3443	2.904	1.058	.9455	1.2392	**71° 00′**
10	345	283	046	476	877	059	446	363	50
20	374	311	021	508	850	060	436	334	40
30	.3403	.3338	2.996	.3541	2.824	1.061	.9426	1.2305	30
40	432	365	971	574	798	062	417	275	20
50	462	393	947	607	773	063	407	246	10
20° 00′	.3491	.3420	2.924	.3640	2.747	1.064	.9397	1.2217	**70° 00′**
		cos θ	**sec θ**	**cot θ**	**tan θ**	**csc θ**	**sin θ**	**Radians**	**Degrees**
								Angle θ	

Table I—*continued*

Angle θ		sin θ	csc θ	tan θ	cot θ	sec θ	cos θ		
Degrees	Radians								
20° 00′	.3491	.3420	2.924	.3640	2.747	1.064	.9397	1.2217	**70° 00′**
10	520	448	901	673	723	065	387	188	50
20	549	475	878	706	699	066	377	159	40
30	.3578	.3502	2.855	.3739	2.675	1.068	.9367	1.2130	30
40	607	529	833	772	651	069	356	101	20
50	636	557	812	805	628	070	346	072	10
21° 00′	.3665	.3584	2.790	.3839	2.605	1.071	.9336	1.2043	**69° 00′**
10	694	611	769	872	583	072	325	1.2014	50
20	723	638	749	906	560	074	315	985	40
30	.3752	.3665	2.729	.3939	2.539	1.075	.9304	1.1956	30
40	782	692	709	973	517	076	293	926	20
50	811	719	689	.4006	496	077	283	897	10
22° 00′	.3840	.3746	2.669	.4040	2.475	1.079	.9272	1.1868	**68° 00′**
10	869	773	650	074	455	080	261	839	50
20	898	800	632	108	434	081	250	810	40
30	.3927	.3827	2.613	.4142	2.414	1.082	.9239	1.1781	30
40	956	854	595	176	394	084	228	752	20
50	985	881	577	210	375	085	216	723	10
23° 00′	.4014	.3907	2.559	.4245	2.356	1.086	.9205	1.1694	**67° 00′**
10	043	934	542	279	337	088	194	665	50
20	072	961	525	314	318	089	182	636	40
30	.4102	.3987	2.508	.4348	2.300	1.090	.9171	1.1606	30
40	131	.4014	491	383	282	092	159	577	20
50	160	041	475	417	264	093	147	548	10
24° 00′	.4189	.4067	2.459	.4452	2.246	1.095	.9135	1.1519	**66° 00′**
10	218	094	443	487	229	096	124	490	50
20	247	120	427	522	211	097	112	461	40
30	.4276	.4147	2.411	.4557	2.194	1.099	.9100	1.1432	30
40	305	173	396	592	177	100	088	403	20
50	334	200	381	628	161	102	075	374	10
25° 00′	.4363	.4226	2.366	.4663	2.145	1.103	.9063	1.1345	**65° 00′**
10	392	253	352	699	128	105	051	316	50
20	422	279	337	734	112	106	038	286	40
30	.4451	.4305	2.323	.4770	2.097	1.108	.9026	1.1257	30
40	480	331	309	806	081	109	013	228	20
50	509	358	295	841	066	111	001	199	10
26° 00′	.4538	.4384	2.281	.4877	2.050	1.113	.8988	1.1170	**64° 00′**
10	567	410	268	913	035	114	975	141	50
20	596	436	254	950	020	116	962	112	40
30	.4625	.4462	2.241	.4986	2.006	1.117	.8949	1.1083	30
40	654	488	228	.5022	1.991	119	936	054	20
50	683	514	215	059	977	121	923	1.1025	10
27° 00′	.4712	.4540	2.203	.5095	1.963	1.122	.8910	1.0996	**63° 00′**
		cos θ	sec θ	cot θ	tan θ	csc θ	sin θ	Radians	Degrees
								Angle θ	

Table I—*continued*

Angle θ									
Degrees	**Radians**	**sin θ**	**csc θ**	**tan θ**	**cot θ**	**sec θ**	**cos θ**		
27° 00′	.4712	.4540	2.203	.5095	1.963	1.122	.8910	1.0996	**63° 00′**
10	741	566	190	132	949	124	897	966	50
20	771	592	178	169	935	126	884	937	40
30	.4800	.4617	2.166	.5206	1.921	1.127	.8870	1.0908	30
40	829	643	154	243	907	129	857	879	20
50	858	669	142	280	894	131	843	850	10
28° 00′	.4887	.4695	2.130	.5317	1.881	1.133	.8829	1.0821	**62° 00′**
10	916	720	118	354	868	134	816	792	50
20	945	746	107	392	855	136	802	763	40
30	.4974	.4772	2.096	.5430	1.842	1.138	.8788	1.0734	30
40	.5003	797	085	467	829	140	774	705	20
50	032	823	074	505	816	142	760	676	10
29° 00′	.5061	.4848	2.063	.5543	1.804	1.143	.8746	1.0647	**61° 00′**
10	091	874	052	581	792	145	732	617	50
20	120	899	041	619	780	147	718	588	40
30	.5149	.4924	2.031	.5658	1.767	1.149	.8704	1.0559	30
40	178	950	020	696	756	151	689	530	20
50	207	975	010	735	744	153	675	501	10
30° 00′	.5236	.5000	2.000	.5774	1.732	1.155	.8660	1.0472	**60° 00′**
10	265	025	1.990	812	720	157	646	443	50
20	294	050	980	851	709	159	631	414	40
30	.5323	.5075	1.970	.5890	1.698	1.161	.8616	1.0385	30
40	352	100	961	930	686	163	601	356	20
50	381	125	951	969	675	165	587	327	10
31° 00′	.5411	.5150	1.942	.6009	1.664	1.167	.8572	1.0297	**59° 00′**
10	440	175	932	048	653	169	557	268	50
20	469	200	923	088	643	171	542	239	40
30	.5498	.5225	1.914	.6128	1.632	1.173	.8526	1.0210	30
40	527	250	905	168	621	175	511	181	20
50	556	275	896	208	611	177	496	152	10
32° 00′	.5585	.5299	1.887	.6249	1.600	1.179	.8480	1.0123	**58° 00′**
10	614	324	878	289	590	181	465	094	50
20	643	348	870	330	580	184	450	065	40
30	.5672	.5373	1.861	.6371	1.570	1.186	.8434	1.0036	30
40	701	398	853	412	560	188	418	1.0007	20
50	730	422	844	453	550	190	403	977	10
33° 00′	.5760	.5446	1.836	.6494	1.540	1.192	.8387	.9948	**57° 00′**
10	789	471	828	536	530	195	371	919	50
20	818	495	820	577	520	197	355	890	40
30	.5847	.5519	1.812	.6619	1.511	1.199	.8339	.9861	30
40	876	544	804	661	501	202	323	832	20
50	905	568	796	703	1.492	204	307	803	10
34° 00′	.5934	.5592	1.788	.6745	1.483	1.206	.8290	.9774	**56° 00′**
		cos θ	**sec θ**	**cot θ**	**tan θ**	**csc θ**	**sin θ**	**Radians**	**Degrees**
									Angle θ

Table I—*continued*

Angle θ									
Degrees	**Radians**	**sin θ**	**csc θ**	**tan θ**	**cot θ**	**sec θ**	**cos θ**		
34° 00′	.5934	.5592	1.788	.6745	1.483	1.206	.8290	.9774	56° 00′
10	963	616	781	787	473	209	274	745	50
20	992	640	773	830	464	211	258	716	40
30	.6021	.5664	1.766	.6873	1.455	1.213	.8241	.9687	30
40	050	688	758	916	446	216	225	657	20
50	080	712	751	959	437	218	208	628	10
35° 00′	.6109	.5736	1.743	.7002	1.428	1.221	.8192	.9599	55° 00′
10	138	760	736	046	419	223	175	570	50
20	167	783	729	089	411	226	158	541	40
30	.6196	.5807	1.722	.7133	1.402	1.228	.8141	.9512	30
40	225	831	715	177	393	231	124	483	20
50	254	854	708	221	385	233	107	454	10
36° 00′	.6283	.5878	1.701	.7265	1.376	1.236	.8090	.9425	54° 00′
10	312	901	695	310	368	239	073	396	50
20	341	925	688	355	360	241	056	367	40
30	.6370	.5948	1.681	.7400	1.351	1.244	.8039	.9338	30
40	400	972	675	445	343	247	021	308	20
50	429	995	668	490	335	249	004	279	10
37° 00′	.6458	.6018	1.662	.7536	1.327	1.252	.7986	.9250	53° 00′
10	487	041	655	581	319	255	969	221	50
20	516	065	649	627	311	258	951	192	40
30	.6545	.6088	1.643	.7673	1.303	1.260	.7934	.9163	30
40	574	111	636	720	295	263	916	134	20
50	603	134	630	766	288	266	898	105	10
38° 00′	.6632	.6157	1.624	.7813	1.280	1.269	.7880	.9076	52° 00′
10	661	180	618	860	272	272	862	047	50
20	690	202	612	907	265	275	844	.9018	40
30	.6720	.6225	1.606	.7954	1.257	1.278	.7826	.8988	30
40	749	248	601	.8002	250	281	808	959	20
50	778	271	595	050	242	284	790	930	10
39° 00′	.6807	.6293	1.589	.8098	1.235	1.287	.7771	.8901	51° 00′
10	836	316	583	146	228	290	753	872	50
20	865	338	578	195	220	293	735	843	40
30	.6894	.6361	1.572	.8243	1.213	1.296	.7716	.8814	30
40	923	383	567	292	206	299	698	785	20
50	952	406	561	342	199	302	679	756	10
40° 00′	.6981	.6428	1.556	.8391	1.192	1.305	.7660	.8727	50° 00′
10	.7010	450	550	441	185	309	642	698	50
20	039	472	545	491	178	312	623	668	40
30	.7069	.6494	1.540	.8541	1.171	1.315	.7604	.8639	30
40	098	517	535	591	164	318	585	610	20
50	127	539	529	642	157	322	566	581	10
41° 00′	.7156	.6561	1.524	.8693	1.150	1.325	.7547	.8552	49° 00′
		cos θ	**sec θ**	**cot θ**	**tan θ**	**csc θ**	**sin θ**	**Radians**	**Degrees**
								Angle θ	

Table I—*continued*

Angle θ									
Degrees	Radians	sin θ	csc θ	tan θ	cot θ	sec θ	cos θ		
41° 00′	.7156	.6561	1.524	.8693	1.150	1.325	.7547	.8552	49° 00′
10	185	583	519	744	144	328	528	523	50
20	214	604	514	796	137	332	509	494	40
30	.7243	.6626	1.509	.8847	1.130	1.335	.7490	.8465	30
40	272	648	504	899	124	339	470	436	20
50	301	670	499	952	117	342	451	407	10
42° 00′	.7330	.6691	1.494	.9004	1.111	1.346	.7431	.8378	48° 00′
10	359	713	490	057	104	349	412	348	50
20	389	734	485	110	098	353	392	319	40
30	.7418	.6756	1.480	.9163	1.091	1.356	.7373	.8290	30
40	447	777	476	217	085	360	353	261	20
50	476	799	471	271	079	364	333	232	10
43° 00′	.7505	.6820	1.466	.9325	1.072	1.367	.7314	.8203	47° 00′
10	534	841	462	380	066	371	294	174	50
20	563	862	457	435	060	375	274	145	40
30	.7592	.6884	1.453	.9490	1.054	1.379	.7254	.8116	30
40	621	905	448	545	048	382	234	087	20
50	650	926	444	601	042	386	214	058	10
44° 00′	.7679	.6947	1.440	.9657	1.036	1.390	.7193	.8029	46° 00′
10	709	967	435	713	030	394	173	.7999	50
20	738	988	431	770	024	398	153	970	40
30	.7767	.7009	1.427	.9827	1.018	1.402	.7133	.7941	30
40	796	030	423	884	012	406	112	912	20
50	825	050	418	942	006	410	092	883	10
45° 00′	.7854	.7071	1.414	1.000	1.000	1.414	.7071	.7854	45° 00′
		cos θ	sec θ	cot θ	tan θ	csc θ	sin θ	Radians	Degrees
								Angle θ	

Table II Four Place Logarithms of Numbers

N	0	1	2	3	4	5	6	7	8	9
10	0000	0043	0086	0128	0170	0212	0253	0294	0334	0374
11	0414	0453	0492	0531	0569	0607	0645	0682	0719	0755
12	0792	0828	0864	0899	0934	0969	1004	1038	1072	1106
13	1139	1173	1206	1239	1271	1303	1335	1367	1399	1430
14	1461	1492	1523	1553	1584	1614	1644	1673	1703	1732
15	1761	1790	1818	1847	1875	1903	1931	1959	1987	2014
16	2041	2068	2095	2122	2148	2175	2201	2227	2253	2279
17	2304	2330	2355	2380	2405	2430	2455	2480	2504	2529
18	2553	2577	2601	2625	2648	2672	2695	2718	2742	2765
19	2788	2810	2833	2856	2878	2900	2923	2945	2967	2989
20	3010	3032	3054	3075	3096	3118	3139	3160	3181	3201
21	3222	3243	3263	3284	3304	3324	3345	3365	3385	3404
22	3424	3444	3464	3483	3502	3522	3541	3560	3579	3598
23	3617	3636	3655	3674	3692	3711	3729	3747	3766	3784
24	3802	3820	3838	3856	3874	3892	3909	3927	3945	3962
25	3979	3997	4014	4031	4048	4065	4082	4099	4116	4133
26	4150	4166	4183	4200	4216	4232	4249	4265	4281	4298
27	4314	4330	4346	4362	4378	4393	4409	4425	4440	4456
28	4472	4487	4502	4518	4533	4548	4564	4579	4594	4609
29	4624	4639	4654	4669	4683	4698	4713	4728	4742	4757
30	4771	4786	4800	4814	4829	4843	4857	4871	4886	4900
31	4914	4928	4942	4955	4969	4983	4997	5011	5024	5038
32	5051	5065	5079	5092	5105	5119	5132	5145	5159	5172
33	5185	5198	5211	5224	5237	5250	5263	5276	5289	5302
34	5315	5328	5340	5353	5366	5378	5391	5403	5416	5428
35	5441	5453	5465	5478	5490	5502	5514	5527	5539	5551
36	5563	5575	5587	5599	5611	5623	5635	5647	5658	5670
37	5682	5694	5705	5717	5729	5740	5752	5763	5775	5786
38	5798	5809	5821	5832	5843	5855	5866	5877	5888	5899
39	5911	5922	5933	5944	5955	5966	5977	5988	5999	6010
40	6021	6031	6042	6053	6064	6075	6085	6096	6107	6117
41	6128	6138	6149	6160	6170	6180	6191	6201	6212	6222
42	6232	6243	6253	6263	6274	6284	6294	6304	6314	6325
43	6335	6345	6355	6365	6375	6385	6395	6405	6415	6425
44	6435	6444	6454	6464	6474	6484	6493	6503	6513	6522
45	6532	6542	6551	6561	6571	6580	6590	6599	6609	6618
46	6628	6637	6646	6656	6665	6675	6684	6693	6702	6712
47	6721	6730	6739	6749	6758	6767	6776	6785	6794	6803
48	6812	6821	6830	6839	6848	6857	6866	6875	6884	6893
49	6902	6911	6920	6928	6937	6946	6955	6964	6972	6981
50	6990	6998	7007	7016	7024	7033	7042	7050	7059	7067
51	7076	7084	7093	7101	7110	7118	7126	7135	7143	7152
52	7160	7168	7177	7185	7193	7202	7210	7218	7226	7235
53	7243	7251	7259	7267	7275	7284	7292	7300	7308	7316
54	7324	7332	7340	7348	7356	7364	7372	7380	7388	7396

Table II—*continued*

N	0	1	2	3	4	5	6	7	8	9
55	7404	7412	7419	7427	7435	7443	7451	7459	7466	7474
56	7482	7490	7497	7505	7513	7520	7528	7536	7543	7551
57	7559	7566	7574	7582	7589	7597	7604	7612	7619	7627
58	7634	7642	7649	7657	7664	7672	7679	7686	7694	7701
59	7709	7716	7723	7731	7738	7745	7752	7760	7767	7774
60	7782	7789	7796	7803	7810	7818	7825	7832	7839	7846
61	7853	7860	7868	7875	7882	7889	7896	7903	7910	7917
62	7924	7931	7938	7945	7952	7959	7966	7973	7980	7987
63	7993	8000	8007	8014	8021	8028	8035	8041	8048	8055
64	8062	8069	8075	8082	8089	8096	8102	8109	8116	8122
65	8129	8136	8142	8149	8156	8162	8169	8176	8182	8189
66	8195	8202	8209	8215	8222	8228	8235	8241	8248	8254
67	8261	8267	8274	8280	8287	8293	8299	8306	8312	8319
68	8325	8331	8338	8344	8351	8357	8363	8370	8376	8382
69	8388	8395	8401	8407	8414	8420	8426	8432	8439	8445
70	8451	8457	8463	8470	8476	8482	8488	8494	8500	8506
71	8513	8519	8525	8531	8537	8543	8549	8555	8561	8567
72	8573	8579	8585	8591	8597	8603	8609	8615	8621	8627
73	8633	8639	8645	8651	8657	8663	8669	8675	8681	8686
74	8692	8698	8704	8710	8716	8722	8727	8733	8739	8745
75	8751	8756	8762	8768	8774	8779	8785	8791	8797	8802
76	8808	8814	8820	8825	8831	8837	8842	8848	8854	8859
77	8865	8871	8876	8882	8887	8893	8899	8904	8910	8915
78	8921	8927	8932	8938	8943	8949	8954	8960	8965	8971
79	8976	8982	8987	8993	8998	9004	9009	9015	9020	9025
80	9031	9036	9042	9047	9053	9058	9063	9069	9074	9079
81	9085	9090	9096	9101	9106	9112	9117	9122	9128	9133
82	9138	9143	9149	9154	9159	9165	9170	9175	9180	9186
83	9191	9196	9201	9206	9212	9217	9222	9227	9232	9238
84	9243	9248	9253	9258	9263	9269	9274	9279	9284	9289
85	9294	9299	9304	9309	9315	9320	9325	9330	9335	9340
86	9345	9350	9355	9360	9365	9370	9375	9380	9385	9390
87	9395	9400	9405	9410	9415	9420	9425	9430	9435	9440
88	9445	9450	9455	9460	9465	9469	9474	9479	9484	9489
89	9494	9499	9504	9509	9513	9518	9523	9528	9533	9538
90	9542	9547	9552	9557	9562	9566	9571	9576	9581	9586
91	9590	9595	9600	9605	9609	9614	9619	9624	9628	9633
92	9638	9643	9647	9652	9657	9661	9666	9671	9675	9680
93	9685	9689	9694	9699	9703	9708	9713	9717	9722	9727
94	9731	9736	9741	9745	9750	9754	9759	9763	9768	9773
95	9777	9782	9786	9791	9795	9800	9805	9809	9814	9818
96	9823	9827	9832	9836	9841	9845	9850	9854	9859	9863
97	9868	9872	9877	9881	9886	9890	9894	9899	9903	9908
98	9912	9917	9921	9926	9930	9934	9939	9943	9948	9952
99	9956	9961	9965	9969	9974	9978	9983	9987	9991	9996

Table III Four-Place Logarithms of Trigonometric Functions

Angle θ in Degrees

Attach -10 to Logarithms Obtained from This Table

Angle θ	L sin θ	L csc θ	L tan θ	L cot θ	L sec θ	L cos θ	
0° 00′	No value	No value	No value	No value	10.0000	10.0000	90° 00′
10	7.4637	12.5363	7.4637	12.5363	.0000	.0000	50
20	.7648	.2352	.7648	.2352	.0000	.0000	40
30	7.9408	12.0592	7.9409	12.0591	.0000	.0000	30
40	8.0658	11.9342	8.0658	11.9342	.0000	.0000	20
50	.1627	.8373	.1627	.8373	.0000	10.0000	10
1° 00′	8.2419	11.7581	8.2419	11.7581	10.0001	9.9999	89° 00′
10	.3088	.6912	.3089	.6911	.0001	.9999	50
20	.3668	.6332	.3669	.6331	.0001	.9999	40
30	.4179	.5821	.4181	.5819	.0001	.9999	30
40	.4637	.5363	.4638	.5362	.0002	.9998	20
50	.5050	.4950	.5053	.4947	.0002	.9998	10
2° 00′	8.5428	11.4572	8.5431	11.4569	10.0003	9.9997	88° 00′
10	.5776	.4224	.5779	.4221	.0003	.9997	50
20	.6097	.3903	.6101	.3899	.0004	.9996	40
30	.6397	.3603	.6401	.3599	.0004	.9996	30
40	.6677	.3323	.6682	.3318	.0005	.9995	20
50	.6940	.3060	.6945	.3055	.0005	.9995	10
3° 00′	8.7188	11.2812	8.7194	11.2806	10.0006	9.9994	87° 00′
10	.7423	.2577	.7429	.2571	.0007	.9993	50
20	.7645	.2355	.7652	.2348	.0007	.9993	40
30	.7857	.2143	.7865	.2135	.0008	.9992	30
40	.8059	.1941	.8067	.1933	.0009	.9991	20
50	.8251	.1749	.8261	.1739	.0010	.9990	10
4° 00′	8.8436	11.1564	8.8446	11.1554	10.0011	9.9989	86° 00′
10	.8613	.1387	.8624	.1376	.0011	.9989	50
20	.8783	.1217	.8795	.1205	.0012	.9988	40
30	.8946	.1054	.8960	.1040	.0013	.9987	30
40	.9104	.0896	.9118	.0882	.0014	.9986	20
50	.9256	.0744	.9272	.0728	.0015	.9985	10
5° 00′	8.9403	11.0597	8.9420	11.0580	10.0017	9.9983	85° 00′
10	.9545	.0455	.9563	.0437	.0018	.9982	50
20	.9682	.0318	.9701	.0299	.0019	.9981	40
30	.9816	.0184	.9836	.0164	.0020	.9980	30
40	8.9945	11.0055	8.9966	11.0034	.0021	.9979	20
50	9.0070	10.9930	9.0093	10.9907	.0023	.9977	10
6° 00′	9.0192	10.9808	9.0216	10.9784	10.0024	9.9976	84° 00′
	L cos θ	L sec θ	L cot θ	L tan θ	L csc θ	L sin θ	Angle θ

Table III—*continued*

Attach −10 to Logarithms Obtained from This Table

Angle θ	L sin θ	L csc θ	L tan θ	L cot θ	L sec θ	L cos θ	
6° 00′	9.0192	10.9808	9.0216	10.9784	10.0024	9.9976	84° 00′
10	.0311	.9689	.0336	.9664	.0025	.9975	50
20	.0426	.9574	.0453	.9547	.0027	.9973	40
30	.0539	.9461	.0567	.9433	.0028	.9972	30
40	.0648	.9352	.0678	.9322	.0029	.9971	20
50	.0755	.9245	.0786	.9214	.0031	.9969	10
7° 00′	9.0859	10.9141	9.0891	10.9109	10.0032	9.9968	83° 00′
10	.0961	.9039	.0995	.9005	.0034	.9966	50
20	.1060	.8940	.1096	.8904	.0036	.9964	40
30	.1157	.8843	.1194	.8806	.0037	.9963	30
40	.1252	.8748	.1291	.8709	.0039	.9961	20
50	.1345	.8655	.1385	.8615	.0041	.9959	10
8° 00′	9.1436	10.8564	9.1478	10.8522	10.0042	9.9958	82° 00′
10	.1525	.8475	.1569	.8431	.0044	.9956	50
20	.1612	.8388	.1658	.8342	.0046	.9954	40
30	.1697	.8303	.1745	.8255	.0048	.9952	30
40	.1781	.8219	.1831	.8169	.0050	.9950	20
50	.1863	.8137	.1915	.8085	.0052	.9948	10
9° 00′	9.1943	10.8057	9.1997	10.8003	10.0054	9.9946	81° 00′
10	.2022	.7978	.2078	.7922	.0056	.9944	50
20	.2100	.7900	.2158	.7842	.0058	.9942	40
30	.2176	.7824	.2236	.7764	.0060	.9940	30
40	.2251	.7749	.2313	.7687	.0062	.9938	20
50	.2324	.7676	.2389	.7611	.0064	.9936	10
10° 00′	9.2397	10.7603	9.2463	10.7537	10.0066	9.9934	80° 00′
10	.2468	.7532	.2536	.7464	.0069	.9931	50
20	.2538	.7462	.2609	.7391	.0071	.9929	40
30	.2606	.7394	.2680	.7320	.0073	.9927	30
40	.2674	.7326	.2750	.7250	.0076	.9924	20
50	.2740	.7260	.2819	.7181	.0078	.9922	10
11° 00′	9.2806	10.7194	9.2887	10.7113	10.0081	9.9919	79° 00′
10	.2870	.7130	.2953	.7047	.0083	.9917	50
20	.2934	.7066	.3020	.6980	.0086	.9914	40
30	.2997	.7003	.3085	.6915	.0088	.9912	30
40	.3058	.6942	.3149	.6851	.0091	.9909	20
50	.3119	.6881	.3212	.6788	.0093	.9907	10
12° 00′	9.3179	10.6821	9.3275	10.6725	10.0096	9.9904	78° 00′
10	.3238	.6762	.3336	.6664	.0099	.9901	50
20	.3296	.6704	.3397	.6603	.0101	.9899	40
30	.3353	.6647	.3458	.6542	.0104	.9896	30
40	.3410	.6590	.3517	.6483	.0107	.9893	20
50	.3466	.6534	.3576	.6424	.0110	.9890	10
13° 00′	9.3521	10.6479	9.3634	10.6366	10.0113	9.9887	77° 00′
	L cos θ	L sec θ	L cot θ	L tan θ	L csc θ	L sin θ	Angle θ

Table III—*continued*

Attach -10 to Logarithms Obtained from This Table

Angle θ	L sin θ	L csc θ	L tan θ	L cot θ	L sec θ	L cos θ	
13° 00′	9.3521	10.6479	9.3634	10.6366	10.0113	9.9887	77° 00′
10	.3575	.6425	.3691	.6309	.0116	.9884	50
20	.3629	.6371	.3748	.6252	.0119	.9881	40
30	.3682	.6318	.3804	.6196	.0122	.9878	30
40	.3734	.6266	.3859	.6141	.0125	.9875	20
50	.3786	.6214	.3914	.6086	.0128	.9872	10
14° 00′	9.3837	10.6163	9.3968	10.6032	10.0131	9.9869	76° 00′
10	.3887	.6113	.4021	.5979	.0134	.9866	50
20	.3937	.6063	.4074	.5926	.0137	.9863	40
30	.3986	.6014	.4127	.5873	.0141	.9859	30
40	.4035	.5965	.4178	.5822	.0144	.9856	20
50	.4083	.5917	.4230	.5770	.0147	.9853	10
15° 00′	9.4130	10.5870	9.4281	10.5719	10.0151	9.9849	75° 00′
10	.4177	.5823	.4331	.5669	.0154	.9846	50
20	.4223	.5777	.4381	.5619	.0157	.9843	40
30	.4269	.5731	.4430	.5570	.0161	.9839	30
40	.4314	.5686	.4479	.5521	.0164	.9836	20
50	.4359	.5641	.4527	.5473	.0168	.9832	10
16° 00′	9.4403	10.5597	9.4575	10.5425	10.0172	9.9828	74° 00′
10	.4447	.5553	.4622	.5378	.0175	.9825	50
20	.4491	.5509	.4669	.5331	.0179	.9821	40
30	.4533	.5467	.4716	.5284	.0183	.9817	30
40	.4576	.5424	.4762	.5238	.0186	.9814	20
50	.4618	.5382	.4808	.5192	.0190	.9810	10
17° 00′	9.4659	10.5341	9.4853	10.5147	10.0194	9.9806	73° 00′
10	.4700	.5300	.4898	.5102	.0198	.9802	50
20	.4741	.5259	.4943	.5057	.0202	.9798	40
30	.4781	.5219	.4987	.5013	.0206	.9794	30
40	.4821	.5179	.5031	.4969	.0210	.9790	20
50	.4861	.5139	.5075	.4925	.0214	.9786	10
18° 00′	9.4900	10.5100	9.5118	10.4882	10.0218	9.9782	72° 00′
10	.4939	.5061	.5161	.4839	.0222	.9778	50
20	.4977	.5023	.5203	.4797	.0226	.9774	40
30	.5015	.4985	.5245	.4755	.0230	.9770	30
40	.5052	.4948	.5287	.4713	.0235	.9765	20
50	.5090	.4910	.5329	.4671	.0239	.9761	10
19° 00′	9.5126	10.4874	9.5370	10.4630	10.0243	9.9757	71° 00′
10	.5163	.4837	.5411	.4589	.0248	.9752	50
20	.5199	.4801	.5451	.4549	.0252	.9748	40
30	.5235	.4765	.5491	.4509	.0257	.9743	30
40	.5270	.4730	.5531	.4469	.0261	.9739	20
50	.5306	.4694	.5571	.4429	.0266	.9734	10
20° 00′	9.5341	10.4659	9.5611	10.4389	10.0270	9.9730	70° 00′
	L cos θ	L sec θ	L cot θ	L tan θ	L csc θ	L sin θ	Angle θ

Table III—*continued*

Attach −10 to Logarithms Obtained from This Table

Angle θ	L sin θ	L csc θ	L tan θ	L cot θ	L sec θ	L cos θ	
20° 00′	9.5341	10.4659	9.5611	10.4389	10.0270	9.9730	**70° 00′**
10	.5375	.4625	.5650	.4350	.0275	.9725	50
20	.5409	.4591	.5689	.4311	.0279	.9721	40
30	.5443	.4557	.5727	.4273	.0284	.9716	30
40	.5477	.4523	.5766	.4234	.0289	.9711	20
50	.5510	.4490	.5804	.4196	.0294	.9706	10
21° 00′	9.5543	10.4457	9.5842	10.4158	10.0298	9.9702	**69° 00′**
10	.5576	.4424	5879	.4121	.0303	.9797	50
20	.5609	.4391	.5917	.4083	.0308	.9692	40
30	.5641	.4359	.5954	.4046	.0313	.9687	30
40	.5673	.4327	.5991	.4009	.0318	.9682	20
50	.5704	.4296	.6028	.3972	.0323	.9677	10
22° 00′	9.5736	10.4264	9.6064	10.3936	10.0328	9.9672	**68° 00′**
10	.5767	.4233	.6100	.3900	.0333	.9667	50
20	.5798	.4202	.6136	.3864	.0339	.9661	40
30	.5828	.4172	.6172	.3828	.0344	.9656	30
40	.5859	.4141	.6208	.3792	.0349	.9651	20
50	.5889	.4111	.6243	.3757	.0354	.9646	10
23° 00′	9.5919	10.4081	9.6279	10.3721	10.0360	9.9640	**67° 00′**
10	.5984	.4052	.6314	.3686	.0365	.9635	50
20	.5978	.4022	.6348	.3652	.0371	.9629	40
30	.6007	.3993	.6383	.3617	.0376	.9624	30
40	.6036	.3964	.6417	.3583	.0382	.9618	20
50	.6065	.3935	.6452	.3548	.0387	.9613	10
24° 00′	9.6093	10.3907	9.6486	10.3514	10.0393	9.9607	**66° 00′**
10	.6121	.3879	.6520	.3480	.0398	.9602	50
20	.6149	.3851	.6553	.3447	.0404	.9596	40
30	.6177	.3823	.6587	.3413	.0410	.9590	30
40	.6205	.3795	.6620	.3380	.0416	.9584	20
50	.6232	.3768	.6654	.3346	.0421	.9579	10
25° 00′	9.6259	10.3741	9.6687	10.3313	10.0427	9.9573	**65°00′**
10	.6286	.3714	.6720	.3280	.0433	.9567	50
20	.6313	.3687	.6752	.3248	.0439	.9561	40
30	.6340	.3660	.6785	.3215	.0445	.9555	30
40	.6366	.3634	.6817	.3183	.0451	.9549	20
50	.6392	.3608	.6850	.3150	.0457	.9543	10
26° 00′	9.6418	10.3582	9.6882	10.3118	10.0463	9.9537	**64° 00′**
10	.6444	.3556	.6914	.3086	.0470	.9530	50
20	.6470	.3530	.6946	.3054	.0476	.9524	40
30	.6495	.3505	.6977	.3023	.0482	.9518	30
40	.6521	.3479	.7009	.2991	.0488	.9512	20
50	.6546	.3454	.7040	.2960	.0495	.9505	10
27° 00′	9.6570	10.3430	9.7072	10.2928	10.0501	9.9499	**63° 00′**
	L cos θ	L sec θ	L cot θ	L tan θ	L csc θ	L sin θ	Angle θ

Table III—*continued*

Attach −10 to Logarithms Obtained from This Table

Angle θ	L sin θ	L csc θ	L tan θ	L cot θ	L sec θ	L cos θ	
27° 00′	9.6570	10.3430	9.7072	10.2928	10.0501	9.9499	**63° 00′**
10	.6595	.3405	.7103	.2897	.0508	.9492	50
20	.6620	.3380	.7134	.2866	.0514	.9486	40
30	.6644	.3356	.7165	.2835	.0521	.9479	30
40	.6668	.3332	.7196	.2804	.0527	.9473	20
50	.6692	.3308	.7226	.2774	.0534	.9466	10
28° 00′	9.6716	10.3284	9.7257	10.2743	10.0541	9.9459	**62° 00′**
10	.6740	.3260	.7287	.2713	.0547	.9453	50
20	.6763	.3237	.7317	.2683	.0554	.9446	40
30	.6787	.3213	.7348	.2652	.0561	.9439	30
40	.6810	.3190	.7378	.2622	.0568	.9432	20
50	.6833	.3167	.7408	.2592	.0575	.9425	10
29° 00′	9.6856	10.3144	9.7438	10.2562	10.0582	9.9418	**61° 00′**
10	.6878	.3122	.7467	.2533	.0589	.9411	50
20	.6901	.3099	.7497	.2503	.0596	.9404	40
30	.6923	.3077	.7526	.2474	.0603	.9397	30
40	.6946	.3054	.7556	.2444	.0610	.9390	20
50	.6968	.3032	.7585	.2415	.0617	.9383	10
30° 00′	9.6990	10.3010	9.7614	10.2386	10.0625	9.9375	**60° 00′**
10	.7012	.2988	.7644	.2356	.0632	.9368	50
20	.7033	.2967	.7673	.2327	.0639	.9361	40
30	.7055	.2945	.7701	.2299	.0647	.9353	30
40	.7076	.2924	.7730	.2270	.0654	.9346	20
50	.7097	.2903	.7759	.2241	.0662	.9338	10
31° 00′	9.7118	10.2882	9.7788	10.2212	10.0669	9.9331	**59° 00′**
10	.7139	.2861	.7816	.2184	.0677	.9323	50
20	.7160	.2840	.7845	.2155	.0685	.9315	40
30	.7181	.2819	.7873	.2127	.0692	.9308	30
40	.7201	.2799	.7902	.2098	.0700	.9300	20
50	.7222	.2778	.7930	.2070	.0708	.9292	10
32° 00′	9.7242	10.2758	9.7958	10.2042	10.0716	9.9284	**58° 00′**
10	.7262	.2738	.7986	.2014	.0724	.9276	50
20	.7282	.2718	.8014	.1986	.0732	.9268	40
30	.7302	.2698	.8042	.1958	.0740	.9260	30
40	.7322	.2678	.8070	.1930	.0748	.9252	20
50	.7342	.2658	.8097	.1903	.0756	.9244	10
33° 00′	9.7361	10.2639	9.8125	10.1875	10.0764	9.9236	**57° 00′**
10	.7380	.2620	.8153	.1847	.0772	.9228	50
20	.7400	.2600	.8180	.1820	.0781	.9219	40
30	.7419	.2581	.8208	.1792	.0789	.9211	30
40	.7438	.2562	.8235	.1765	.0797	.9203	20
50	.7457	.2543	.8263	.1737	.0806	.9194	10
34° 00′	9.7476	10.2524	9.8290	10.1710	10.0814	9.9186	**56° 00′**
	L cos θ	L sec θ	L cot θ	L tan θ	L csc θ	L sin θ	Angle θ

Table III—*continued*

Attach −10 to Logarithms Obtained from This Table

Angle θ	L sin θ	L csc θ	L tan θ	L cot θ	L sec θ	L cos θ	
34° 00′	9.7476	10.2524	9.8290	10.1710	10.0814	9.9186	**56° 00′**
10	.7494	.2506	.8317	.1683	.0823	.9177	50
20	.7513	.2487	.8344	.1656	.0831	.9169	40
30	.7531	.2469	.8371	.1629	.0840	.9160	30
40	.7550	.2450	.8398	.1602	.0849	.9151	20
50	.7568	.2432	.8425	.1575	.0858	.9142	10
35° 00′	9.7586	10.2414	9.8452	10.1548	10.0866	9.9134	**55° 00′**
10	.7604	.2396	.8479	.1521	.0875	.9125	50
20	.7622	.2378	.8506	.1494	.0884	.9116	40
30	.7640	.2360	.8533	.1467	.0893	.9107	30
40	.7657	.2343	.8559	.1441	.0902	.9098	20
50	.7675	.2325	.8586	.1414	.0911	.9089	10
36° 00′	9.7692	10.2308	9.8613	10.1387	10.0920	9.9080	**54° 00′**
10	.7710	.2290	.8639	.1361	.0930	.9070	50
20	.7727	.2273	.8666	.1334	.0939	.9061	40
30	.7744	.2256	.8692	.1308	.0948	.9052	30
40	.7761	.2239	.8718	.1282	.0958	.9042	20
50	.7778	.2222	.8745	.1255	.0967	.9033	10
37° 00′	9.7795	10.2205	9.8771	10.1229	10.0977	9.9023	**53° 00′**
10	.7811	.2189	.8797	.1203	.0986	.9014	50
20	.7828	.2172	.8824	.1176	.0996	.9004	40
30	.7844	.2156	.8850	.1150	.1005	.8995	30
40	.7861	.2139	.8876	.1124	.1015	.8985	20
50	.7877	.2123	.8902	.1098	.1025	.8975	10
38° 00′	9.7893	10.2107	9.8928	10.1072	10.1035	9.8965	**52° 00′**
10	.7910	.2090	.8954	.1046	.1045	.8955	50
20	.7926	.2074	.8980	.1020	.1055	.8945	40
30	.7941	.2059	.9006	.0994	.1065	.8935	30
40	.7957	.2043	.9032	.0968	.1075	.8925	20
50	.7973	.2027	.9058	.0942	.1085	.8915	10
39° 00′	9.7989	10.2011	9.9084	10.0916	10.1095	9.8905	**51° 00′**
10	.8004	.1996	.9110	.0890	.1105	.8895	50
20	.8020	.1980	.9135	.0865	.1116	.8884	40
30	.8035	.1965	.9161	.0839	.1126	.8874	30
40	.8050	.1950	.9187	.0813	.1136	.8864	20
50	.8066	.1934	.9212	.0788	.1147	.8853	10
40° 00′	9.8081	10.1919	9.9238	10.0762	10.1157	9.8843	**50° 00′**
10	.8096	.1904	.9264	.0736	.1168	.8832	50
20	.8111	.1889	.9289	.0711	.1179	.8821	40
30	.8125	.1875	.9315	.0685	.1190	.8810	30
40	.8140	.1860	.9341	.0659	.1200	.8800	20
50	.8155	.1845	.9366	.0634	.1211	.8789	10
41° 00′	9.8169	10.1831	9.9392	10.0608	10.1222	9.8778	**49° 00′**
	L cos θ	L sec θ	L cot θ	L tan θ	L csc θ	L sin θ	Angle θ

Table III—*continued*

Attach −10 to Logarithms Obtained from This Table

Angle θ	L sin θ	L csc θ	L tan θ	L cot θ	L sec θ	L cos θ	
41° 00′	9.8168	10.1831	9.9392	10.0608	10.1222	9.8778	**49° 00′**
10	.8184	.1816	.9417	.0583	.1233	.8767	50
20	.8198	.1802	.9443	.0557	.1244	.8756	40
30	.8213	.1787	.9468	.0532	.1255	.8745	30
40	.8227	.1773	.9494	.0506	.1267	.8733	20
50	.8241	.1759	.9519	.0481	.1278	.8722	10
42° 00′	9.8255	10.1745	9.9544	10.0456	10.1289	9.8711	**48° 00′**
10	.8269	.1731	.9570	.0430	.1301	.8699	50
20	.8283	.1717	.9595	.0405	.1312	.8688	40
30	.8297	.1703	.9621	.0379	.1324	.8676	30
40	.8311	.1689	.9646	.0354	.1335	.8665	20
50	.8324	.1676	.9671	.0329	.1347	.8653	10
43° 00′	9.8338	10.1662	9.9697	10.0303	10.1359	9.8641	**47° 00′**
10	.8351	.1649	.9722	.0278	.1371	.8629	50
20	.8365	.1635	.9747	.0253	.1382	.8618	40
30	.8378	.1622	.9772	.0228	.1394	.8606	30
40	.8391	.1609	.9798	.0202	.1406	.8594	20
50	.8405	.1595	.9823	.0177	.1418	.8582	10
44° 00′	9.8418	10.1582	9.9848	10.0152	10.1431	9.8569	**46° 00′**
10	.8431	.1569	.9874	.0126	.1443	.8557	50
20	.8444	.1556	.9899	.0101	.1455	.8545	40
30	.8457	.1543	.9924	.0076	.1468	.8532	30
40	.8469	.1531	.9949	.0051	.1480	.8520	20
50	.8482	.1518	9.9975	.0025	.1493	.8507	10
45° 00′	9.8495	10.1505	10.0000	10.0000	10.1505	9.8495	**45° 00′**
	L cos θ	L sec θ	L cot θ	L tan θ	L csc θ	L sin θ	Angle θ

INDEX